terra australis 33

Terra Australis reports the results of archaeological and related research within the south and east of Asia, though mainly Australia, New Guinea and island Melanesia — lands that remained terra australis incognita to generations of prehistorians. Its subject is the settlement of the diverse environments in this isolated quarter of the globe by peoples who have maintained their discrete and traditional ways of life into the recent recorded or remembered past and at times into the observable present.

List of volumes in Terra Australis

Volume 1: Burrill Lake and Currarong: Coastal Sites in Southern New South Wales. R.J. Lampert (1971)
Volume 2: Ol Tumbuna: Archaeological Excavations in the Eastern Central Highlands, Papua New Guinea. J.P. White (1972)
Volume 3: New Guinea Stone Age Trade: The Geography and Ecology of Traffic in the Interior. I. Hughes (1977)
Volume 4: Recent Prehistory in Southeast Papua. B. Egloff (1979)
Volume 5: The Great Kartan Mystery. R. Lampert (1981)
Volume 6: Early Man in North Queensland: Art and Archaeology in the Laura Area. A. Rosenfeld, D. Horton and J. Winter (1981)
Volume 7: The Alligator Rivers: Prehistory and Ecology in Western Arnhem Land. C. Schrire (1982)
Volume 8: Hunter Hill, Hunter Island: Archaeological Investigations of a Prehistoric Tasmanian Site. S. Bowdler (1984)
Volume 9: Coastal South-West Tasmania: The Prehistory of Louisa Bay and Maatsuyker Island. R. Vanderwal and D. Horton (1984)
Volume 10: The Emergence of Mailu. G. Irwin (1985)
Volume 11: Archaeology in Eastern Timor, 1966–67. I. Glover (1986)
Volume 12: Early Tongan Prehistory: The Lapita Period on Tongatapu and its Relationships. J. Poulsen (1987)
Volume 13: Coobool Creek. P. Brown (1989)
Volume 14: 30,000 Years of Aboriginal Occupation: Kimberley, North-West Australia. S. O'Connor (1999)
Volume 15: Lapita Interaction. G. Summerhayes (2000)
Volume 16: The Prehistory of Buka: A Stepping Stone Island in the Northern Solomons. S. Wickler (2001)
Volume 17: The Archaeology of Lapita Dispersal in Oceania. G.R. Clark, A.J. Anderson and T. Vunidilo (2001)
Volume 18: An Archaeology of West Polynesian Prehistory. A. Smith (2002)
Volume 19: Phytolith and Starch Research in the Australian-Pacific-Asian Regions: The State of the Art. D. Hart and L. Wallis (2003)
Volume 20: The Sea People: Late-Holocene Maritime Specialisation in the Whitsunday Islands, Central Queensland. B. Barker (2004)
Volume 21: What's Changing: Population Size or Land-Use Patterns? The Archaeology of Upper Mangrove Creek, Sydney Basin. V. Attenbrow (2004)
Volume 22: The Archaeology of the Aru Islands, Eastern Indonesia. S. O'Connor, M. Spriggs and P. Veth (2005)
Volume 23: Pieces of the Vanuatu Puzzle: Archaeology of the North, South and Centre. S. Bedford (2006)
Volume 24: Coastal Themes: An Archaeology of the Southern Curtis Coast, Queensland. S. Ulm (2006)
Volume 25: Lithics in the Land of the Lightning Brothers: The Archaeology of Wardaman Country, Northern Territory. C. Clarkson (2007)
Volume 26: Oceanic Explorations: Lapita and Western Pacific Settlement. S. Bedford, C. Sand and S. P. Connaughton (2007)
Volume 27: Dreamtime Superhighway: Sydney Basin Rock Art and Prehistoric Information Exchange. J. McDonald (2008)
Volume 28: New Directions in Archaeological Science. A. Fairbairn, S. O'Connor and B. Marwick (2008)
Volume 29: Islands of Inquiry: Colonisation, Seafaring and the Archaeology of Maritime Landscapes. G. Clark, F. Leach and S. O'Connor (2008)
Volume 30: Archaeological Science Under a Microscope: Studies in Residue and Ancient DNA Analysis in Honour of Thomas H. Loy. M. Haslam, G. Robertson, A. Crowther, S. Nugent and L. Kirkwood (2009)
Volume 31: The Early Prehistory of Fiji. G. Clark and A. Anderson (2009)
Volume 32: Altered Ecologies: Fire, Climate and Human Influence on Terrestrial Landscapes. S. Haberle, J. Stevenson and M. Prebble (2010)

terra australis 33

MAN BAC: THE EXCAVATION OF A NEOLITHIC SITE IN NORTHERN VIETNAM

The Biology

edited by Marc F. Oxenham,
Hirofumi Matsumura and Nguyen Kim Dung

Published by ANU E Press
The Australian National University
Canberra ACT 0200 Australia
Email: anuepress@anu.edu.au
Web: http://epress.anu.edu.au

National Library of Australia Cataloguing-in-Publication entry

Title: Man Bac : the excavation of a neolithic site in northern Vietnam / edited by Marc F. Oxenham, Hirofumi Matsumura and Nguyen Kim Dung.

ISBN: 9781921862229 (pbk.) 9781921862236 (eBook)

Series: Terra Australis ; number 33.

Subjects: Excavations (Archaeology)--Vietnam--Red River Delta.
Man Bac Site (Vietnam).
Vietnam--Antiquities.

Other Authors/Contributors:
Oxenham, Marc.
Matsumura, Hirofumi.
Nguyen, Dung Kim.

Dewey Number: 959.701

Series Editor: Sue O'Connor

Typesetting and design: Anna Willis

Cover image: Excavation trench 2007H2 at the site of Man Bac

Back cover map: Hollandia Nova. Thevenot 1663 by courtesy of the National Library of Australia.
Reprinted with permission of the National Library of Australia.

Contributors
(alphabetical order)

Peter BELLWOOD
School of Archaeology and Anthropology, Australian National University, Canberra, Australia

Yukio DODO
Department of Anatomy, Tohoku University, Sendai, Japan

Kate M. DOMETT
School of Medicine and Dentistry, James Cook University, Australia, Townsville, Australia

Damien G. HUFFER
School of Archaeology and Anthropology, Australian National University, Canberra, Australia

Hirofumi MATSUMURA
Department of Anatomy, Sapporo Medical University, Sapporo, Japan

NGUYEN Lan Cuong
The Vietnamese Institute of Archaeology, Hanoi, Vietnam

NGUYEN Kim Thuy
The Vietnamese Institute of Archaeology, Hanoi, Vietnam

NGUYEN Anh Tuan
The Vietnamese Institute of Archaeology, Hanoi, Vietnam

Marc F. OXENHAM
School of Archaeology and Anthropology, Australian National University, Canberra, Australia

Junmei SAWADA
Department of Anatomy, St. Marianna University School of Medicine, Kawasaki, Japan

Ken-ichi SHINODA
Department of Anthropology, National Museum of Nature and Science, Tokyo, Japan

Takeji TOIZUMI
Institute of Comparative Archaeology, Waseda University, Tokyo, Japan

TRINH Hoang Hiep
The Vietnamese Institute of Archaeology, Hanoi, Vietnam

Wataru TAKIGAWA
Department of Physical Therapy, International University of Health and Welfare, Fukuoka, Japan

Shinya WATANABE
Department of Archaeology, Waseda University, Japan

Mariko YAMAGATA
Department of Archaeology, Waseda University, Japan

Preface

Man Bac is a prehistoric site dated to 3,500-3,800 years BP in Ninh Binh Province, Vietnam, that has revealed a large number of human burials in an excellent state of preservation as well as an enormous assemblage of diverse domestic artefacts. Man Bac represents a large cemetery population of the neolithic period, unprecedented among contemporaneous sites in Vietnam. Biological analysis of the remains provides crucial insights into past demography, life style and behaviour, health, food resources and subsistence as well as mortuary practices. This volume is a product of multi-disciplinary and multi-national efforts over the past several years. We hope that this book will provide an invaluable source of information concerning prehistoric peoples in Southeast Asia. The editors would like to thank all of the authors for their contributions to this volume.

Marc F. Oxenham
Hirofumi Matsumura
Nguyen Kim Dung

Acknowledgements

In addition to those that have contributed chapters and appendices to this volume, we are grateful to the additional collaborators, listed below, of the multi-national excavation projects at the site of Man Bac during 2005-2007 for their fieldwork, laboratory work, analyses, assistance, official support and/or helpful comments. Further, special thanks are owed to Anna Willis for the enormous amount of time and effort she has spent in proofing and formatting this monograph. Any errors or omissions are the sole responsibility of the editors.

Vietnam
Ha Van Phung, Former Director, the Vietnamese Institute of Archaeology
Nguyen Giang Hai, Vice Director, the Vietnamese Institute of Archaeology
Vu The Long, the Vietnamese Institute of Archaeology
Bui Thu Phuong, the Vietnamese Institute of Archaeology
Ha Manh Thang, the Vietnamese Institute of Archaeology
Nguyen Ngoc Quy, the Vietnamese Institute of Archaeology
Nguyen Dang Cuong, the Vietnamese Institute of Archaeology
Tran Thi Thuy Ha, the Vietnamese Institute of Archaeology
Nguyen Hann Khang, Ninh Binh Provincial Museum
Nguyen Cao Tan, Ninh Binh Provincial Museum
Nguyen Van Sai, landowner of the Man Bac site
Community of Chung Village
Local workers in excavation at the site of Man Bac

Australia
Bennet A. Mosig, School of Archaeology and Anthropology, ANU
Amy Way, School of Archaeology and Anthropology, ANU
Brad Smith, School of Archaeology and Anthropology, ANU
Lorna Tilley, School of Archaeology and Anthropology, ANU
Janelle Stevenson, School of Culture, History & Language, ANU

Japan
Minoru Yoneda, Department of Integrated Biosciences, the University of Tokyo
Masanari Nishimura, Institute for Cultural Interaction Studies, Kansai University

Financial Assistance
This excavation project was supported by a Grant-in-Aid (B) in 2003-2005 (No.15405018) and 2008-2010 (No.20370096) from the Japan Society for the Promotion of Science, the Toyota Foundation 2006-2007 (No. D06-R-0035) and an Australian Research Council Grant (DP 077 4 079).

Contents

Figures

Tables

1
Introduction: Man Bac Biological Research Objectives

Hirofumi Matsumura[1] and Marc F. Oxenham[2]

[1]*Department of Anatomy, Sapporo Medical University, Japan.*
[2] *School of Archaeology and Anthropology, Australian National University*

The principle aim of this volume is the examination and elucidation of the human biology of the Man Bac cemetery population and associated faunal assemblages, in order to reveal the micro-evolutionary history, palaeohealth, local palaeo-environmental conditions, subsistence strategies and general life-ways of this ancient community. Building on previous Man Bac research we wish to provide a wealth of new information about population history, colonisation, diet, nutrition, adaptive shifts, and specific and general aspects of health in the current volume.

Quantitative and qualitative cranio-dental analyses will speak to a complex population history involving both migration and *in situ* development. Ancient mitochondrial DNA investigations will reveal at least one piece of the genetic landscape during this period, helping to shed light on Man Bac's place in debates over the origins of present day Southeast Asian populations.

An investigation of long bone morphology will provide important information on antemortem life styles related to physical activities, while explorations of ancient patterns of health and past demographic trends at Man Bac will sketch a picture of the history of human wellbeing and behaviour in the region.

Work on the faunal remains will reveal a subsistence base rich in both terrestrial and aquatic resources with the differential targeting of certain terrestrial mammals and marine resources, alluding to the complexity of their hunting and gathering behaviours and abilities. Further, the analysis of age distributions for *Sus* remains will provide evidence for the initial stages of the domestication of pigs in the region.

The human and non-humans of Man Bac specifically dealt with in this volume will provide an informative database in developing, reconstructing and interpreting the lives of the first food producers in this region.

Man Bac is one of the best cemetery/habitation sites in Vietnam for investigating questions regarding genetic history, health, disease, environment, social systems and identity, as well as a wide range of other factors, in detail. Previous publications have already revealed evidence for an age-based social hierarchy (Oxenham et al., 2008a), sophisticated system of palliative care (Oxenham et al., 2009) and the expression of social identity, including that of children, through complex mortuary practices (Oxenham et al., 2008b). Further, preliminary analysis of cranial and dental morphometric data recorded from skeletons of the earlier excavation seasons (1999-2005) suggests the existence within Man Bac of immigrants, which may prove to be of crucial importance in debates concerning the population history of this region (Matsumura et al., 2008).

MAN BAC: THE SITE

An exhaustive account of the palaeoenvironment and archaeology of Man Bac will be detailed in a subsequent publication. Here, the environment, geography and archaeology of the site are summarised. The site of Man Bac is located in Yen Mo district, Ninh Binh province, northern Vietnam (109^0 59' 17" East and 20^0 08' 00" North) (Figure 1.1). Man Bac is approximately 25km from the coast and surrounded by karst limestone mountains. The following summary of the current climate and environment of Vietnam, while not necessarily representative of the situation two to three thousand years ago, does, however, serve as an approximation to conditions in the past.

> Topographically, while about three-quarters of Vietnam can be described as mountainous, 85% of these are below 1000m in elevation. Vietnam has three plains systems that are still in the process of expansion. [Man Bac] situated in the low lying northern Bac Bo plain and fall[s] between latitude 18^0 and 22^0 north. Vietnam presents two distinct climatic zones, a northern and a southern. Northern Vietnam has two seasons, cold and hot but with high levels of humidity occurring during both periods. Further, the north experiences marked climatic variability or instability that has restricted levels of ecological variation in comparison to southern Vietnam. The country is prone to typhoons and has experienced over 400 during the last one hundred years. Storms are also very frequent with winds reaching up to 50 m/sec. This in combination with rainfall in excess of 600mm in a 24 hour period can lead to extensive agricultural and human disruption. Because the Bac Bo region is essentially subtropical, tropical forests are found only below about 500m while more tropical flora is only found below some 300m. Around the coastal regions mangroves still predominate, while dense bamboo forests are common all over the northern plains. The north is home to a diverse range of bird, riverine and marine life. Some 900 species of fish are recognized in the Gulf of Bac Bo alone. Terrestrial animals such as sambar deer, muntjac, chamois and numerous arboreal primates are still common in the region. In the past elephant, rhinoceros, tiger, and panther were also common.
> (Oxenham 2006:212-213)

The excavation history of this approximately 2m deep deposit is as follows: Vietnam Institute of Archaeology and Ninh Binh Museum in 1999 ($30m^2$, 6 burials); the same group in 2001 ($30m^2$, 11 burials); the same group in collaboration with the Sapporo Medical University and Australian National University in 2004-5 ($36m^2$, 35 burials); and finally by the same multi-national team in 2007 (pit H1 $12m^2$ 15 burials, pit H2 $24m^2$ 32 burials) (Cuong, 2001; Phung, 2001; Hiep and Phung, 2004; Nishimura, 2006, Oxenham et al. 2008a, Matsumura et al. 2008). Figure 1.2 shows both the 2005 excavation and the history of excavation at the site. Figure 1.3 indicates the high concentration of burials in the western part of the site during the 2007 excavations. Figure 1.4 provides a schematic burial plan for all four seasons. Colour coding provides information on the age structure and distribution (within

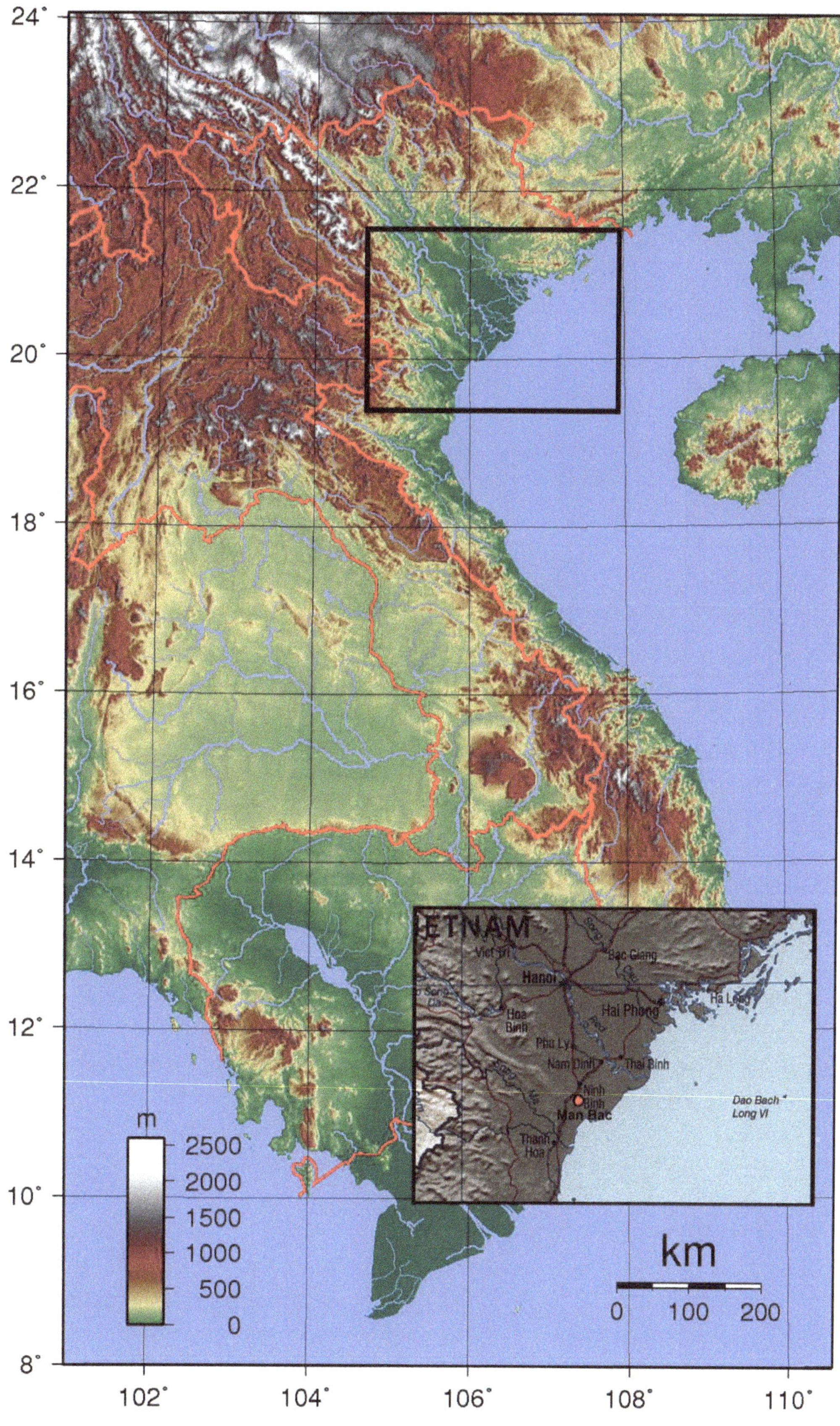

Figure 1.1 Topographic map of Vietnam and location of Man Bac (inset lower right) on the southern edge of the Red River Delta (map source: Sadalmelik; inset source: Wiki.verkata).

the cemetery) of the sample (see Chapters 2 and 3 for more details), while the orientation of the burials are represented by the long axes of the ellipses (the majority of burials were oriented approximating east-west with the head in the east). Regarding the actual size of the original settlement,

> [i]t is difficult to determine the extent of the site, primarily due to subsequent terracing and the development of a catholic cemetery to the east of the site in the historic period, but it likely approximates 200-300m^2. Preliminary analyses suggest that two distinct cultural phases are associated with three stratigraphic levels: the upper two units being occupation phases and the third (bottom) layer being almost exclusively burials in otherwise sterile silt. Material cultural similarities between the occupation layers and grave inclusions in the third level suggest the burials are associated with the occupation level(s).
> Oxenham et al. (2008b:191)

A number of C14 dates on charcoal [2 sigma range calibrations (INTCAL04) after Reimer et al. (2004)] sandwich the occupation and burial layers quite narrowly: 3,341±38 BP (1,737-1,524 BC); 3,393±36 BP (1,775–1,608 BC); 3,560±30 BP (2,066-1,775 BC) (Oxenham et al. 2008b:192; see Appendix 2 in this monograph for a complete list of dates). In terms of the local cultural chronologies, the material culture displays many similarities with the Phung Nguyen period (4,000–3,500 years BP), whose culture flourished in the north of Vietnam along the upper reaches of the Red River Delta. The Phung Nguyen culture period, and indeed Man Bac to a greater or lesser extent, is generally associated with evidence for agriculture, land clearance, ceramic manufacture, hunting, marine resource gathering, extensive and far reaching trade networks, and local food production (although Man Bac lacks hard evidence for rice agriculture) (Hiep and Phung, 2004; Nishimura, 2006).

TERMINOLOGY

Following the convention established in Oxenham and Tayles (2006) the term 'neolithic' is rendered in lower case. The principle reason for this is the general unsuitability of a term which has a raft of connotations specific to archaeological culture chronologies in Europe and the Near East. Additionally, 'neolithic' implies a definitional precision lacking in Southeast Asian archaeology. How the term 'neolithic' may be tentatively used in a Southeast Asian archaeological context is discussed in more detail in Chapter 12. For the purposes of the following chapters, 'neolithic' refers to Southeast Asian food-producing communities that lacked evidence for metal.

Geographic terminology is crucial to the issues discussed in this volume because many researchers who argue for the *in situ* evolution of indigenous Southeast Asians include south China in their definition of prehistoric Southeast Asia (e.g., Turner, 1995). Although such a definition does have certain advantages (Solheim, 1985), in this study we use the separate designations of East and Southeast Asia: "East Asia" refers to modern China, Taiwan, North and South Korea, Japan, Mongolia, and the Russian Far East; "Southeast Asia" includes modern Myanmar (Burma), Thailand, Vietnam, Laos, Cambodia, Malaysia, Singapore, Indonesia, Brunei, the Philippines, and the Andaman and Nicobar Islands. In choosing this

Figure 1.2 View of Man Bac looking south. Note modern catholic cemetery to the east (left in picture) of the excavated area. Arrows show excavation history: red 1999, yellow 2001, green 2004/5 (open excavation square seen), blue 2007.

Figure 1.3 View of 2007 pit H2, looking east. There was a high concentration of burials of all ages from neonate, to infant to adult in this 30m^2 trench.

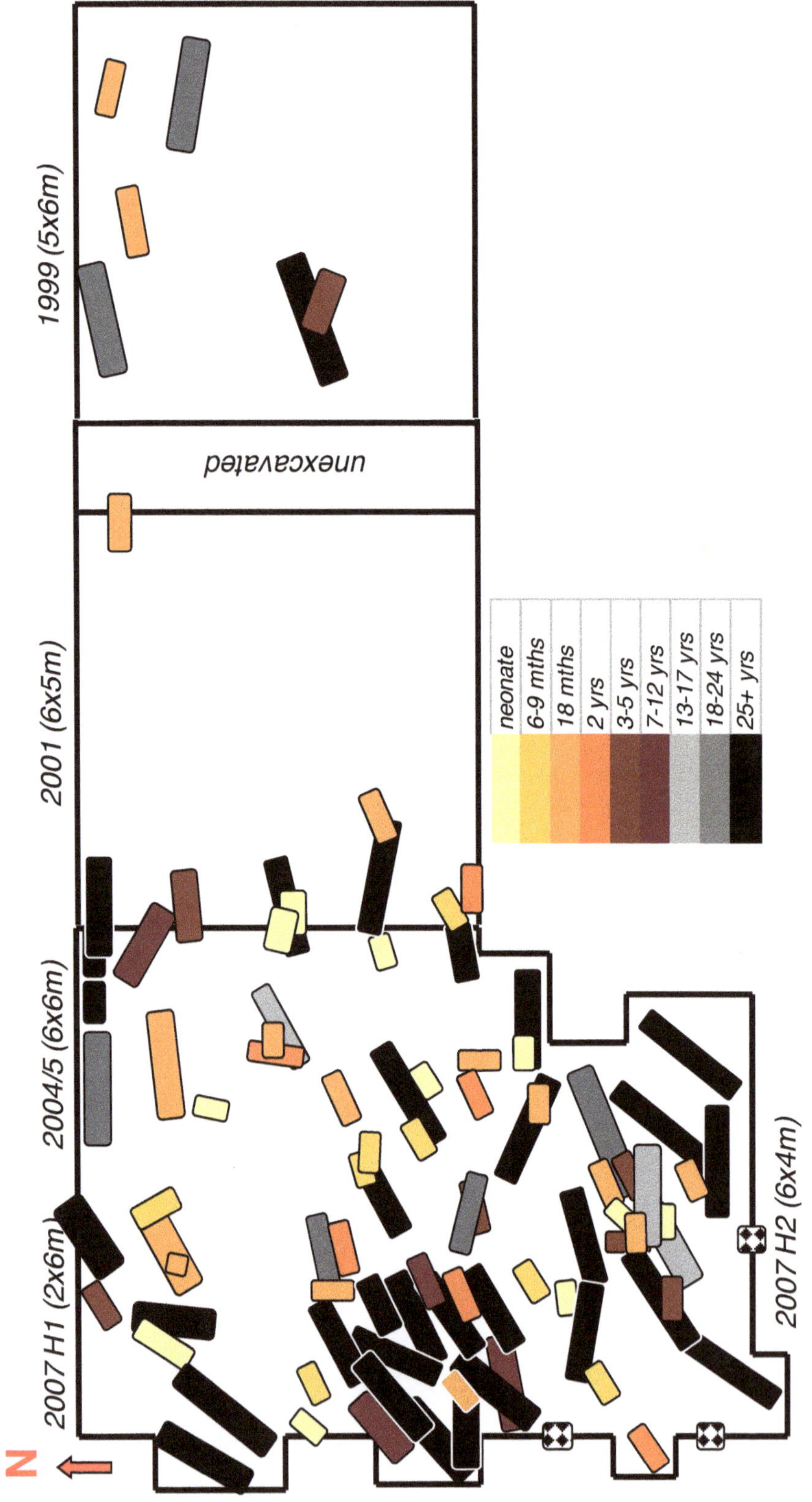

Figure 1.4 Schematic burial plan for all seasons of excavation at Man Bac (after Oxenham 2006, 2009).

terminology, we are aware that the concept of "Asia" itself is a Western one which maps that which is "non-European" and thus lacks geographic or historical precision (Chaudhuri, 1990, p.22-23), nonetheless it has a heuristic value.

MONOGRAPH STRUCTURE

The following Chapter (2) explores the demography of the sample as well as explaining the chief methodological protocols employed for age-at-death and sex assessment of the human sample. Chapters 3 through 5 analyse the cranio-dental morphology of the sample using both quantitative and qualitative techniques with the aim of elucidating the micro-evolutionary history and phenotypic relationships of the Man Bac population within the local and larger region. Chapter 6 explores stature at Man Bac and regionally, as well as looking at long bone cross-sectional morphology. Chapter 7 explores certain aspects of the health of the Man Bac inhabitants, particularly with respect to oral disease and two signatures of physiological well-being (enamel hypoplasia and cribra orbitalia). In Chapter 8 an analysis of ancient mtDNA haplotypes helps elucidate the genetic relationships of the Man Bac population with local and more distant populations. Chapter 9 assesses the mammalian vertebrate assemblage and discusses the evidence for pig domestication and the contribution of hunting to the economy. Following this, Chapter 10 examines the fish assemblage and discusses the habitats and species targeted by the ancient Man Bac community. Finally, Chapter 11 synthesises the finding of the preceding chapters and places Man Bac into a broader interpretive context. Three large appendices are included; the first describes the mortuary context of each burial, in addition to age-at-death and the sex of each burial; the second describes and illustrates the mortuary pottery and the third appendix describes the state of preservation of each burial and details the cranial and postcranial morphology of the adult remains.

LITERATURE CITED

Chaudhuri KN. 1990. Asia before Europe: economy and civilization of the Indian Ocean from the rise of Islam to 1750. Cambridge: Cambridge University Press.

Cuong NL. 2001. About human remains at Man Bac site. Khao Co Hoc (Vietnamese Archaeology) 1-2001: 17-46 (in Vietnamese with English title and summary).

Hiep TH, Phung HV. 2004. Man Bac location and its relationship through ceramic data. Khao Co Hoc (Vietnamese Archaeology) 6-2004: 13-48 (in Vietnamese with English summary).

Matsumura H, Oxenham MF, Dodo Y, Domett K, Cuong NL, Thuy NK, Dung NK, Huffer D, Yamagata M. 2008. Morphometric affinity of the late Neolithic human remains from Man Bac, Ninh Binh Province, Vietnam: key skeletons with which to debate the 'two layer' hypothesis. Anthropological Science 116: 135-148.

Nishimura M. 2006. Archaeological Study in the Deltas of Red River Dong Nai River. Tokyo: PhD Dissertation, the University of Tokyo.

Oxenham MF. 2006. Biological responses to change in prehistoric Viet Nam. Asian perspectives. 45(2):212-239.

Oxenham MF. 2009. The Social and Biological Construction of Childhood in Ancient Vietnam. A paper presented at the 19th Congress of the Indo-Pacific Prehistory

Association, Hanoi, Vietnam 29th November-5th December 2009.

Oxenham MF, and Tayles N. 2006. Bioarchaeology of Southeast Asia. Cambridge. Cambridge University Press.

Oxenham MF, Matsumura H, Domett K, Thuy NK., Dung NK, Cuong NL, Huffer D, Muller S. 2008a. Childhood in late Neolithic Vietnam: bio-mortuary insights into an ambiguous life stage. In: Bacvarov K, editor. Babies Reborn: Infant/Child Burials in Pre- and Protohistory. Oxford: BAR International Series Archaeopress. p 123-136.

Oxenham MF, Matsumura H, Domett K, Nguyen KT, Dung NK, Cuong NL, Huffer D, Muller S. 2008b. Health and the experience of childhood in late Neolithic Vietnam. Asian Perspectives 47:190-209.

Oxenham MF, Tilley L, Matsumura H, Cuong NL. Thuy NK, Dung NK, Domett K, Huffer D. 2009. Paralysis and severe disability requiring intensive care in Neolithic Asia. Anthropological Science 2: 107-112.

Phung HV. 2001. Man Bac site –data and perception. Khao Co Hoc (Vietnamese Archaeology) 1-2001: 17-46 (in Vietnamese with English title and summary).

Sadalmelik. nd. http://commons.wikimedia.org/wiki/User:Sadalmelik [public domain]

Solheim WG. 1985. 'Southeast Asia': what's in a name?", another point of view. Journal of the Southeast Asian Studies 16:141-147.

Turner CG II. 1995. Shifting continuity: modern human origin. In: Brenner S, Hanihara K, editors. The Origin and Past of Modern Humans as Viewed from DNA. Singapore: World Scientific. p 216-243.

Wiki.verkata. nd. http:// wiki.verkata.com/en/wiki/File:Laos_2003_CIA_map.jpg [public domain]

2
The Demographic Profile of the Man Bac Cemetery Sample

Kate M. Domett[1] and Marc F. Oxenham[2]

[1]*School of Medicine and Dentistry, James Cook University, Australia*
[2]*School of Archaeology and Anthropology, Australian National University*

The chief aims of this chapter are to describe the Man Bac human skeletal sample in terms of its sex and age-at-death distributions. Moreover, the preservation of the sample will be discussed in the context of a demographic reconstruction of the past population, which will include a range of measures of fertility. Inferences regarding the demographic 'health' of the population will be made with reference to major social and behavioural changes seen in the region some 3,500 years ago.

MATERIALS

Preservation, Completeness and Disturbance

Over the course of three excavations undertaken from 2004/5 to 2007 at Man Bac, 84 individuals with a range of skeletal preservation were observed. Some skeletons were extremely incomplete and/or highly disturbed with only a few bone fragments and grave goods. Fortunately, these were in the minority as many skeletons were fully articulated and complete. Some of the subadults were in fact remarkably well preserved with many preserving separate epiphyses and the small developing bones of the hands and feet (see Appendix 1).

Six of the 84 burials are not included in any of the following calculations (MB05M35, MB07H2M11, MB07H2M23 and MB07H2M25 were not excavated, while MB07H1M13b and MB05M33 were represented by a few isolated teeth only), leaving a total sample for demographic analysis of 78. Of the assessable sample 47/78 (60.3%) were complete or near complete, and in many of these the bone quality was very good. A further 22/78 (28.2%) were classed as incomplete where they were missing a skull and/or some major limb bones, usually from some disturbance in prehistory, or they were not within the bounds of the excavation. Only 9/78 (11.5%) were considered highly incomplete and fragmented. These data indicates that excellent preservation existed at the Man Bac cemetery compared with many other Southeast Asian skeletal samples, with many showing significantly less well preserved material. For example, less than a third of burials were deemed to be near complete at Noen U-Loke (Tayles et al., 2007); only 18% at Ban Lum Khao where, although bone quality was good, there was a lot of disturbance (Domett, 2004). Nong Nor was poorly preserved, with only 19% complete (Tayles et al., 1998); and at Ban Chiang, more than half of the skulls and postcranial skeletons were incomplete or fragmented (Pietrusewsky and Douglas, 2002a). Khok Phanom Di is one of the few sites that is on a par with Man Bac in terms of preservation with less than 10% missing any major element (Domett, 2001). The well preserved nature of the Man Bac skeletal sample has enabled a thorough assessment of demography and health to be undertaken. A complete set of burial descriptions is provided in Appendices 1 and 3.

METHODS

Age-At-Death Estimation

Subadults

The most reliable method of the age-at-death estimation of children (up to approximately 12 years) is through observations of the development (calcification) and eruption of the dentition. The dentition has been found to be more strongly controlled by genetic factors and less influenced by environmental factors compared to skeletal growth (Ubelaker, 1987; Saunders, 1992) and is therefore a more reliable indicator of *biological* age and provides a close approximation of *chronological* age (Saunders, 1992).

All mandibular and maxillary elements from subadults were radiographed in order to provide accurate evidence of calcification (or formation), rather than relying on the stage of eruption of the deciduous and permanent dentition; the latter approach is thought to be less accurate (Halcrow et al., 2007). These results were then compared with published standards (Buikstra and Ubelaker, 1994; White, 2000). These standards are not derived from Asian populations, but they are used in the absence of more population-specific information. There is very limited information regarding the development of teeth in Southeast Asian children and that which is published (eg. Kamalanathan, 1960) is based on modern populations. A full review of the effect of using non-population specific standards on prehistoric Southeast Asian children has recently been reviewed by Halcrow et al. (2007).

A number of subadults at Man Bac did not have any dentition preserved. In these cases long bone development (either complete or partial diaphyses) were compared with those Man Bac subadults that had been aged from their dentition. This method establishes a population specific set of standards rather than relying on individuals from other Southeast Asian populations as the comparison. There is also considerable information provided by Scheur and Black (2000) on the development of individual skeletal elements, this was used in conjunction with the population standards. There were only two subadults (MB05M6 and MB05M22) who had neither dentition nor complete diaphyseal lengths with which to estimate age-at-death. For these individuals, sections of their long bones were compared to similar bone sections in individuals aged by their dentition.

Once the second permanent molar has erupted around the age of 12 years, the dentition is no longer the most reliable indicator of age-at-death. Skeletons of these older children at Man Bac were aged through observation of epiphyseal fusion, predominantly based on Scheur and Black (2000). Again these published standards are not based on prehistoric Asian children but provide an excellent summary of the information that is available.

The categorisation of 'subadults' is often different in different studies. The issues that arise from this have recently been reviewed (Halcrow and Tayles, 2008). Biologically, most growth and development of a skeleton has been completed by the late teenage years and into the early 20s. However, socially, particularly in prehistory, a person of this age has likely been contributing to the economy and life of the community for some years. In addition there are also issues involved if comparisons want to be made to previous studies. In this way, the allocation of a specific age range to subadults necessitates changes depending on the questions asked. This will be made clear in the following discussion.

Adults

The estimation of age-at-death in those over 15 years is most reliably estimated through observation of the pubic symphyseal face. The Suchey Brooks standards (Buikstra and Ubelaker, 1994) were used on the Man Bac skeletons where the pubic symphysis was preserved. For those adults with no pubic symphyseal face preserved, some of the younger adults were able to be aged through observation of late fusing epiphyses. For those adults with neither of these observations possible (N=7), age-at-death has been estimated using functions, developed by Oxenham (2000) on a Da But period (c. 5,500 years BP) sample from northern Vietnam. These functions were originally developed by regressing Scott's (1979) molar wear scores on age-at-death determinations based on either symphyseal or auricular morphology. In order to test the validity of using these functions on a different ancient Vietnamese population, age-at-death was estimated for the 26 Man Bac individuals for which there existed independent (for the most part based on pubic symphyseal morphology and epiphyseal fusion) age estimates. Of the independently aged sub-sample (N=26), 20 individuals (76.9%) were found to fall within the same 10 year age bracket as provided by independent age estimates. Four individuals (15.4%) fell into an adjacent age category (all were aged by tooth wear as 30-39 years, whereas symphyseal morphology in each instance indicated an age of 40-49 years), while the final 2 cases had pathological wear patterns. In total, 24/26 individuals (92.3%) had their symphyseal and/or epiphyseal fusion age estimates confirmed to within a single decade of tooth wear age estimates. On this basis, some confidence was placed in estimating the age-at-death, using tooth wear scores, of the 6 individuals without other forms of independent age estimation.

Sex Estimation

Standard morphological analyses of the pelvis and skull were the primary sources of information for estimating the sex of an adult skeleton at Man Bac (Buikstra and Ubelaker, 1994). Those without the pelvis or skull have not been estimated for sex.

Sex estimation of subadults has not been proven to be particularly reliable (e.g. Cardaso and Saunders, 2008; Vlak et al., 2008) so results are not presented here.

RESULTS AND DISCUSSION

Overview

Table 2.1 details the age-at-death and sex estimate of each individual where possible. Table 2.2 provides a summary of the age-at-death of all individuals and the sex estimate of adults where possible. From this information it is possible to state that the skeletal remains comprising the excavated sample probably provide a representative sample of the entire cemetery population. This is primarily based on the results of the subadult mortality rate and the sex ratio of adults. Waldron (1994) suggests at least 30% of any pre-industrial sample should be subadult (less than 15 years); anything much lower could lead to inaccurate epidemiological calculations. The Man Bac subadult sample comprises 59% of the total sample, well above the suggested 30%. This, however, is a quite high subadult mortality, reasons for which will be discussed later, but does suggest a good retrieval rate for these fragile skeletons. The adult sex ratio of males to females from Man Bac is 1:0.8 (15:12). This is reasonably close to parity, to indicate a non sex-biased cemetery.

Subadults

The subadult section of the demographic profile can be very useful in providing a picture of health and quality of life in a prehistoric sample. This group is

Table 2.1 The age-at-death and sex estimation of each individual from Man Bac (2004/5 to 2007).

Excavation ID	Burial No.	Sex	Age-at-death	Notes
MB05	1		18 mths +/- 5 mths	incomplete, fragmented
MB05	2		neonate	near complete
MB05	3		6 mths +/- 2 mths	near complete
MB05	4		2 yrs +/- 6 mths	incomplete
MB05	5		18 mths +/- 3 mths	near complete
MB05	6		~18 mths	incomplete
MB05	7		neonate	incomplete
MB05	8		6 mths	incomplete
MB05	9	Female	40-49 yrs	near complete
MB05	10		9 yrs +/- 9 mths	near complete
MB05	11	Male	18-25 yrs	near complete
MB05	12		2 yrs +/- 6 mths	near complete
MB05	13	?	16 yrs	near complete
MB05	14		2.5 yrs	near complete
MB05	15	Female	17-18 yrs	incomplete
MB05	16a	Female	40-49 yrs	incomplete
MB05	16b		neonate	incomplete, fragmented
MB05	17	?	Adult	incomplete, fragmented
MB05	18		18 mths +/- 3 mths	near complete
MB05	19	?	Adult	incomplete, fragmented
MB05	20	Male?	15-29 yrs	near complete
MB05	21		6 mths	near complete
MB05	22		18 mths	incomplete, fragmented
MB05	23		15 mths	incomplete
MB05	24		8 yrs +/- 9 mths	near complete
MB05	25		5 yrs +/- 9 mths	near complete
MB05	26		4-5 yrs +/- 1 yr	incomplete, fragmented
MB05	27	?	Adult	incomplete
MB05	28	Female?	15-29 yrs	incomplete
MB05	29	Male	30-30 yrs	near complete
MB05	30		6 months	near complete
MB05	31	Male	20-29 yrs	near complete
MB05	32	Male?	15-29 yrs	incomplete, fragmented
MB05	33			teeth only
MB05	34	Female	40-40 yrs	near complete
MB05	35			not excavated
MB05	36		3 yrs +/- 6 mths	incomplete

Table 2.1 continued next page.

Table 2.1 (continued).

Excavation ID	Burial No.	Sex	Age-at-death	Notes
MB07H1	1		12 yrs +/- 6 mths	incomplete
MB07H1	2		neonate	incomplete,fragmented
MB07H1	3		12-18 yrs	near complete
MB07H1	4	Female?	30+ yrs	incomplete
MB07H1	5	Male	40-49 yrs	near complete
MB07H1	6		6 mths +/- 2 mths / 9 mths +/-2 mth	near complete
MB07H1	7		1 yr +/- 3 mths	incomplete
MB07H1	8	Male	30-39 yrs	near complete
MB07H1	9	Male?	15-29 yrs	incomplete
MB07H1	10	Male?	40-49 yrs	near complete
MB07H1	11	Female	50+ yrs	near complete
MB07H1	12		neonate	incomplete,fragmented
MB07H1	13a	?	30+ yrs	incomplete
MB07H1	13b		~10 yrs	incomplete
MB07H1	14	?	30+ yrs	incomplete
MB07H2	1	Male	40-49 yrs	near complete
MB07H2	2		12-18 yrs	near complete
MB07H2	3		neonate	incomplete
MB07H2	4		18 mths	incomplete
MB07H2	5	Female	20-25 yrs	near complete
MB07H2	6		2 yrs +/- 6 mths	near complete
MB07H2	7		18 mths +/- 5 mths	near complete
MB07H2	8		1 yr +/- 3 mths / 18 mths +/- 5 mths	near complete
MB07H2	9		neonate	incomplete
MB07H2	10	Male	30-39 yrs	near complete
MB07H2	11	?	Adult	not excavated
MB07H2	12	Female	50+ yrs	near complete
MB07H2	13		4 yrs +/- 9 mths	near complete
MB07H2	14		neonate	near complete
MB07H2	15		4 yrs +/- 9 mths	near complete
MB07H2	16		1 yr +/- 3 mths / 18 mths +/- 5 mths	incomplete
MB07H2	17		13-18 yrs	near complete
MB07H2	18	Female	18-24 yrs	near complete
MB07H2	19	Male	20-24 yrs	near complete
MB07H2	20		6 mths +/- 2 mths	near complete
MB07H2	21		9 mths +/- 2 mths	incomplete
MB07H2	22	Female	30-39 yrs	near complete
MB07H2	23	?	Adult	not excavated
MB07H2	24	Female?	40-49 yrs	near complete
MB07H2	25	?	Adult	not excavated
MB07H2	26		18 mths +/- 5 mths	near complete
MB07H2	27	Male	30-39 yrs	near complete
MB07H2	28		neonate	near complete
MB07H2	29		7 yrs +/- 9 mths	incomplete
MB07H2	30	Male	30-39 yrs	near complete
MB07H2	31		4 yrs +/- 9 mths	near complete
MB07H2	32	Male	<25 yrs	near complete

particularly vulnerable to environmental and cultural pressures, and their response to such pressures can indicate how robust the whole population may have been in buffering against these pressures (Saunders, 2008; Lewis and Gowland, 2007; Bogin 1999; Goodman and Armelagos, 1989).

The subadult section of Man Bac is also worthy of further detailed examination given their high mortality (46/78, 59%). Figure 2.1 (and Table 2.2) shows the breakdown of the subadults into specific age classes. The 1-4 year age group shows the highest mortality (21/78, 26.9%), with the infant age class at 20.5% (16/78). It would perhaps be more typical to see the highest mortality in the infant age class representing the most vulnerable period in the first year (typically the first month) of life (Goodman and Armelagos, 1989). Table 2.3 and Figure 2.2 provide a demographical comparison between Man Bac and other Southeast Asian samples.

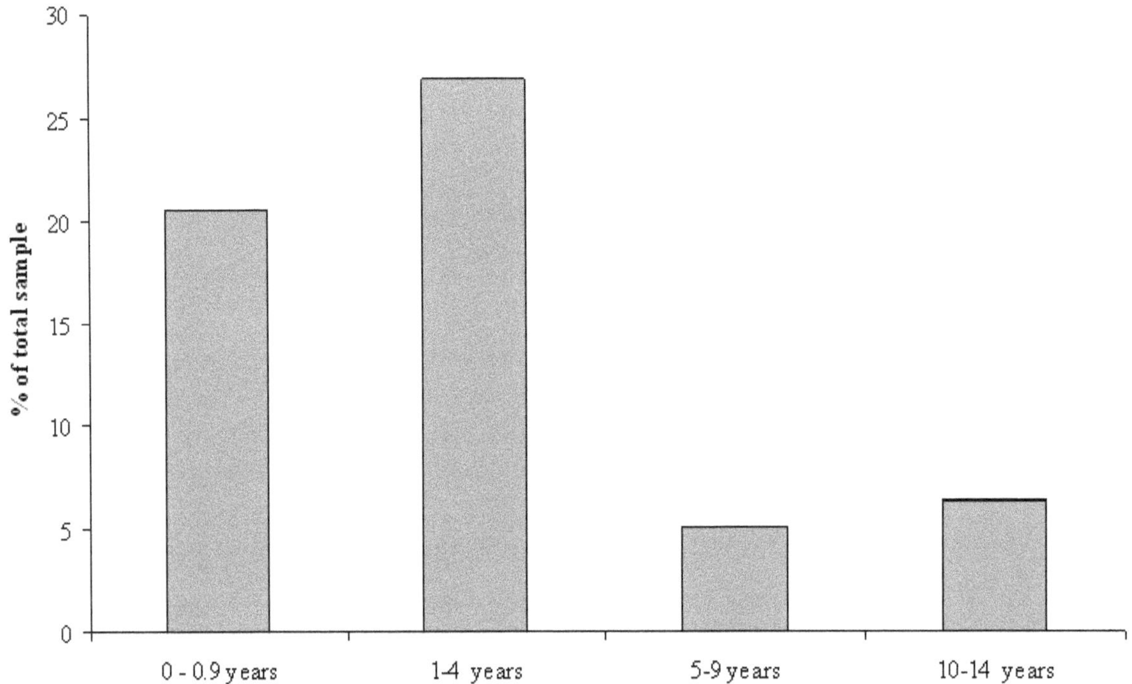

Figure 2.1 Age specific mortality for Man Bac subadults as a percentage of the total sample (n=78).

Table 2.2 Demographic profile of the skeletal remains from Man Bac.

Age	Number	%							Notes
0 - 0.9	16	20.5							
1-4	21	26.9							
5-9 years	4	5.1							
10-18	5	6.4							1
Subtotal	46	59.0							
			Female	%	Male	%	?sex	%	
18-29	11	14.1	4		7				2
30-39	9	11.5	2		5		2		
40-49+	7	9.0	4		3				
50+	2	2.6	2		0				
Unknown	3	3.8	0		0		3		
Subtotal	32	41.0	12	37.5	15	46.9	5	15.6	
Total	78								3

Notes:

1: Includes those aged 12-18 years and 13-18 years

2: Includes adults aged 15-29 years, 17-18 years, 18-24 years, 18-25 years

3: Excludes MB05 M33 and M35; MB07H2 M25, M23 and M11

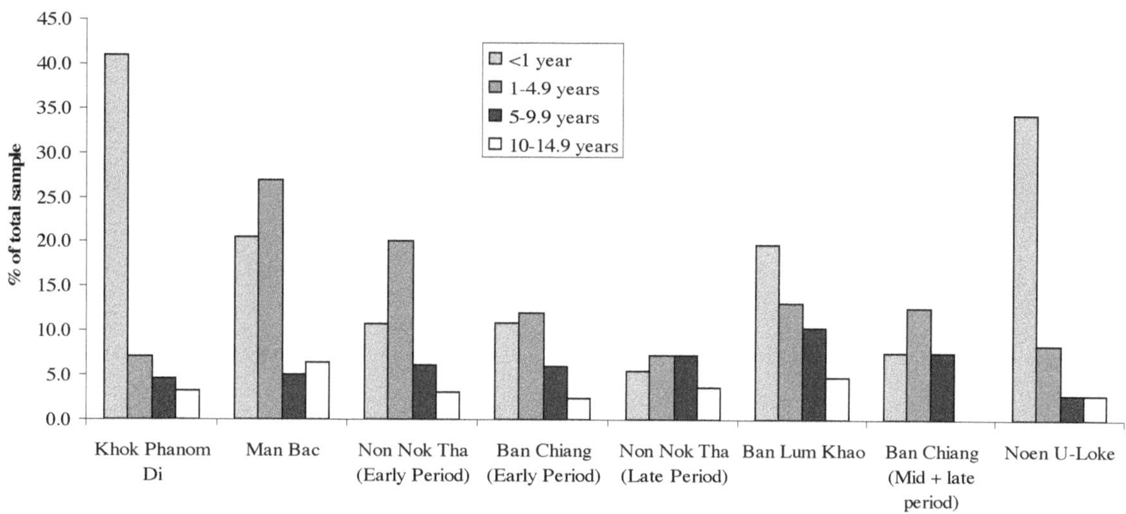

Figure 2.2 Subadult mortality across prehistoric Southeast Asia.

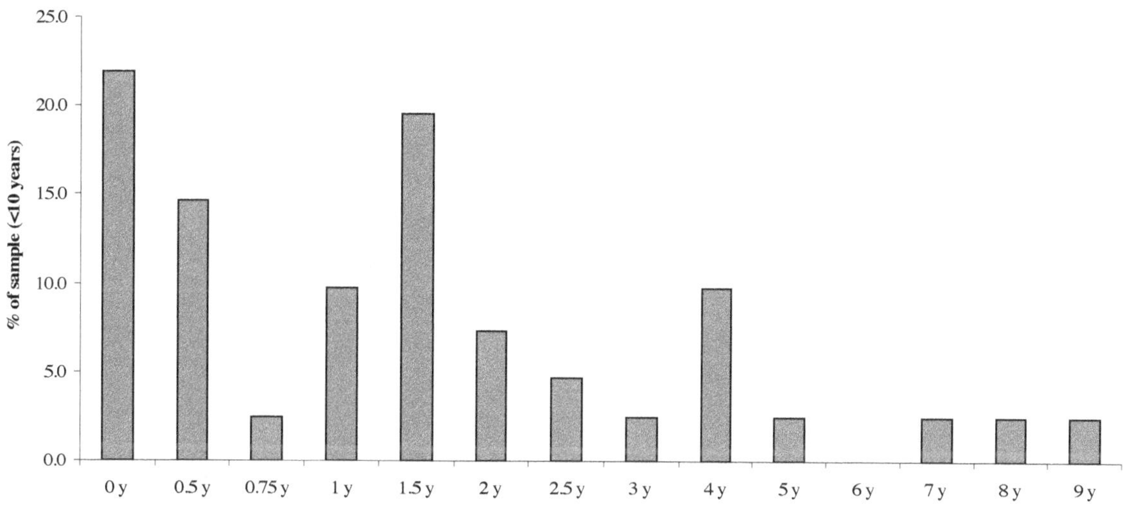

Figure 2.3 Subadult mortality by age at Man Bac (as a proportion of the total of 0-9 year olds, N= 41).

Evidence from three sites, Khok Phanom Di, Ban Lum Khao, and Noen U-Loke, that span the neolithic, Bronze and Iron Ages respectively, indicate that all have higher mortality rates in the less than 1 year of age class, with a decrease with age. Other sites, such as Ban Chiang and Non Nok Tha, have higher mortality in the 1-4 year olds. However, Pietrusewsky and Douglas (2002b) state that subadults in the Ban Chiang sample are underrepresented, although, they are just within the 30% cut-off suggested by Waldron (1994), but at Late Non Nok Tha only 12.5% were subadults (less than 15 years) (Douglas, 1996). All samples of skeletons in Figure 2.2 have particular sampling issues but perhaps Khok Phanom Di, Ban Lum Khao and possibly Noen U-Loke are somewhat more representative.

Given that the sample size is quite adequate for Man Bac, with a good retrieval rate for subadult skeletons, it is likely that taphonomic reasons are not the explanation for the discrepancy in mortality between the first year of life and 1-4 years of age. In order to delve into this further it is useful to break down the data into more specific age ranges. Figure 2.3 shows the breakdown of those subadults less than 10 years of age (N=41). There appear to be peaks at 0 years (N=9) and 1.5

Table 2.3 Palaeodemographic values for Man Bac and other prehistoric Southeast Asian communities (from Oxenham et al., 2008b, but with new data for Man Bac).

Sample	Time period	<5 years	5-9.9 years	10-14.9 years	15-19.9 years	20+ years	JA Ratio	D20+/ D5+	MCM	DR
Khok Phanom Di	Neolithic	48.1	4.5	3.2	5.2	39.0	0.20	0.750	0.091	1.30
Man Bac	Neolithic - Early Bronze	47.4	5.1	3.8	3.8	35.9	0.25	0.737	0.097	1.48
Non Nok Tha (Early)	Neolithic - Bronze Age	27.7	4.8	2.4	2.4	62.6	0.12	0.867	0.047	1.27
Ban Chiang (Early)	Neolithic - Bronze Age	20.6	5.4	2.2	7.9	64.5	0.12	0.851	0.052	0.52
Non Nok Tha (Late)	Bronze Age	5.0	5.0	2.5	1.3	86.3	0.09	0.908	0.032	0.30
Ban Lum Khao	Bronze Age	32.7	10.3	4.7	4.7	47.7	0.30	0.708	0.108	1.38
Ban Chiang (Mid + Late)	Iron Age	17.4	6.5	0.0	10.9	65.3	0.10	0.842	0.055	0.48
Noen U-Loke	Iron Age	43.0	2.8	2.8	3.7	47.7	0.11	0.709	0.058	1.45

Subadults <15 years; JA Ratio – Juvenile Adult Ratio; D20+/D5+ - proportion of those aged over 20years compared to those aged over 5years; MCM - Mean childhood mortality; DR – dependency ratio

Note: In order to make comparisons with these Southeast Asian samples, the Man Bac sample age ranges (Table 2.2) have been modified to fit this format.

years (N=8) in particular, but also a lesser peak at 6 months (N=6) (Figure 2.3). This is similar to the pattern reported by Oxenham et al. (2008a) where it was suggested the effects of weaning could be responsible for the second mortality peak at 18 months. It is well accepted that the risk of dying is highest at birth and within the first week or so of birth (see discussion in Halcrow et al., 2008: 388), therefore a peak at the neonate period is not surprising and conforms to expectations. Peaks at 6 months and 1.5 years may well relate to other well known phenomena, those of the introduction of solid foods at 6 months and weaning around 2-4 years (Lewis and Roberts, 1997). The weaning period is known to be associated with significant risk of morbidity and mortality for a number of factors. This is the time period where breast milk is removed from the infant diet and replaced with foodstuffs likely to contain a higher pathogen load and lead to gastrointestinal infection and diarrhoea which can be a serious disease in young children. The infant is also now required to develop its own antibodies to new diseases and can no longer rely on antibodies from breast-milk. In addition, the infant's immature gastrointestinal system is required to adapt to digesting larger amounts of these new foodstuffs and can result in calorie deficiencies and also induce diarrhoea (Lewis and Roberts, 1997; Goodman and Armelagos, 1989).

If the peak at 1.5 years indicates the period of weaning at Man Bac then it may be considered to have occurred quite early in comparison with developing countries today (Lewis and Roberts, 1997). However, it is also possible that the age-at-death estimation methods for Southeast Asian children are inappropriate, as many are based on American or European children. Halcrow et al. (2007) suggest Southeast Asian subadults (in this case older children) are being overaged by 1 or sometimes 2 years when European standards for the permanent dentition, such as Moorrees et al. (1963), are used to estimate age-at-death. Therefore it is quite possible younger Southeast Asian children and infants are also being overaged; at the very least age-at-death estimation methods need to be considered as a factor in the interpretation of the timing of significant events during childhood (Halcrow et al., 2007). It may be possible to combine demographic evidence with enamel hypoplasia presented later (Chapter 7) in order to further investigate a possible peak of stress.

Adults

Thirty two (41%) of the 78 burials excavated were those of adult individuals. As mentioned above there were both males and females identified at a ratio of 1:0.8 (15:12) and this difference is not statistically significant (FET p-value = 0.5867). There were five adult individuals that were not able to be assigned a sex estimate. Three of these could also not be assessed for age. This small number of unknowns is indicative of the excellent preservation of the material.

Table 2.2 indicates the distribution of males and females in each of the major age ranges. Although there are some disparities within the age groups, there were no statistically significant differences in the proportion of males and females within each age range (18-29 years FET p-value = 0.395; 30-39 years FET p-value = 0.286; 40-49+ years FET p-value = 0.347). This is probably at least partly due to the small sample sizes within each class. There were slightly more young adults than middle or older aged adults but the differences were not statistically significant (18-29 years vs 30-39 years FET p-value = 0.783; 18-29 years vs 40-49+years FET p-value

= 0.783; 30-39 years vs 40-49+ years FET p-value = 1.000).

Palaeodemographic Calculations

Palaeodemographic calculations can provide measures of fertility within a population (Jackes 1992). Table 2.3 provides the results of the calculations for the juvenile/adult ratio (JA ratio = ratio of children aged between 5 and 15 years to adults aged 20 years and over) and the mean childhood mortality (average of probability measures 5q5, 5q10, 5q15 from a life table), both of which have increasing values with increasing fertility. The D20+/D5+ ratio (the proportion of individuals living over 20 years to all the individuals that survived to at least 5 years of age) decreases with increasing fertility.

The values for Man Bac in comparison with other Southeast Asian samples indicate that Man Bac has a high JA (0.25), a high MCM (0.097) and a lower D20+/D5+ (0.737) (Table 2.3). This would indicate that Man Bac has a high level of fertility in comparison with other samples. On a broader scale, for example in comparison with worldwide values for JA and MCM indicated in Jackes (1994), the Man Bac values are still high and, like Ban Lum Khao, indicate a population that was rapidly growing. Khok Phanom Di also shows high fertility values (high JA and MCM and low D20+/D5+), and like Man Bac, has a very high sub 5 year old mortality (Table 2.3). The sample from Noen U-Loke also showed a high proportion of individuals dying before 5 years of age, but the JA and MCM values (which do not take this into account) are much lower (0.11 and 0.058 respectively) although the D20+/D5+ value (0.709) is low, similar to Man Bac. These results tend to show that Noen U-Loke is still growing but at a much slower rate than Man Bac (Domett and Tayles, 2006).

The dependency ratio (DR, Table 2.3) shows that Man Bac had the highest value (1.48). This means that were a high number of children per adult. An earlier report on the smaller 2005 Man Bac sample alone indicated an extremely high DR of 4.48 (Oxenham et al., 2008b). After subsequent excavations and an enlargement of the skeletal sample that has been reduced to 1.48 which, although still high, indicates the value of excavating as much of a prehistoric site as possible. Other sites with high DR values include Noen U-Loke (1.45), Ban Lum Khao (1.38), and Khok Phanom Di (1.30).

SUMMARY

The state of preservation of the Man Bac human skeletal material is excellent by Southeast Asian standards with nearly 60% of individuals complete or near complete. The age-at-death of the vast majority of subadult and adult remains was determined using a range of age-appropriate techniques. An unexpected bimodal subadult mortality distribution was noted, with an expected peak among the neonates and an unexpected peak at approximately 18 months of age. If the 18 months peak is not an artefact of the age-at-death determination methods employed, it may be correlated with weaning behaviours. While the majority of adults were aged using either epiphyseal fusion timing or pubic symphyseal morphology, a small subset were aged using dental wear functions developed on a temporally earlier northern Vietnamese sample. Testing of the accuracy of these equations was carried out on known age Man Bac individuals. That these equations could be accurately applied to the Man Bac assemblage suggests that similar tooth wear trajectories existed in both populations. Why this should be the case may be due to the observation that both the Man Bac and mid Holocene Da But communities were primarily hunter-gatherers in very similar environments, rather than agriculturalists. Adult sex estimation suggests an expected ratio of males to females, again supporting the demographic representativeness of the sample.

The demographic reconstruction of the Man Bac sample suggests a community experiencing elevated levels of fertility. The conclusions of an earlier assessment of a much smaller Man Bac sample (see Bellwood and Oxenham, 2008) are reconfirmed by this study, although the extreme values for each of the fertility measures in the earlier study have been substantially revised with a much larger sample. Nonetheless, Man Bac, along with Khok Phanom Di, essentially contemporaneous populations, show levels of fertility consistent with a major economic and/or behavioural shift in the region. These demographic findings are entirely consistent with the elevated levels of physiological disruption and oral disease, and the evidence for population shifts discussed in later chapters.

LITERATURE CITED

Bellwood P, Oxenham M. 2008. The expansions of farming societies and the role of the Neolithic Demographic Transition. In: Bocquet-Appel J-P, Bar-Yosef O, editors. The Neolithic Demographic Transition and its Consequences. Dordrecht: Springer.

Bogin B. 1999. Patterns of Human Growth. Cambridge: Cambridge University Press.

Buikstra JE, Ubelaker DH. 1994. Standards for Data Collection from Human Skeletal Remains. Fayetteville, Arkansas: Arkansas Archaeological Survey.

Cardoso HFV, and Saunders SR. 2008. Two arch criteria of the ilium for sex determination of immature skeletal remains: A test of their accuracy and an assessment of intra- and inter-observer error. Forensic Science International 178:24–29

Domett K. 2001. Health in late prehistoric Thailand. BAR International Series 946. Oxford: Archaeopress.

Domett KM. 2004. The People of Ban Lum Khao. In: Higham CFW, and Thosarat R, editors. The Origins of The Civilization of Angkor Volume I: The Excavation of Ban Lum Khao. Bangkok: The Thai Fine Arts Department. p 113-151.

Domett K, Tayles N. 2006. Human Biology from the Bronze Age to the Iron Age in the Mun River valley of Northeast Thailand. In: Oxenham MF, and Tayles N, editors. Bioarchaeology of Southeast Asia: Human Skeletal Biology of the Late Prehistoric Inhabitants of Tropical Southeast Asia and Island Southwest Pacific. Cambridge: Cambridge University Press. p 220-240.

Douglas MT. 1996. Paleopathology in Human Skeletal Remains from the Pre-Metal, Bronze and Iron Ages, Northeastern Thailand. PhD dissertation. Hawaii: University of Hawaii.

Goodman AH, and Armelagos GJ. 1989. Infant and childhood morbidity and mortality risks in archaeological populations. World Arch 21:225-243.

Halcrow S, Tayles N, Buckley HR. 2007. Age estimation of children from prehistoric Southeast Asia: are the dental formation methods used appropriate? J Anthropol Sci 34:1158-1168.

Halcrow S, and Tayles N. 2008. The bioarchaeological investigation of childhood and social age: problems and prospects. J Arch Meth Theory 15:190-215.

Halcrow SE, Tayles N, Livingstone V. 2008. Infant death in late prehistoric Southeast Asia. Asian Perspectives 47:371-404.

Jackes M. 1992. Palaeodemography: Problems and Techniques. In: Saunders SR, Katzenberg MA, editors. Skeletal Biology of Past Peoples: Research Methods. New York: Wiley-Liss, Inc. p 189-224.

Jackes M. 1994. Birth rates and bones. In: Herring A, Chan L, editors. Strength in Diversity: A Reader in Physical Anthropology. Toronto: Canadian Scholars' Press. p 155-185.

Kamalanathan GS, Hauck HM, Kittiveja C. 1960. Dental development of children in a Siamese village, Bang Chan, 1953. J Dent Res 39(3):455-461.

Lewis M, Roberts C. 1997. Growing pains: the interpretation of stress indicators. International Journal of Osteoarchaeology 7: 581-586.

Lewis M, Gowland RL. 2007. Brief and precarious lives: Infant mortality in contrasting sites from Medieval and Post-Medieval England (AD 850-1859). Am J Phys Anthropol 134:117-129.

Moorrees CFA, Fanning EA, Hunt EE. 1963. Formation and resorption of three deciduous teeth in children. Am J Phys Anthropol 21:205-213.

Oxenham MF. 2000. Health and Behaviour During the Mid-Holocene and Metal Period of Northern Viet Nam. PhD thesis, Northern Territory University, NTU, (Charles Darwin University), Australia.

Oxenham MF, Ross KW, Nguyen KD, Matsumura H. 2008a. Children through adult eyes: subadult identity in neolithic Vietnam. A paper presented at the 6th World Archaeological Congress. Theme: Peopling the past, individualizing the present: bioarchaeological contributions in a global context. Session: a cast of thousands: children in the archaeological record, 29th June-4th July, Dublin, Ireland.

Oxenham MF, Matsumura H, Domett K, Nguyen KT, Nguyen KD, Nguyen LC, Huffer D, and Muller S. 2008b. Health and the experience of childhood in Late Neolithic Viet Nam. Asian Perspectives 47:190-209.

Pietrusewsky M, Douglas MT. 2002a. Ban Chiang: a prehistoric village site in northeast Thailand. I: The human skeletal remains. Philadelphia: Museum of Archaeology and Anthropology, University of Pennsylvania.

Pietrusewsky M, Douglas MT. 2002b. Intensification of agriculture at Ban Chiang: Is there evidence from the skeletons? Asian Perspectives 40:157-178.

Saunders SR. 1992. Subadult skeletons and growth related studies. In: Saunders SR, Katzenberg MA, editors. Skeletal Biology of Past Peoples: Research Methods. New York: Wiley-Liss. p 1-20.

Saunders SR. 2008. Juvenile skeletons and growth-related studies. In: Katzenberg MA, Saunders SR, editors. Biological Anthropology of the Human Skeleton. Hoboken. New Jersey: John Wiley & Sons. p 117-147.

Scheur L, Black S. 2000. Developmental Juvenile Osteology. San Diego: Academic Press.

Scott EC 1979. Dental wear scoring technique. Am J Phys Anthropol 51:213-218.

Tayles N, Domett K, Hunt V. 1998. The People of Nong Nor. In: Higham CFW, Thosarat R, editors. The Excavation of Nong Nor: A Prehistoric Site in Central Thailand. Dunedin: Department of Anthropology, University of Otago. p. 321-368.

Tayles N, Halcrow S, Domett K. 2007. The People of Noen U-Loke. In: Higham CFW, Kijngam A, Talbot S, editors. The Origins of the Civilization of Angkor Volume II: The Excavation of Noen U-Loke and Non Muang Kao. Bangkok: Thai Fine Arts Department. p 244-304.

Waldron T. 1994. Counting the dead. The epidemiology of skeletal populations. Chichester: John Wiley and Sons.

White TD. 2000. Human Osteology. San Diego: Academic Press.

Ubelaker DH. 1987. Estimating age at death from immature human skeletons: an overview. J For Sci 32(5):1254-1263.

Vlak D, Roksandic M, Schillaci MA. 2008. Greater sciatic notch as a sex indicator in juveniles. Am J Phys Anthropol 137:309-31.

3
Quantitative Cranio-Morphology at Man Bac

Hirofumi Matsumura

Department of Anatomy, Sapporo Medical University, Japan

The aim of this chapter is to quantitatively assess cranial morphology of the Man Bac assemblage and explore any evidence for biological relationships between Man Bac and surrounding populations dating from prehistoric through to more recent times. An assessment of the morphometric affinities presented here addresses the issue of the origin of this group of neolithic people in northern Vietnam.

MATERIALS AND METHODS

Of the human remains excavated from Man Bac between 1999 and 2007, comprehensive sets of cranial and mandibular measurements were available for 17 adult males and 13 females. A maximum of 32 measurements, and five indices, were recorded for each cranium and mandible (a complete set was not always available) based on Martin's definitions (see Bräuer, 1988) and are presented in the Appendix to this chapter. Male skulls representative of the sample are shown in Figure 3.1 and Table 3.1 provides a basic statistical description of the cranial and mandibular series based on these measurements. Note that Table 3.1 provides summary data for two recognised Man Bac male subgroups (this is discussed below) as well as the total male sample and complete (both sexes) sample.

Of the cranial assemblage, 17 male skulls are utilised for craniometric analysis. The craniometric affinities among the comparison samples are assessed using Q-mode correlation coefficients (Sneath and Sokal, 1973). The comparative samples are listed in Table 3.2, which includes both archaeological and modern samples from East/Southeast Asia and the Pacific. To aid interpretation of phenotypic affinities between the samples, un-rooted tree diagrams were generated using the Neighbour Joining method (Saitou and Nei, 1987), applied to the distance (1-r) matrix of Q-mode correlation coefficients (r). This procedure was undertaken using the software package "Splits Tree Version 4.0" provided by Huson and Bryant (2006).

RESULTS

The majority of the Man Bac cranial series can be characterised as having a relatively narrow and flat face with round orbits (Figure 3.1 left). However, some individuals present quite different features, such as a dolichocephalic cranium with a prominent glabella and a low and wide face (Figure 3.1, right). Since such visually clear morphological variation among the Man Bac cranial series implies the

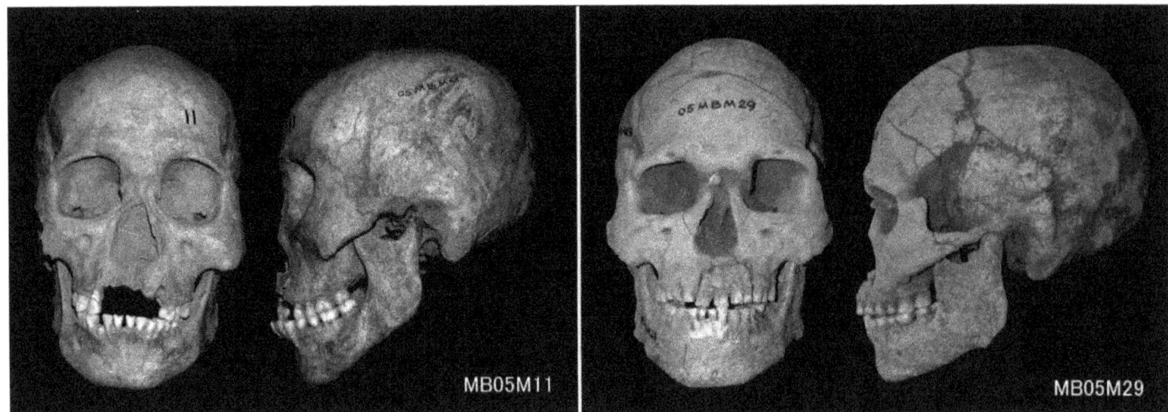

Figure 3.1 Views of representative skulls from the site of Man Bac.

possibility of genetic heterogeneity, multivariate craniometric comparisons were carried out in order to both confirm and assess the degree of phenotypic variation within the Man Bac adult male series.

In order to utilise the greatest number of individual specimens, given differential preservation and consequent availability of measurements, nine cranial measurements (Martin's 1, 8, 9, 45, 48, 51, 52, 54 and 55) were selected for calculating the Q-mode correlation coefficient. A complete data set without any missing values was derived for 14/17 of the Man Bac male adult series. Table 3.3 (upper right triangle) provides the distance matrix (1-r) transformed from the Q-mode correlation coefficients (r) thus computed.

The result of the Neighbour Joining analysis applied to the distance matrix of Q-mode correlation coefficients is presented in Figure 3.2. The non-rooted tree in this figure depicts an apparent divergence into two major clusters. One consists of the majority of Man Bac individuals (n = 9) which have branched from neolithic, Metal Period and modern samples from East and Southeast Asia. In this clustering pattern seven Man Bac specimens are more closely associated with the Metal Period Dong Son Vietnamese among the range of modern Southeast Asian samples. This sub-cluster contains the neolithic and Metal Period sample of Ban Chiang crania from northern Thailand. The modern Vietnamese sample, together with another Man Bac specimen, branches from this sub-cluster. The Hoabinhian, Australian and Melanesian samples form another major cluster, which includes five Man Bac individuals, together with the Jomon.

In the next step of analysis cranial affinities were assessed using group-average measurements for the two identified Man Bac samples, utilising 17 male cranial specimens. As the analysis using the individual dataset yielded a clear dichotomisation of the Man Bac adult male series, five individuals were treated as a separate group designated 'Man Bac 1', while the remaining specimens were combined with other incomplete male crania and labelled 'Man Bac 2'. The descriptive statistics for these two sub-samples is given in Table 3.2. Q-mode correlation coefficients were calculated using the new cranial data set consisting of 16 measurements (Martin's: M1, M8, M9, M17, M43(1), M43c, M45, M46b, M46c,

M57, M57a, M48, M51, M52, M54, M55). A distance matrix (1-r) transformed from the Q-mode correlation coefficients thus calculated are given in Table 3.3 (left lower triangle). Figure 3.3 depicts an un-rooted tree using the Neighbour Joining method based on the distance matrix of Q-mode correlation coefficients. 'Man Bac 1' is tightly connected with the early Vietnamese samples including the Hoabinhian, Bac Son and Da But (Con Co Ngua) series. These samples form a mega cluster together with the Australo-Melanesian, Gua Cha (Malaysia) and Jomon (Japan) samples. On the other hand, the 'Man Bac 2' and Dong Son Vietnamese, forming a sub-cluster together with Ban Chiang (Thailand) and modern Vietnamese, are linked closely with another major cluster consisting of the neolithic to modern period samples from East/Southeast Asia, as well as the Neolithic Weidun sample from China.

Table 3.1 Cranial and mandibular measurements (mm) and indices for the Man Bac people.

Martin's No and measurement	Man Bac 1 (males)			Man Bac 2 (males)			Man Bac all (males)			Man Bac all (females)		
	n	M	SD	n	M	SD	n	M	SD	n	M	SD
1 Max. cranial length	5	185.6	4.0	11	179.7	5.7	16	181.6	5.8	12	174.5	6.0
5 Basion-nasion length	5	102.6	4.5	5	99.4	6.7	10	101.0	5.6	5	95.0	4.8
8 Max. cranial breadth	5	143.6	6.8	11	142.4	5.7	16	142.8	5.8	12	134.8	3.3
9 Min. frontal breadth	5	100.4	5.9	11	98.5	4.0	16	99.1	4.6	12	94.0	5.0
10 Max. frontal breadth	5	114.6	9.0	8	118.8	5.8	13	117.2	7.2	10	113.8	4.6
12 Max. occipital breadth	5	113.4	5.9	8	109.4	4.9	13	110.9	5.4	8	108.5	3.9
17 Basion-bregma height	5	143.8	6.3	6	140.0	8.9	11	141.7	7.7	5	134.2	10.7
29 Frontal chord	5	116.4	2.7	8	111.5	3.5	13	113.4	3.9	10	107.3	5.6
30 Parietal chord	5	118.6	3.8	8	114.5	6.7	13	116.1	6.0	10	111.3	7.8
31 Occipital chord	5	105.0	2.8	7	106.0	5.7	12	105.6	4.6	6	100.8	5.8
40 Basion-prosthion length	5	102.2	0.8	5	99.0	2.7	10	100.6	2.5	5	90.8	3.3
43 Upper facial breadth	5	113.0	5.1	11	110.5	5.7	16	111.3	5.5	8	105.5	4.1
45 Bizygomatic breadth	5	142.0	6.2	10	141.6	6.9	15	141.7	6.5	8	132.5	7.1
46 Bimaxillary breadth	5	109.0	4.5	10	108.2	6.1	15	108.5	5.5	6	103.3	4.1
48 Upper facial height	5	68.8	3.1	10	71.0	4.0	15	70.3	3.8	7	68.0	3.4
51 Orbital breadth	5	44.2	1.9	9	41.6	2.1	14	42.5	2.3	8	41.0	1.3
52 Orbital height	5	33.0	2.0	11	35.5	1.8	16	34.7	2.1	9	34.6	1.3
54 Nasal breadth	5	28.8	3.1	11	28.1	1.4	16	28.3	2.0	8	25.9	2.6
55 Nasal height	5	51.0	2.5	10	54.1	3.5	15	53.1	3.5	7	51.6	3.1
60 Upper alveolar length	5	54.2	0.8	10	53.0	3.0	15	53.4	2.5	7	52.4	3.7
61 Upper alveolar breadth	5	66.6	2.1	10	67.0	3.5	15	66.9	3.0	6	63.5	3.5
8:1 Cranial index	5	77.5	5.2	11	79.3	4.2	16	78.7	4.4	10	77.9	4.4
48:45 Upper facial index	5	48.6	4.2	10	45.7	16.3	15	46.6	13.4	6	52.4	1.3
43(1) Frontal chord	5	104.2	5.1	8	103.5	6.6	13	103.8	5.9			
43c Frontal subtense	5	16.3	5.1	8	14.6	3.2	13	15.2	3.9			
57 Simotic chord	5	9.6	1.6	5	10.2	2.6	10	9.9	2.1			
57a Simotic subtense	4	3.2	0.7	5	3.0	1.9	9	3.1	1.4			
46b Zygomaxillary chord	5	107.6	6.1	6	107.0	6.8	11	107.3	6.2			
46c Zygomaxillary subtense	5	22.6	4.1	6	22.5	4.7	11	22.5	4.2			
43c:43(1) Frontal index	5	15.8	5.5	8	14.1	2.8	13	14.7	3.9			
57a:57 Simotic index	4	33.6	1.7	5	26.6	12.4	9	29.7	9.6			
46c:46b Zygomaxillary index	5	21.1	4.2	6	20.9	3.8	11	21.0	3.8			
66 Bigonial breadth	5	104.6	7.5	9	105.9	9.4	14	105.4	8.5	8	96.3	6.3
68 Mandibular length	5	83.2	1.6	9	80.8	4.1	14	81.6	3.5	8	74.5	5.5
69 Symphyseal height	5	33.2	4.1	9	33.3	3.4	14	33.3	3.5	7	32.9	2.3
70 Ramus height	5	68.0	6.3	8	67.6	5.6	13	67.8	5.6	8	59.1	6.8
71 Ramus breadth	5	39.2	1.6	9	37.0	3.2	14	37.8	2.9	8	35.5	1.8

Man Bac 1 consists of 05M29, 07H1M8, 07H2M27, 07H2M30 and 07H2M32.
Man Bac 2 consists of the other 12 individuals given in Appendix 3.1.

Table 3.2 Comparative prehistoric cranial samples from East/Southeast Asia.

	Period	Data Source	Storage	Remark
Hoabinhian Vietnam	Late Pleistocene - Early Holocene	H.M.	IAH, MHO	Sites of Mai Da Nuoc, Mai Da Dieu, Lang Gao, Lang Bon in northern Vietnam (Cuong, 1986, 2007)
Bac Son	Early Holocene (c. 8,000-7,000 BP)	H.M.	MHO	Sites of Pho Binh Gia, Lang Cuom, Cua Gi, Dong Thuoc in northern Vietnam
Con Co Ngua	Early Neolithic Da But culture (c.5,000 BP)	Cuong, 2003	IAH, MHO	Sites of Con Co Ngua 'Thuy, 1990) and Da But in Than Hoa Prov. northern Vietnam; M43(1),43c,46b,46c,57,57a by H.M.
Dong Son	Early Metal Age (3,000-1,700 BP)	Cuong, 1996, and H.M.	IAH, CSPH	Sites of Vinh Quang, Chau Son, Doi Son, Quy Chu, Thieu Duong, Nui Nap, Dong Mom, Minh Duc, Dong Xa in northern Vietnam
Gua Cha	Hoabinhian (c.8,000-6,000 BP)	H.M.	UCB	Site in Kelantan Prov., Malaysia; specimen No. H12; Sieveking (1954)
Ban Chiang	Neolithic-Bronze Age (c. 3,500-1,800 BP)	Pietrusewsky and Douglas, 2002	UHW, SAC	Site in Udon Thani Prov., Thailand; M51by Hanihara, 1993; M43(1),43c,46b,46c,57,57a by H.M.
Weidun	Neolithic (c.7,000-5,000 BP)	Nakahashi et al., 2002.		Nakahashi and Li, 2002
Anyang	Bronze - Iron Age (c. 3,300 BP)	IHIA 1982; Han and Qi, 1985	AST	M43(1),43c,46b,46c,57,57a by H.M.
Jiangnan	Zhou-Western Han (2,770-1,992 BP)	Nakahashi et al., 2002		Jiangnan Region, Sth China; Nakahashi and Li, 2002
Jomon	Late Jomon (c. 5,000-2,300 BP)	Hanihara, 1993, 2000		Sites in Japan; Yamaguchi, 1982
Yayoi	Early Metal Age (2,800-1,700 BP)	Ishida, 1992; Nakahashi, 1993		
Bunun	Modern	Pietrusewsky and Chang, 2003	NTW	M43(1),43c,45,46b,46c,48,51,55,57,57a by H.M.
Cambodia	Modern	H.M.	MHO	-
Celebes	Modern	Pietrusewsky, 1981; Hanihara, 2000	BMNH	M17,45,48,51 by H.M.
Dayak	Modern	Yokoh, 1940; Hanihara 2000		
Hainan	Modern	Howells, 1989	NTW	M43(1),43c,46b,46c,48,51,55,57,57a by H.M
Java	Modern	Pietrusewsky, 1981; Hanihara, 2000	BMNH	M17,45,48,51 by H.M.
Laos	Modern	Cuong, 1996	MHO	M43(1),43c,46b,46c,57,57a by H.M.
Myanmar	Modern	Pietrusewsky, 1981; Hanihara, 2000	BMNH	M17,45,48,51 by H.M.
North China	Modern	Hanihara, 1993, 2000		
Philippines	Modern	Suzuki et al., 1993	NMP	M43(1),43c,46b,46c,57,57a by H.M.
Sumatra	Modern	Pietrusewsky, 1981; Hanihara, 2000	BMNH	M17,45,48,51 by H.M.
Thai	Modern	Sangvichien, 1971; Hanihara, 2000		
Vietnam	Modern	H.M.	MHO	
Australia	Modern	Hanihara, 1993	BMNH, UCB	M43(1),43c,46b,46c,57,57a by H.M.
Melanesia	Modern	Hanihara, 1993, 2000		
Loyalty	Modern	H.M.	MHO	

In 'Remarks': M=Martin's cranial measurment number, In 'Storage': institutions of materials studied by H.M. (H. Matsumura) AST=Academia Sinica of the Republic of China in Taipei; BMNH=Department of Palaeontology, Natural History Museum, London; CSPH=Center for South East Asian Prehistory, Hanoi; IAH=Department of Anthropology, Institute of Archaeology, Hanoi; MHO=Laboratoire d'Anthropologie Biologique, Musée de l'Homme, Paris; NMP=Department of Archaeology, National Museum of the Philippines, Manila; NTW=Department of Anatomy, National Taiwan University, SAC=Princess Maha Chakri Sirindhorn Anthropology Centre, Bangkok; UCB=Department of Biological Anthropology, University of Cambridge; UHW=Department of Anthropology, University of Hawaii.

Figure 3.2 An un-rooted tree of neighbour joining analysis applied to the distances of Q-mode correlation coefficients between the Man Bac individuals and comparative samples (based on 9 male cranial measurements).

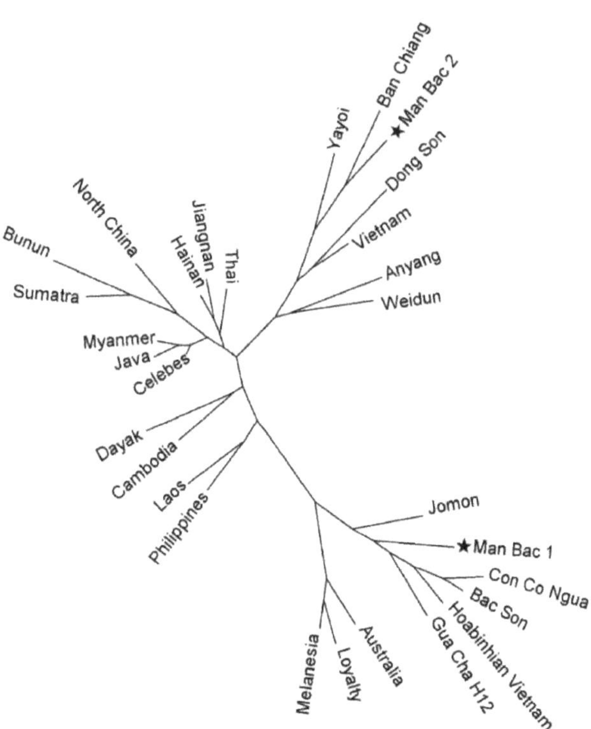

Figure 3.3 An un-rooted tree of neighbour joining analysis applied to the distances of Q-mode correlation coefficients between the two groups of Man Bac individuals and comparative samples (based on 16 male cranial measurements).

DISCUSSION

Archaeological and linguistic research has linked the dispersal of Austronesian and Austroasiatic language families with the demographic expansion of rice cultivating people during the Neolithic period, and have sought the ultimate homeland of these language and population dispersals in southern China and Taiwan (Renfrew, 1987, 1989, 1992; Bellwood, 1991, 1993, 1996, 1997; Bellwood et al, 1992; Blust, 1996a, b; Glover and Higham, 1996; Higham, 1998, 2001; Bellwood and Renfrew, 2003; Diamond and Bellwood, 2003). With respect to analyses of human skeletal data, the 'Two Layer' model, is instrumental in understanding the population history of mainland Southeast Asia (e.g. Callenfels, 1936; Mijsberg, 1940; Von Koenigswald, 1952; Coon, 1962; Jacob, 1967; Brace et al, 1991, Matsumura and Hudson, 2005; Matsumura, 2006). This model hypothesises that Southeast Asia was initially occupied by indigenous populations, akin to modern Australo-Melanesians, that later exchanged genes with immigrants from North and/or East Asia, leading to the formation of present day Southeast Asian populations. However, some recent cranial and dental studies question this model, alternatively advocating regional continuity or local evolutionary scenarios in order to account for the region's population history (e.g. Turner, 1990; Hanihara, 1994; 2006; Pietrusewsky, 1994, 2005, 2006, 2008). The question arises as to whether these opposing models address the timing and scale of the population dispersal under debate with regard to the expansion of Austroasiatic and Austronesian languages and rice farming cultures, and whether there was a resultant mixture with replacement of extant populations.

Table 3.3 Distance (1-r) matrices of Q-mode correlation coefficients (r), based on 9 cranial measurements (upper right triangle), and on 16 cranial measurements (lower left triangle).

	2	3	4	5	6	7	8	9	10	11	12	13	14	15	16	17
1 MB99M3	0.86	1.50	1.40	0.77	1.32	1.15	1.18	1.65	1.56	0.93	0.80	1.21	1.08	1.72	0.98	1.09
2 MB01M5		0.55	0.43	1.12	0.78	0.37	1.28	1.08	1.32	0.49	0.90	0.86	1.12	1.07	0.98	1.43
3 MB01M9			0.47	0.75	1.13	0.47	0.75	0.62	0.69	0.65	1.59	0.61	0.80	0.67	0.96	1.37
4 MB01M10				1.15	0.83	0.45	1.30	0.55	0.73	0.55	1.22	0.64	1.09	0.88	1.18	1.17
5 MB05M11					1.92	1.00	1.23	1.27	0.80	0.87	1.55	1.30	1.36	1.25	1.31	1.23
6 MB05M29						0.97	0.96	0.60	1.36	1.30	0.47	0.67	0.63	0.55	0.49	0.65
7 MB05M31							1.07	0.67	0.65	0.83	1.49	0.56	0.91	0.83	1.36	1.11
8 MB07H1M5								0.95	0.76	0.96	1.21	0.69	0.40	0.92	0.84	1.16
9 MB07H1M8									0.73	1.40	1.48	0.21	0.44	0.50	0.77	0.53
10 MB07H2M1										0.86	1.56	0.97	1.16	0.77	1.68	1.08
11 MB07H2M10											0.93	1.25	1.44	1.52	1.46	1.92
12 MB07H2M27												1.53	1.34	1.08	0.78	1.05
13 MB07H2M30													0.15	0.83	0.64	0.68
14 MB07H2M32														0.78	0.37	0.59
15 Hoabinhian Mai Da Dieu 16															0.66	0.48
16 Hoabinhian Mai Da Nuoc																0.56
17 Hoabinhian Lang Bon																

Table 3.3 (Continued).

	18	19	20	21	22	23	24	25	26	27	28	29	30	31	32	33	34	35	36	37	38	39	40	41	42	43	44	45	46
1 MB99M3			1.01	0.89	1.27	0.88	0.91	0.69	0.87	1.26	0.76	1.03	1.13	0.64	1.23	1.04	0.96	1.50	0.80	1.31	0.74	0.82	0.93	1.08	0.36	0.67	1.15	0.20	1.44
2 MB01M5			0.75	1.39	1.22	0.85	1.00	1.00	0.97	1.08	1.08	0.20	0.70	0.60	1.13	1.14	1.00	1.16	1.18	1.33	1.48	0.91	1.28	0.74	0.89	0.85	0.78	0.79	1.02
3 MB01M9			1.20	1.26	1.05	1.12	1.39	0.83	1.05	0.88	0.55	0.41	0.74	0.93	1.21	1.01	0.83	0.92	1.03	1.17	1.56	1.18	1.39	0.52	1.68	0.96	0.35	1.34	1.16
4 MB01M10			1.11	1.50	0.92	0.76	0.92	1.19	1.09	0.77	1.27	0.54	0.78	1.27	0.94	0.92	1.39	0.66	1.06	1.06	1.31	0.96	1.17	0.68	0.98	1.14	0.65	1.47	1.04
5 MB05M11			1.32	1.25	1.15	1.09	0.74	0.76	1.03	1.08	0.25	0.91	1.20	0.76	1.35	0.99	0.56	1.64	0.89	1.05	1.08	1.02	0.84	1.18	0.93	0.77	0.45	1.06	1.37
6 MB05M29			0.71	0.72	0.60	1.00	1.10	1.40	1.28	0.69	1.67	0.92	0.49	1.30	0.43	1.37	1.50	0.38	1.39	0.76	0.89	1.26	1.39	1.00	1.17	1.42	1.63	1.04	0.67
7 MB05M31			0.68	1.80	1.32	1.23	1.11	0.48	0.67	1.34	0.94	0.80	0.87	0.64	1.21	0.80	0.77	0.81	0.58	1.70	1.84	0.70	1.35	0.43	1.16	0.41	0.34	0.86	0.90
8 MB07H1M5			1.07	0.65	1.29	1.18	1.91	0.53	0.58	1.26	0.73	1.21	1.21	0.90	1.34	0.60	0.75	0.72	0.59	1.33	1.19	0.90	1.08	0.39	1.67	0.77	0.83	0.80	1.01
9 MB07H1M8			1.34	1.05	0.49	1.05	1.18	0.99	1.38	0.43	1.12	1.16	0.49	1.65	0.59	1.25	1.50	0.15	0.96	0.89	1.10	1.31	1.63	0.85	1.52	1.31	0.94	1.42	0.97
10 MB07H2M1			0.90	1.59	1.31	1.21	1.09	0.69	0.50	1.39	0.96	1.38	1.50	0.99	1.32	0.31	0.62	0.80	0.45	1.30	1.50	0.59	0.65	0.61	1.27	0.59	0.31	1.46	0.71
11 MB07H2M10			1.07	1.30	1.66	0.43	1.18	0.99	0.52	1.28	1.07	0.44	1.51	0.61	1.71	0.46	0.77	1.39	0.90	1.29	1.38	0.46	0.53	0.78	0.80	0.81	0.55	1.02	0.89
12 MB07H2M27			0.59	0.62	0.95	0.76	0.71	1.63	1.07	1.02	1.66	0.85	1.05	0.91	0.74	1.17	1.16	1.20	1.54	0.59	0.61	1.00	0.64	1.45	0.56	1.35	1.77	0.89	0.63
13 MB07H2M30			1.41	0.95	0.72	1.03	1.55	0.60	1.19	0.60	0.87	1.07	0.48	1.44	0.83	1.17	1.50	0.21	0.70	1.29	1.13	1.21	1.81	0.49	1.47	1.05	0.86	0.91	1.31
14 MB07H2M32			1.34	0.55	0.67	1.22	1.71	0.61	1.21	0.67	0.74	1.19	0.49	1.38	0.76	1.26	1.39	0.32	0.79	1.16	0.90	1.38	1.78	0.52	1.60	1.10	1.13	0.74	1.39
15 Hoabin. Mai Da Dieu 16			0.64	1.02	0.45	1.69	0.92	1.30	1.38	0.83	1.04	0.97	0.47	1.25	0.36	1.37	1.03	0.61	1.38	0.65	1.10	1.58	1.37	0.87	1.61	1.25	1.10	1.54	0.86
16 Hoabin. Mai Da Nuoc			1.25	0.22	0.28	1.19	1.22	1.32	1.74	0.36	0.83	0.75	0.20	1.45	0.35	1.80	1.54	0.73	1.58	0.51	0.44	1.85	1.69	1.07	1.36	1.63	1.64	0.96	1.47
17 Hoabin. Lang Bon			0.95	0.79	0.24	1.54	0.73	1.12	1.54	0.66	1.07	1.53	0.44	1.59	0.17	1.50	1.48	0.47	1.11	0.67	0.53	1.56	1.46	1.11	1.04	1.26	1.46	1.11	1.24
18 Man Bac 1																													
19 Man Bac 2	0.78																												
20 Anyang	0.97	1.12		1.42	1.28	1.51	0.75	1.01	0.61	1.69	1.35	1.03	1.13	0.42	0.92	0.82	0.53	1.18	1.01	1.25	1.40	0.76	0.74	0.85	0.86	0.55	1.04	0.83	0.51
21 Australia	1.51	0.94	1.54		0.54	0.87	1.30	1.34	1.47	0.48	0.92	1.03	0.72	1.42	0.67	1.45	1.40	0.89	1.44	0.44	0.25	1.54	1.20	1.28	1.22	1.64	1.71	0.93	1.28
22 Bac Son	1.00	0.59	1.30	0.78		1.25	0.69	1.56	1.93	0.20	1.02	1.00	0.20	1.80	0.08	1.81	1.71	0.58	1.62	0.25	0.35	1.89	1.55	1.32	1.17	1.77	1.57	1.46	1.37
23 Ban Chiang	0.44	0.73	0.78	1.42	0.98		1.04	1.21	0.92	0.69	1.39	0.80	1.31	1.12	1.40	0.83	1.29	1.03	1.04	0.92	0.88	0.60	0.72	1.37	0.64	1.29	1.10	0.99	0.78
24 Bunun	1.52	1.66	0.81	0.91	1.07	1.37		1.54	1.36	0.90	1.28	0.99	0.94	1.07	0.60	1.30	1.08	1.35	1.43	0.55	0.74	1.12	0.71	1.67	0.36	1.21	1.24	1.32	0.88
25 Cambodia	0.99	1.01	1.19	1.14	1.34	1.37	1.30		0.40	1.53	0.51	1.34	1.28	0.57	1.58	0.54	0.57	0.95	0.11	1.89	1.53	0.53	1.18	0.39	1.18	0.17	0.38	0.37	1.16
26 Celebes	1.23	1.46	0.69	1.25	1.75	1.25	0.76	0.52		1.87	1.10	1.26	1.82	0.34	1.80	0.08	0.35	1.17	0.25	1.77	1.63	0.10	0.45	0.55	0.91	0.19	0.47	0.61	0.56
27 Da But	0.74	0.49	1.47	0.85	0.20	0.62	1.33	1.35	1.81		1.03	0.79	0.34	1.80	0.43	1.71	1.78	0.60	1.58	0.30	0.43	1.69	1.52	1.42	1.18	1.88	1.45	1.43	1.31
28 Dayak	1.29	1.38	1.18	0.85	1.04	1.52	0.81	0.52	0.78	1.29		0.87	0.87	0.82	1.23	1.14	0.66	1.36	0.82	1.20	1.06	1.30	1.30	0.73	1.37	0.75	0.53	0.82	1.67
29 Dong Son	0.62	0.87	1.09	1.12	1.05	0.89	1.14	1.15	0.96	0.98	1.11		0.65	0.78	1.05	1.33	1.02	1.35	1.57	0.86	1.19	1.26	1.20	0.97	1.10	1.26	0.90	1.13	1.16
30 Gua Cha	1.00	0.51	1.21	0.61	0.51	1.26	1.01	1.18	1.65	0.68	0.89	0.91		1.45	0.23	1.88	1.57	0.62	1.54	0.64	0.74	1.84	1.88	1.01	1.30	1.51	1.37	1.11	1.43
31 Hainan	1.34	1.56	0.71	1.13	1.55	1.50	0.69	0.54	0.29	1.68	0.50	0.93	1.34		1.64	0.62	0.15	1.66	0.73	1.61	1.65	0.45	0.67	0.84	0.89	0.25	0.59	0.46	0.69
32 Hoabinhian	1.13	0.63	0.99	0.76	0.15	1.11	0.85	1.44	1.61	0.54	1.09	1.07	0.36	1.48		1.75	1.63	0.59	1.61	0.34	0.44	1.79	1.47	1.26	1.04	1.63	1.66	1.35	1.20
33 Java	1.22	1.46	0.95	1.18	1.61	1.30	0.88	0.51	0.13	1.63	0.79	1.14	1.70	0.44	1.62		0.52	1.06	0.27	1.60	1.48	0.15	0.34	0.56	0.90	0.36	0.44	0.92	0.64
34 Jiangnan	1.11	1.65	0.56	1.33	1.47	1.21	0.74	0.65	0.34	1.54	0.59	1.10	1.60	0.21	1.50	0.48		1.61	0.67	1.46	1.63	0.53	0.55	0.87	1.15	0.30	0.45	0.78	0.63
35 Jomon	1.09	0.43	1.28	0.68	0.89	1.19	1.07	0.99	1.12	0.88	1.34	1.06	0.58	1.46	0.78	1.05	1.57		0.83	0.99	0.96	1.17	1.51	0.67	1.39	1.25	1.16	1.22	0.94
36 Laos	1.00	1.14	1.29	0.94	1.34	1.37	1.03	0.37	0.57	1.27	1.01	1.22	1.23	0.85	1.35	0.55	0.85	0.69		1.84	1.46	0.31	0.87	0.44	0.98	0.20	0.40	0.56	0.98
37 Loyalty	1.55	1.07	1.30	0.42	0.59	1.42	0.70	1.45	1.42	0.79	0.79	1.20	0.55	1.12	0.52	1.24	1.29	0.96	1.34		0.28	1.68	0.97	1.67	1.00	1.91	1.65	1.64	1.05
38 Melanesia	1.72	1.18	1.42	0.26	0.67	1.53	0.61	1.15	1.24	0.89	0.66	1.44	0.69	1.01	0.66	1.04	1.26	0.90	1.07	0.22		1.55	1.01	1.49	0.70	1.70	1.76	1.12	1.43
39 Myanmar	1.41	1.43	0.90	1.14	1.72	1.36	0.72	0.54	0.15	1.74	0.76	1.06	1.49	0.29	1.60	0.15	0.55	1.06	0.66	1.12	0.97		0.39	0.83	0.64	0.32	0.55	0.66	0.46
40 North China	1.43	1.65	0.77	1.09	1.44	1.11	0.60	1.08	0.46	1.49	0.85	1.16	1.66	0.50	1.43	0.32	0.51	1.39	1.18	0.87	0.77	0.35		1.26	0.52	0.79	0.91	1.14	0.54
41 Philippines	0.93	1.11	0.90	0.94	1.37	1.36	0.84	0.64	0.52	1.47	0.81	1.05	0.95	0.81	1.18	0.65	0.83	0.61	0.40	1.31	1.11	0.79	1.20		1.40	0.48	0.47	0.80	1.26
42 Sumatra	1.60	1.47	1.16	0.83	1.16	1.57	0.40	0.86	0.53	1.39	0.73	0.81	1.16	0.46	1.06	0.55	0.85	1.08	0.85	0.79	0.58	0.37	0.51	0.97		0.91	1.24	0.74	1.01
43 Thai	1.05	1.43	0.57	1.45	1.69	1.26	0.85	0.30	0.22	1.76	0.66	1.21	1.46	0.29	1.58	0.37	0.28	1.23	0.50	1.56	1.32	0.36	0.71	0.51	0.79		0.35	0.44	0.87
44 Vietnam	0.39	1.06	1.03	1.63	1.29	0.86	1.32	0.55	0.63	1.18	0.89	0.47	1.36	0.83	1.44	0.66	0.70	1.15	0.74	1.62	1.75	0.81	1.06	0.83	1.04	0.62		1.05	1.03
45 Weidun	0.98	0.94	0.56	1.42	1.16	0.65	1.13	0.74	0.64	1.24	0.80	0.82	1.27	0.72	1.08	0.95	0.72	1.43	1.18	1.46	1.40	0.83	0.99	1.04	0.95	0.63	0.74		1.15
46 Yayoi	0.80	0.86	0.63	1.48	0.94	0.44	1.20	1.17	0.93	0.80	1.39	0.88	1.48	1.00	1.12	0.91	0.77	1.21	1.44	1.40	1.46	1.02	0.68	1.50	1.19	1.04	0.79	0.73	

With regard to qualitative cranial morphology, the late Pleistocene/early Holocene Hoabinhian and Bac Son samples, in addition to the mid-Holocene Da But individuals, share dolichocephalic calvaria, large zygomatic bones, a remarkably prominent glabella, a concave nasal root and a low and wide face with prominent prognathism. On the other hand, the majority of Metal Period Dong Son individuals tend to possess an array of distinctive cranial features represented by relatively narrow and long faces, flat glabella and nasal roots, and round orbits. Such a remarkable discontinuity in cranial morphology between the pre- and early historic populations suggests that the neolithic period may be regarded as a turning point in terms of the micro-evolutionary history of northern Vietnam, at least. Multivariate analysis using the data set of the craniomorphometric dataset supports the view that Bac Son and Da But populations are direct descendants of Hoabinhian settlers, while much later Dong Son populations owe a significant proportion of their genetic heritage to immigrant populations from the northern peripheral areas of Vietnam, including southern China. In the current analysis it can be seen that the neolithic Man Bac sample is not a genetically homogeneous group. Many Man Bac individuals display cranial features common in the later Dong Son sample, whereas some individuals exhibit characteristics possibly inherited or retained from earlier mid-Holocene and even late Pleistocene Hoabinhian populations. This suggests an initial appearance of immigrants during the Phung Nguyen culture phase currently best characterised, in terms of human biology, by Man Bac, and the coexistence of different population lineages in a single site. The Man Bac specimens lend strong support to the 'Two Layer' model.

SUMMARY

This chapter has described the quantitative morphology of the cranial series from the Man Bac site. Multivariate comparisons using craniometric data demonstrates that the Man Bac series is clearly not a monophyletic group. Some individuals closely resemble the earlier pre-neolithic settlers of the region, while others show a close affinity to the later Dong Son, or Metal Period, inhabitants. This remarkable intra-group variation in cranial morphology suggests an initial appearance of immigrants at Man Bac with a genetic heritage located in the northern peripheral region of Vietnam, which includes the area currently encompassed by southern China.

ACKNOWLEDGMENTS

I am grateful for the collaborations of Drs. Ha Van Phung, Nguyen Giang Hai, the Vietnamese Institute of Archaeology, and Dr. Peter Bellwood, Australian National University for the Man Bac excavation projects in Vietnam.

Thanks are also due to Dr. Chris Stringer, Department of Palaeontology, the Natural History Museum, London, Mr. Korakot Boonlop, the Princess Maha Chakri Sirindhorn Anthropology Centre, Bangkok, Dr. Michael Pietrusewsky, the University of Hawai'i, Dr. Nguyen Viet, the Centre for Southeast Asian Prehistory, Hanoi, Dr. Philippe Mennecier, Department Hommes, Musee de l'Homme, Paris, Dr. Robert Foley, Department of Biological Anthropology, the University of Cambridge, Dr. Tsai

Hsi-Kue, National Taiwan University, College of Medicine, Dr. Wang Daw-Hwan, IHP, Academia Sinica, Taipei, and Dr. Wilfredo Ronquillio, Archaeology Division, National Museum of the Philippines for permission to study the comparative cranial specimens.

This study was supported in part by a Grant-in-Aid in 2003–2011 (No.15405018, No.20370096, No.18520593, No. 20520666) from the Japan Society for the Promotion of Science, and by the Toyota Foundation (No. D06–R–0035)

LITERATURE CITED

Bellwood P. 1991. The Austronesian dispersal and the origin of languages. Scientific America 265: 88-93.

Bellwood P. 1993. An archaeologist's view of language macrofamily relationships. Bulletin of the Indo-Pacific Prehistory Association 13: 46-60.

Bellwood P. 1996. Early agriculture and the dispersal of the southern Mongoloids. In: Akazawa T, Szathmàry EJE, editors. Prehistoric Mongoloid Dispersals. Oxford: Oxford University Press. p 287-302.

Bellwood P. 1997. Prehistory of the Indo-Malaysian archipelago, revised edition. Honolulu: University of Hawai'i Press.

Bellwood P, Renfrew C, editors. 2003. Examining the Farming/Language Dispersal Hypothesis. , Cambridge: McDonald Institute for Archaeological Research.

Bellwood P, Gillespie R, Thompson GB, Vogel JS, Ardika IW, Datan I. 1992. New dates for prehistoric Asian rice. Asian Perspectives 31:161-170.

Blust RA. 1996a. Austronesian culture history: the window of language. In: Goodenough WH, editor. Prehistoric Settlement of the Pacific. Philadelphia: American Philosophical Society. p 28-35.

Blust RA. 1996b. Beyond the Austronesian homeland: the Austric hypothesis and its implications for archaeology. In: Goodenough WH, editor. Prehistoric Settlement of the Pacific. Philadelphia American Philosophical Society. p 117-140.

Brace CL, Tracer DP, Hunt KD. 1991. Human craniofacial form and the evidence for the peopling of the Pacific. Bulletin of the Indo-Pacific Prehistory Association 12: 247-269.

Bräuer G. 1988. Osteometrie. In: Martin R, Knussmann K, editors. Anthropologie. Stuttgart: Gustav Fisher. p 160-232.

Callenfels VS. 1936. The Melanesoid civilizations of Eastern Asia. Bulletin of the Raffles Museum Series B,1: 41-51.

Coon CS. 1962. The Origin of Races. New York: Alfred A Knoph.

Cuong NL. 1986. Two early Hoabinhian crania from Thanh Hoa province, Vietnam. Zeitschrift für Morphologie und Anthropologie 77: 11-17.

Cuong NL. 1996. Anthropological Characteristics of Dong Son Population in Vietnam. Hanoi: Social Sciences Publishing House (in Vietnamese with English title and summary).

Cuong NL. 2003. Ancient human bones in Da But Culture - Thanh Hoa Province. Khao Co Hoc (Vietnamese Archaeology) 3-2003: 66-79 (in Vietnamese with English title and summary).

Cuong NL. 2007. Paleoanthropology in Vietnam. Vietnam Archaeology 2-2007: 23-41.

Diamond J, Bellwood P. 2003. Farmers and their languages: the first expansions. Science 300: 597-603.

Glover IC, Higham CFW. 1996. New evidence for early rice cultivation in South, Southeast and East Asia. In: Harris DR, editor. The Origins and Spread of Agriculture and Pastoralism in Eurasia. London: UCL Press. p 413-441.

Han KX, Qi PF. 1985. The study of the human bones of the middle and small cemeteries of

Yin sites, Anyang. In: Institute of History and Institute of Archaeology IHIA, editor. Contributions to The Study on Human Skulls from the Shang Sites at Anyang Cultural Relics. Beijing: Publishing House. p 50–81 (in Chinese).

Hanihara T. 1993. Craniofacial features of Southeast Asians and Jomonese: a reconsideration of their microevolution since the late Pleistocene. Anthropol Sci 101: 25-46.

Hanihara T. 1994. Craniofacial continuity and discontinuity of Far Easterners in the late Pleistocene and Holocene. J Hum Evol 27: 417-441.

Hanihara T. 2000. Frontal and facial flatness of major human populations. Am J Phys Anthropol 111: 105-134.

Hanihara T. 2006. Interpretation of craniofacial variation and diversification of East and Southeast Asia. In: Oxenham MF, Tayles N, editors. Bioarchaeology of Southeast Asia. Cambridge: Cambridge University Press. p 91-111..

Higham CFW. 1998. Archaeology, linguistics and the expansion of the East and Southeast Asian Neolithic. In: Blench R, Spriggs M, editors. Archaeology and Language II: Archaeological Data and Linguistic Hypotheses. London: Rutledge. p 103-114.

Higham CFW. 2001. Prehistory, language and human biology: is there a consensus in East and Southeast Asia? In: Jin L, Seielstad M, Xiao CJ, editors. Genetic, Linguistic and Archaeological Perspectives on Human Diversity in Southeast Asia. Singapore: World Scientific. p 3-16.

Howells WW. 1989. Skull Shapes and the Map: Cranio-Metric Analysis in the Dispersion of Modern Homo. Papers of the Peabody Museum of Archaeology and Ethnology, Volume 79. Cambridge: Harvard University Press.

Huson DH, Bryant D. 2006. Application of phylogenetic networks in evolutionary studies. Molecular Biology and Evolution 23:254-267.

Institute of History and Institute of Archaeology IHIA, Chinese Academy of Social Science CASS, editors. 1982. Contributions to the Study on Human Skulls from the Shang Sites at Anyang. Beijing: Cultural Relics Publishing House (in Chinese with English summary).

Ishida, H. 1992. Flatness of facial skeletons in Siberian and other Circum-Pacific populations. Z. Morph. Anthrop 79: 53-67.

Jacob T. 1967. Some Problems Pertaining to the Racial History of the Indonesian Region. Utrecht: Ph.D. Dissertation of University of Utrecht,.

Matsumura H. 2006. The population history of Southeast Asia viewed from morphometric analyses of human skeletal and dental remains. In: Oxenham M, Nancy T, editors. Bioarchaeology of Southeast Asia. Cambridge: Cambridge University Press. p 33-58.

Matsumura H, Hudson MJ. 2005. Dental perspectives on the population history of Southeast Asia. Am J Phys Anthropol 127: 182-209.

Mijsberg WA. 1940. On a Neolithic Paleo-Melanesian Lower Jaw Found in Kitchen Midden at Guar Kepah, Province Wellesley, Straits Settlements. Singapore: Proceedings of 3rd Congress of Prehistorians of the Far East. p 100-118.

Nakahashi T. 1993. Temporal cranio changes from the Jomon to the modern period in western Japan. Am J Phys Anthropol 90: 409-425.

Nakahashi T, Li M, editors. 2002. Ancient People in the Jiangnan Region, China. Fukuoka: Kyushu University Press.

Nakahashi T, Li M, Yamaguchi B. 2002. Anthropological study on the cranial measurements of the human remains from Jiangnan region, China. In: Nakahashi T, Li M, editors. Ancient People in The Jiangnan Region, China. Fukuoka: Kyushu University Press. p17-33.

Pietrusewsky M. 1981. Cranial variation in early metal age Thailand and Southeast Asia studied by multivariate procedures. Homo 32: 1–26.

Pietrusewsky M. 1994. Pacific-Asian relationships: a physical anthropological perspective. Oceanic Linguistics 33: 407-429.

Pietrusewsky M. 2005. The physical anthropology of the Pacific, East Asia: A multivariate craniometric analysis. . In: Sagart L, Blench R, Sanchez-Mazos A, editors. The Peopling of East Asia Putting Together Archaeology, Linguistics and Genetics. Curzon: Rutledge. p 201-229.

Pietrusewsky M. 2006. A multivariate craniometric study of the prehistoric and modern inhabitants of Southeast Asia, East Asia and surrounding regions: a human kaleidoscope ?. In: Oxenham MF, Tayles N, editors. Bioarchaeology of Southeast Asia. Cambridge: Cambridge University Press. p 59-90.

Pietrusewsky M. 2008. Craniometric variation in Southeast Asia and neighbouring regions: a multivariate analysis of cranial measurements. Human Evolution 23: 49-86.

Pietrusewsky M, Douglas MT. 2002. Ban Chiang, a Prehistoric Village Site in Northeast Thailand I: the Human Skeletal Remains. Philadelphia: University of Pennsylvania, Museum of Archaeology and Anthropology.

Pietrusewsky M, Chang C. 2003. Taiwan aboriginals and peoples of the Pacific-Asia region: multivariate craniometric comparisons. Anthropological Science 111: 293-332.

Renfrew C. 1987. Archaeology and Language: the Puzzle of Indo-European Origins. London: Jonathan Cape.

Renfrew C. 1989. Models of change in language and archaeology. Transactions of the Philological Society 87: 103-155.

Renfrew C. 1992. World languages and human dispersals: a minimalist view. In: Hall JA, Jarvie IC, editors. Transition to Modernity: essays on power, wealth and belief. Cambridge: Cambridge University Press. p 11-68.

Saitou N, Nei M. 1987. The neighbour-joining method: A new method for reconstructing phylogenetic trees. Molecular Biology and Evolution 4:406-425.

Sangvichien S. 1971. Physical Anthropology of the Skull of Thai. Bangkok: Ph.D. Dissertation, Faculty of Medicine, Siriraj Hospital, Mahidol University, No.2514.

Sieveking GG. 1954. Excavations at Gua Cha, Kelantan, Part 1. Federation Museums Journal 1: 75-143.

Sneath PH, Sokal RR. 1973. Numerical Taxonomy. San Francisco: WH Freeman and Co.

Suzuki H, Mizoguchi Y, Conese E. 1993. Craniofacial measurement of artificially deformed skulls from the Philippines. Anthropological Science 101: 111-127.

Thuy NK. 1990. Ancient human skeletons at Con Co Ngua. Khao Co Hoc (Vietnamese Archaeology) 3-1990:37-48 (in Vietnamese with English summary).

Turner CGII 1990. Major features of Sundadonty and Sinodonty, including suggestions about East Asian microevolution, population history and late Pleistocene relationships with Australian Aborigines. Am J Phys Anthropol 82: 295-317.

Von Koenigswald GHR. 1952. Evidence of a prehistoric Australo-Melanesoid population in Malaya and Indonesia. Southwestern Journal of Anthropology 8: 92-96.

Yamaguchi B. 1982. A review of the osteological characteristics of the Jomon population in prehistoric Japan. J Anthropol Soc Nippon 90(Supplement): 77-90.

Yokoh Y. 1940. Beiträge zur kraniologie der Dajak. Japanese Journal of Medical Science, Part I Anatomy 8: 1-354.

Appendix 3.1 Cranial and mandibular measurements (mm) and indices for the Man Bac specimens (Sex M=male, F=female).

Sample No.	99 M3	01 M5	01 M9	01 M10	05 M11	05 M20	05 M29	05 M31	07H1 M5	07H1 M8	07H1 M9	07H2 M1	07H2 M10	07H2 M19	07H2 M27	07H2 M30	07H2 M32	99 M2	99 M5b	05 M9	05 M15	05 M16	05 M34	07H1 M4	07H1 M10	07H1 M11	07H2 M5	07H2 M12	07H2 M22	07H2 M24
Martin's No. / Sex	M	M	M	M	M	M	M	M	M	M	M	M	M	M	M	M	M	F	F	F	F	F	F	F	F	F	F	F	F	F
1 Max. cranial length	180	185	174	184	173	180	186	175	177	185	192	176	181		192	181	184	176	169	169	183	163	174	171	177	180		180	171	181
5 Basion-nasion length	99	93		104	93		96			105			108		105	100	107	91	94	90	100								100	
8 Max. cranial breadth	139	143	137	145	136	140	138	155	140	147	140	150	141		135	151	147	137	131	136	129	137		139	137	137	135	130	137	133
9 Min. frontal breadth	96	103	98	102	93	98	95	101	96	101	97	94	106		94	108	104	93	91	93	101	85	97	99	96	97		85	95	96
10 Max. frontal breadth	118				116	113	110	130	117	123	117	125	114		120	119	101	106	114	119		107	119	115		113		111	116	118
12 Max. occipital breadth	104				104	110	105	107	116	110	109	108	117		115	118	119	112	107	105	109					107		103	115	110
17 Basion-bregma height	142	134		138	128		136			147		154	144		138	150	148	121	132	128	147								143	
29 Frontal chord	117				111	113	115	110	107	116	113	114	107		114	116	121	99	102	100	116	110	106			106		111	111	112
30 Parietal chord	115				102	116	120	112	118	119	126	112	115		114	124	116	108	123	105	121	97	108			109		113	111	118
31 Occipital chord	101				108		102	98	104	104	115	110	106		108	108	103	103	90	107	100								102	103
40 Basion-prosthion length	96	100		99	97		103			102			103		102	101	103	92	93	85	92								92	
43 Upper facial breadth	102	118	110	110	106	107	109	111	115	114	104	113	120		107	120	115	99	106	107	109	108				109		99	107	
45 Bizygomatic breadth	128	142	147	144	130		138	146	142	146	144	147	146		133	147	146	120	128	133	133	144		137		136			129	
46 Bimaxillary breadth	102	108	109	102	113		111	100	110	105		109	121	108	107	116	106	97		103	109	106						101	104	
48 Upper facial height	70	75	67	68	70	66	67	74	70	69		71	79		74	66	68	62	65	69	71					71		68	70	
51 Orbital breadth	42	40	40	41	40		42	39	45	44		44	43		43	45	47	41		39	40	43		42		41		40	42	
52 Orbital height	34	34	35	35	39	37	30	34	34	35	34	37	37		32	34	34	33	35	35	37	34		34		33		34	36	
54 Nasal breadth	28	29	28	26	28	28	26	29	29	28		26	27	31	26	31	33	22	25	25	27	28				27		30	23	
55 Nasal height	53	56	52	54	50	52	49	53	54	49		54	63		55	50	52	47	53	49	56					52		54	50	
60 Upper alveolar length	49	57		50	54	50	54	53	55	53	52	52	58		55	55	54	51	55	46	53				57	55			50	
61 Upper alveolar breadth	62	67		66	71	62	68	66	67	64	72	71		66	65	67	69	60	60	65	69				65				62	
8:1 Cranial index	77.2	77.3	78.7	78.8	78.6	77.8	74.2	88.6	79.1	79.5	72.9	85.2	77.9		70.3	83.4	79.9	77.8	77.5	80.5	70.5	85.6		81.3	77.4	76.1		72.2	80.1	
48:45 Upper facial index	54.7	52.8	45.6	47.2	53.8		48.6	50.7	49.3	47.3	0.0	48.3	54.1		55.6	44.9	46.6	51.7	50.8	51.9	53.4					52.2			54.3	
43(1) Frontal chord		109	110		95		101	100	107	107	94	103	110		97	110	105													
43c Frontal subtense		18	16		12		17	10	14	8	16	12	19		23	17	16													
57 Simotic chord		11.4			6.5		10.1		8.6	7.6		11.7	12.9		9.5	11.9	9.0													
57a Simotic subtense		2.8			0.8				1.5	2.6		4.3	5.4		3.3	4.1	2.8													
46b Zygomaxillary chord		109	112				110	95	108	100		104	114		104	116	107													
46c Zygomaxillary subtense		23	25				22	19	30	18		16	23		28	19	26													
43c:43(1) Frontal index		16.5	14.9		12.2		16.6	10.1	13.3	7.9	17.5	11.2	16.9		23.3	15.7	15.2													
57a:57 Simotic index		24.1			12.5				17.9	33.8		36.5	42.0		34.6	34.7	31.1													
46c:46b Zygomaxillary index		20.8	21.9				20.2	19.5	27.3	18.0		15.6	20.6		26.8	16.5	23.9													
66 Bigonial breadth	94				103	112	106	113	108	106	95	115	118	95	93	114	104	86	90	94	104				102	94		101	99	
68 Mandibular length	83				75	82	82	74	78	81	85	82	84	84	85	84	84	67	69	75	71				83	75		75	81	
69 Symphyseal height	32				38	31	38	36	27	34	36	32	36	32	36	29	29	37	30	34	32				33	31			33	30
70 Ramus height	63				56	70	69	71	69	66		69	69	74	66	78	61	46	58	56	55				65	64		63	66	
71 Ramus breadth	39				32	39	40	34	35	37	41	41	35	37	40	41	38	35	35	35	37				36	38		36	32	

4
Qualitative Cranio-Morphology at Man Bac

Yukio Dodo

Tohoku University School of Medicine, Japan

Cranial nonmetric traits are widely accepted to be effective for reconstructing population histories, not only within limited regions but also globally (Ossenberg, 1986, 1994; Dodo and Ishida, 1990; Dodo and Kawakubo, 2002; Hanihara et al., 2003; Dodo and Sawada, 2010). In this chapter, the occurrence of cranial nonmetric traits is assessed for the Man Bac series, and the origins and affinities of the Man Bac people are discussed in the context of local and regional populations in East and Southeast Asia.

MATERIALS AND METHODS

The presence/absence of 22 nonmetric traits was examined for 33 adult and near-adult crania from the Man Bac site: 4 from the 1999-2001 season, 11 from the 2005 season, and 18 from the 2007 season. The criteria employed here for scoring nonmetric traits are given in Dodo (1974) and Dodo and Ishida (1990). The following 6 traits were used for comparison of the frequencies among cranial samples:

Supraorbital foramen (SOF)
Hypoglossal canal bridging (HGCB)
Transverse zygomatic suture vestige (TZS)
Ossicle at the lambda (OL)
Mylohyoid nerve groove bridging (MHB)
Medial palatine canal (MPC)

These 6 nonmetric traits are little affected by interobserver error in scoring (Ishida and Dodo, 1990) and have been noted as good measures for population relationships in the Japanese Islands (Dodo and Ishida, 1990). Furthermore, the supraorbital foramen and hypoglossal canal bridging are believed to be highly effective in discriminating amongst major human groupings globally (Dodo, 1986; Dodo and Sawada, 2010).

Table 4.1 provides summary information on the 6 cranial samples compared: Neolithic Weidun and early historic Eastern Zhou/Western Han on the lower reaches of the Yangtze River, Jiangsu, southern China; modern mainland Southeast Asians including inhabitants of Vietnam, Laos, Cambodia, and Thailand; modern southern Chinese derived from south of the Yangtze River; and modern Australian

Aborigines.

Biological distances among the samples were assessed via Smith's Mean Measure of Divergence statistic (MMD) defined as follows:

$$MMD = 1/r \sum [(\theta_1 - \theta_2)^2 - (1/n_1 + 1/n_2)]$$

where r is the number of traits; θ_1 and θ_2 are angular transformations in radian of the trait frequencies p_1 and p_2 in two samples, obtained by the formula θ = arcsine (1 - 2p); and n_1 and n_2 are the numbers of observations in the two samples (Sjøvold, 1973).

Two statistical methods for graphic representation were applied to the matrix of MMDs to depict the relationships of the samples. One is group average clustering analysis and the other is the multi-dimensional scaling method. The procedures of these statistical analyses were kindly carried out by Professor H. Matsumura of Sapporo Medical University, using data analysis software "STATISTICA Version 06J" produced by StatSoft Japan Inc., Tokyo.

Table 4.1 Cranial samples used for nonmetric analyses.

Sample name	Provenance	Period	Reference
Man Bac	Ninh Binh, northern Vietnam	neolithic (3,300 - 3,500 uncal.BP)	Present study
Weidun	Lower reaches of the Yangtze River, Jiangsu, China	Neolithic (6,000 - 5,000 BP)	Wakebe, 2002
Zhou/Han (Eastern Zhou-Western Han)	Lower reaches of the Yangtze River, Jiangsu, China	Early Historic (2,800 -2,000 BP)	Wakebe, 2002
SE-Asia (Mainland Southeast Asians)	Vietnam, Laos, Cambodia, and Thailand	Modern	Hanihara and Ishida, 2001a,b,c,d,e
S China (Southern Chinese)	South of the Yangtze River, China	Modern	Hanihara and Ishida, 2001a,b,c,d,e
Australia (Australian Aborigines)	New South Wales, Queensland, and Victoria	Modern	Hanihara and Ishida, 2001a,b,c,d,e

Table 4.2 Comparison of side-incidences of 6 cranial nonmetric traits.

	Man Bac		Weidun		Zhou/Han	
Trait	z	p	n	p	n	p
1 SOF	57	0.439	69	0.406	52	0.519
2 HGCB	32	0.156	45	0.044	44	0.136
3 TZS	39	0.026	37	0.081	42	0.006*
4 OL	24	0.010*	22	0.045	23	0.217
5 MHB	57	0.070	81	0.099	44	0.023
6 MPC	50	0.020	74	0.041	50	0.060

* 1/4n (Bartlett's adjustment)

RESULTS

The presence/absence of the 22 nonmetric traits in each cranium of the Man Bac series is shown in the Appendix of this chapter. In Table 4.2, sex- and side-pooled incidences of the 6 nonmetric traits are given for the 6 cranial samples, with the zero proportions being replaced by 1/4n as recommended by Bartlett (Snedecor and Cochran, 1980). Although an anthroposcopic impression suggested the mingling of two types of crania in the Man Bac series (see Chapter 3), i.e., a gracile one and a robust one, no such distinction was noticed in the patterning of cranial nonmetric traits. For this reason the Man Bac individuals were treated here as a single

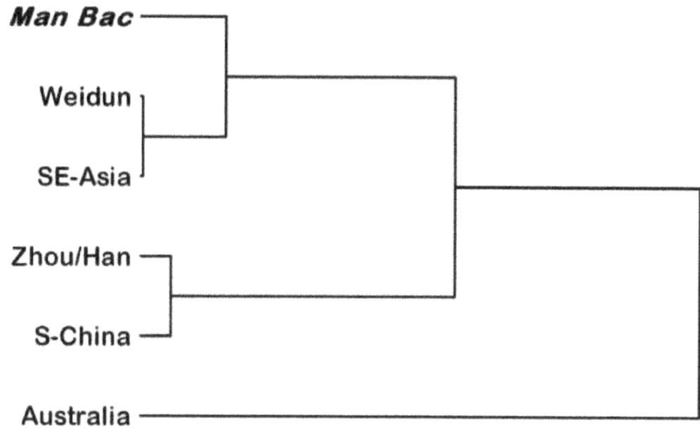

Figure 4.1 Dendrogram of a cluster analysis applied to the MMDs among the samples from mainland Southeast Asia, southern China, and Australia.

Table 4.3 MMDs among the 6 cranial samples compared.

	Man Bac	Weidun	Zhou/Han	SE-Asia	S China	Australia
Man Bac		0.0016	0.0735	0.0270	0.0959	0.1068
Weidun	0.0016		0.0778	0.0005	0.0359	0.0652
Zhou/Han	0.0735	0.0778		0.0233	0.0052	0.1476
SE-Asia	0.0270	0.0005	0.0233		0.0084	0.0614
S China	0.0959	0.0359	0.0052	0.0084		0.0880
Australia	0.1068	0.0652	0.1476	0.0614	0.0880	

population sample. Table 4.3 gives MMDs based on the frequencies of the 6 nonmetric traits among the 6 cranial samples compared.

The Man Bac cranial series (3800–3500 years BP) was compared with that of the Neolithic Weidun site (6,000–5,000 years BP) and that of the early historic Eastern Zhou/Western Han (2,800–2,000 years BP) in the Yangtze Basin, southern China. Moreover, comparisons were made with modern cranial samples from mainland Southeast Asia and southern China. Australian aboriginal crania were also used for comparison.

In the MMD matrix of Table 4.3, the closest sample to the Man Bac population is the Neolithic Weidun series from in the Yangtze Basin, and the next closest are modern mainland Southeast Asians. The Australian aboriginal population is the furthest away, and the samples of Eastern Zhou/Western Han and southern Chinese are in-between.

A dendrogram of cluster analysis and a two-dimensional display of multidimensional scaling are depicted in Figure 4.1 and Figure 4.2, respectively. Both figures show a relatively tight cluster of Man Bac, Weidun, and mainland Southeast Asians. Another cluster is seen between the Eastern Zhou/Western Han and southern Chinese.

DISCUSSION

It was noticed that there is a close relationship between Man Bac and the Weidun sample, and that these two series are also close to the crania of mainland Southeast Asians (Table 4.3, Figures 4.1 and 4.2). The cranial and dental metric study of the Man Bac specimens from the 1999-2001 and 2004-2005

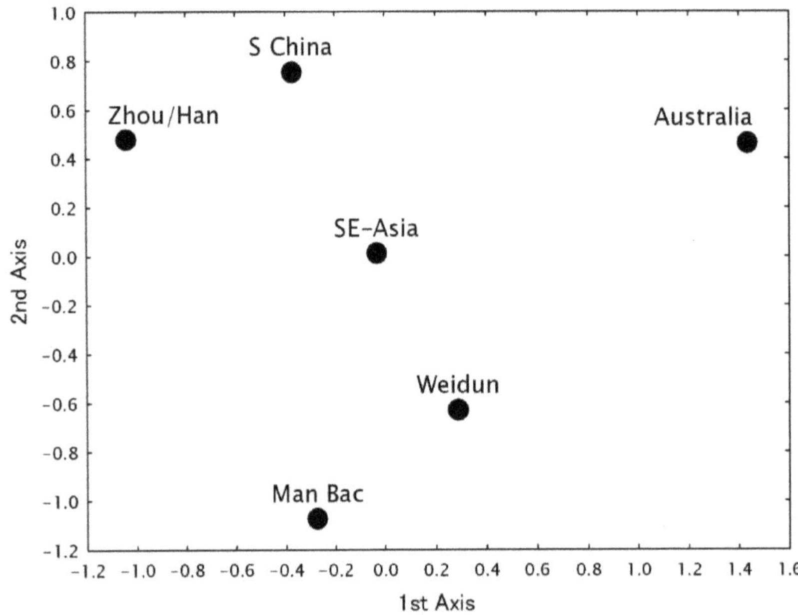

Figure 4.2 Two-dimensional display of the multidimensional scaling method applied to the MMDs among the samples from mainland Southeast Asia, southern China and Australia.

seasons revealed that the Man Bac sample is grouped with the early Metal Age to modern East/Southeast Asian and the Neolithic Weidun people (Matsumura et al., 2008a). Taking into account the findings of the cranial nonmetric and cranial/dental metric analyses, it can be postulated that the inhabitants of Man Bac from northern Vietnam were closely related to the Neolithic Weidun people, essentially a rice-farming culture on the lower reaches of the Yangtze River, and the following schema of population history can be outlined: Neolithic Weidun→neolithic Man Bac→early Iron Age Dong Son→modern mainland Southeast Asians.

Recent studies have disclosed that the late Pleistocene and early Holocene human remains from Southeast Asia, such as Gua Gunung Runtuh in Peninsular Malaysia and Mai Da Nuoc, Mai Da Dieu, and Hang Cho in northern Vietnam, exhibit osteological characteristics shared with 'Australo-Melanesians' (Matsumura and Zuraina, 1999; Cuong, 1986; Matsumura et al., 2008b). These researchers have argued that Southeast Asia was first occupied by an indigenous population, sometimes referred to as 'Australo-Melanesian', before immigrants from East Asia dispersed widely into this region (Matsumura and Hudson, 2005; Matsumura et al., 2008a).

The results of the present nonmetric analysis, however, revealed little affiliation between the Man Bac inhabitants and the Australian aboriginal sample, as shown in Table 4.3 and Figures 4.1-4.2. Most likely, the prototype population ancestral to modern mainland Southeast Asians, which would appear to be quite different to 'Australo-Melanesians', was already established by the time of the neolithic in northern Vietnam.

In order to reconstruct the population history in Vietnam more systematically, samples of the Early to Middle Holocene Hoabinhian, Bacsonian, and Da But cultures, as well as the Early Metal Age Dong Son culture, need to be investigated in terms of cranial nonmetric variation.

SUMMARY

The presence/absence of 22 nonmetric traits was examined for 33 adult and near-adult crania from the Man Bac site. The frequencies of the 6 traits, which are little affected by interobserver error in scoring, were used for comparison among the 6 neolithic to modern cranial samples from mainland Southeast Asia, southern China, and Australia. Biological distances assessed by Smith's Mean Measure of Divergence indicated that the Man Bac series is closest to Neolithic Weidun in the Yangtze Basin in southern China, and next closest to modern mainland Southeast Asians. From these findings, it was inferred that the Man Bac people, genetically influenced by those represented by the Neolithic Weidun rice-farming people in the Yangtze Basin, are a prototype population ancestral to modern mainland Southeast Asians.

ACKNOWLEDGMENTS

I express my sincere gratitude to Professor T. Hanihara of Kitasato University School of Medicine for providing me with cranial nonmetric raw data of the Southeast Asian, southern Chinese, and Australian samples. The present study was supported by Grant-in-Aid in 2003-2005 (No. 15405018) and 2008-2009 (No. 20370096) from the Japan Society for the Promotion of Science.

LITERATURE CITED

Cuong NL. 1986. Two early Hoabinhian crania from Thanh Hoa province, Vietnam. Zeitschrift für Morphologie und Anthropologie 77: 11-17. Dodo Y. 1974. Nonmetrical cranial traits in the Hokkaido Ainu and the northern Japanese of recent times. J Anthropol Soc Nippon 82: 31-51.

Dodo Y. 1986. Supraorbital foramen and hypoglossal canal bridging: the two most suggestive nonmetric traits in discriminating major racial groupings of man. J Anthropol Soc Nippon 95: 19-35.

Dodo Y, Ishida H. 1990. Population history of Japan as viewed from cranial nonmetric variation. J Anthropol Soc Nippon 98: 269-287.

Dodo Y, Kawakubo Y. 2002. Cranial affinities of the Epi-Jomon inhabitants in Hokkaido, Japan. Anthropol Sci 110: 1-32.

Dodo Y, Sawada J. 2010. Supraorbital foramen and hypoglossal canal bridging revisited: their worldwide frequency distribution. Anthropol Sci 118 (in press)

Hanihara T, Ishida H. 2001a. Os incae: variation in frequency in major human population groups. J Anat 198: 137-152.

Hanihara T, Ishida H. 2001b. Frequency variations of discrete cranial traits in major human populations. I. Supernumerary ossicle variations. J Anat 198: 689-706.

Hanihara T, Ishida H. 2001c. Frequency variations of discrete cranial traits in major human populations. II. Hypostotic variations. J Anat 198: 707-725.

Hanihara T, Ishida H. 2001d. Frequency variations of discrete cranial traits in major human populations. III. Hyperostotic variations. J Anat 199: 251-272.

Hanihara T, Ishida H. 2001e. Frequency variations of discrete cranial traits in major human populations. IV. Vessel and nerve related variations. J Anat 199: 273-287.

Hanihara T, Ishida H, Dodo Y. 2003. Characterization of biological diversity through analysis of discrete cranial traits. Am J Phys Anthropol 121: 241-251.

Ishida H, Dodo Y. 1990. Interobserver error in scoring nonmetric cranial traits. J Anthropol Soc Nippon 98: 403-409.

Matsumura H, Zuraina M. 1999. Metric analyses of an early Holocene human skeleton from Gua Gunung Runtuh, Malaysia. Am J Phys Anthropol 109: 327-340.

Matsumura H, Hudson MJ. 2005. Dental perspective on the population history of Southeast Asia. Am J Phys Anthropol 127: 182-209.

Matsumura H, Oxenham MF, Dodo Y, Domett K, Thuy NK, Cuong NL, Dung NK, Huffer D, Yamagata M. 2008a. Morphometric affinity of the late Neolithic human remains from Man Bac, Ninh Binh Province, Vietnam: key skeletons with which to debate the 'two layer' hypothesis. Anthropol Sci 116: 135-148.

Matsumura H, Yoneda M, Dodo Y, Oxenham MF, Cuong NL, Thuy NK, Dung LM, Long VT, Yamagata M, Sawada J, Shinoda K, Takigawa W. 2008b. Terminal Pleistocene human skeleton from Hang Cho Cave, northern Vietnam: implications for the biological affinities of Hoabinhian people. Anthropol Sci 116: 201-217.

Ossenberg NS. 1986. Isolate conservatism and hybridization in the population history of Japan: the evidence of nonmetric cranial traits. In: Akazawa T, Aikens CM, editors. Prehistoric Hunter-Gatherers in Japan. Univ Mus Univ Tokyo Bulletin 27: 199-215.

Ossenberg NS. 1994. Origins and affinities of the native people of northwestern North America: the evidence of cranial nonmetric traits. In: Bonichsen R, Steele DG, editors. Method and Theory for Investigating the Peopling of America. Corvallis, OR: Center for the First Americans, Oregon State University. p 79-115.

Sjøvold T. 1973. The occurrence of minor nonmetrical variants in the skeleton and their quantitative treatment for population comparisons. Homo: 24: 204-233.

Snedecor GW, Cochran WG. 1980. Statistical Methods. 7th edition. Iowa: Iowa State University Press.

Wakebe T. 2002. Human skeletal remains excavated from Jinagnan area in China as viewed from cranial nonmetric variation. In: Nakahashi T, Li M, editors. Ancient People in The Jiangnan Region, China. Fukuoka; Kyushu University Press. p 35-49.

Appendix 4.1 Presence or absence of cranial nonmetric traits in each cranium of the Man Bac series (* median trait; 1 present; 0 absent; / unobservable).

No.	MB99M2		MB01M5		MB01M9		MB01M10		MB05M29		MB05M31		MB05M32		MB05M34		MB07H1M04	
Age/Sex	adult/female		adult/male		adult/male		adult/male		adult/male		adult/male		adult/male		adult/female		adult/male	
	R	L	R	L	R	L	R	L	R	L	R	L	R	L	R	L	R	L
1 Metopism*	1		0		0		0		0		0		/		0		0	
2 Supraorbital nerve groove	0	0	0	0	0	0	0	0	0	0	0	0	/	/	0	1	0	0
3 Supraorbital foramen	1	1	0	0	0	0	1	1	1	1	0	0	/	/	0	1	0	0
4 Ossicle at lambda*	/		0		0		0		0		0		/		/		0	
5 Biasterionic suture (10mm-)	0	0	0	0	0	0	0	0	0	0	0	1	/	/	/	0	0	0
6 Asterionic bone	0	0	0	/	1	/	0	0	0	0	1	0	/	/	/	/	/	0
7 Occipitomastoid bone	/	0	/	/	/	/	0	0	0	/	1	0	/	/	/	/	/	/
8 Parietal notch bone	0	0	0	0	0	0	0	0	0	0	1	0	/	/	/	/	/	0
9 Condylar canal	1	1	/	0	/	/	/	/	1	0	/	1	/	/	/	/	/	/
10 Precondylar tubercle	0	0	0	0	/	/	0	0	0	0	/	/	/	/	/	/	/	/
11 Paracondylar process	0	0	0	0	/	/	/	/	/	0	/	/	/	/	/	/	/	/
12 Hypoglossal canal bridging	0	0	1	0	/	/	0	0	1	1	/	/	/	/	/	/	/	/
13 Foramen of Huschke	0	0	0	0	0	0	1	1	/	0	0	0	0	0	/	/	0	0
14 Ovale-spinosum open	0	0	0	0	/	/	/	/	/	0	0	/	/	/	/	/	0	0
15 Foramen of Vesalius	0	0	/	/	/	/	/	/	/	/	0	/	/	/	/	/	0	0
16 Pterygospinous foramen	0	0	0	0	/	/	0	0	/	0	0	/	/	/	/	/	0	0
17 Medial palatine canal	0	0	0	0	0	0	0	0	0	0	0	0	/	/	/	/	0	0
18 Transv. zygomatic suture (5mm-)	0	0	0	0	0	0	0	0	0	0	0	0	/	/	/	/	0	/
19 Jugular foramen bridging	0	0	0	0	/	/	0	0	/	/	/	/	/	/	/		/	/
20 Sagittal sinus groove left*	1		/		0		/		/		1		0		0		0	
21 Clinoid bridging	/	/	/	/	/	/	/	/	/	/	/	/	/	/	/	/	/	/
22 Mylohyoid bridging	0	0	0	0	0	0	0	0	0	0	0	0	/	/	0	0	0	0

Appendix 4.1 (continued 1).

No.	MB07H1M05		MB07H1M08		MB07H1M09		MB07H1M10		MB07H1M11		MB07H2M01		MB07H1M04		MB07H1M05		MB07H1M08	
Age/Sex	adult/male		adult/male		adult/male		adult/female		adult/female		adult/male		adult/male		adult/male		adult/male	
	R	L	R	L	R	L	R	L	R	L	R	L	R	L	R	L	R	L
1 Metopism*	0		0		0		0		0		0		0		0		0	
2 Supraorbital nerve groove	0	0	0	0	0	0	0	0	0	0	0	1	0	0	0	0	0	0
3 Supraorbital foramen	0	0	0	0	/	1	/	0	0	1	0	0	0	0	0	0	0	0
4 Ossicle at lambda*	0		0		0		/		0		/		0		0		0	
5 Biasterionic suture (10mm-)	0	0	0	0	1	1	/	/	0	1	0	/	0	0	0	0	0	0
6 Asterionic bone	0	0	0	0	0	/	/	/	/	/	1	1	/	0	0	0	0	0
7 Occipitomastoid bone	/	/	0	/	1	/	/	/	/	/	0	0	/	/	/	/	0	/
8 Parietal notch bone	0	1	0	0	0	0	/	/	1	1	0	0	/	0	0	1	0	0
9 Condylar canal	/	/	/	1	/	/	/	/	/	/	/	/	/	/	/	/	/	1
10 Precondylar tubercle	/	/	0	0	/	/	/	/	/	/	/	/	/	/	/	/	0	0
11 Paracondylar process	/	/	/	0	/	/	/	/	/	/	/	/	/	/	/	/	/	0
12 Hypoglossal canal bridging	/	/	0	0	/	/	/	/	/	/	/	/	/	/	/	/	0	0
13 Foramen of Huschke	/	0	0	/	0	/	/	/	0	1	/	0	0	0	/	0	0	/
14 Ovale-spinosum open	0	/	0	0	/	0	/	/	0	/	0	0	0	0	0	/	0	0
15 Foramen of Vesalius	/	/	0	1	/	0	/	/	0	/	/	/	0	0	/	/	0	1
16 Pterygospinous foramen	0	/	0	0	0	0	/	/	0	0	0	0	0	0	0	/	0	0
17 Medial palatine canal	0	0	0	0	0	/	0	/	0	0	1	0	0	0	0	0	0	0
18 Transv. zygomatic suture (5mm-)	0	0	0	0	0	/	/	/	0	0	0	/	0	/	0	0	0	0
19 Jugular foramen bridging	/	/	/	0	/	/	/	/	/	/	/	/	/	/	/	/	/	0
20 Sagittal sinus groove left*	/		0		1		0		0		/		0		/		0	
21 Clinoid bridging	/	/	/	/	/	/	/	/	/	/	/	/	/	/	/	/	/	/
22 Mylohyoid bridging	1	1	0	0	0	0	0	0	0	0	0	0	0	0	1	1	0	0

Appendix 4.1 (continued 2).

No.	MB07H1M09		MB07H1M10		MB07H1M11		MB07H2M01		MB07H2M02		MB07H2M05		MB07H2M10		MB07H2M12		MB07H2M18	
Age/Sex	adult/male		adult/female		adult/female		adult/male		subadult/female		adult/female		adult/male		adult/female		subadult/male	
	R	L	R	L	R	L	R	L	R	L	R	L	R	L	R	L	R	L
1 Metopism*	0		0		0		0		0		0		0		0		0	
2 Supraorbital nerve groove	0	0	0	0	0	0	0	1	/	0	/	/	0	0	0	0	0	0
3 Supraorbital foramen	/	1	/	0	0	1	0	0	1	1	0	/	0	0	1	1	1	0
4 Ossicle at lambda*	0		/		0		/		/		0		0		0		0	
5 Biasterionic suture (10mm-)	1	1	/	/	0	1	0	/	/	0	0	0	0	1	0	0	0	0
6 Asterionic bone	0	/	/	/	/	/	1	1	/	0	0	0	0	0	0	0	0	0
7 Occipitomastoid bone	1	/	/	/	/	/	0	0	/	/	/	0	0	0	/	/	0	0
8 Parietal notch bone	0	0	/	/	1	1	0	0	/	0	0	0	0	0	0	0	0	0
9 Condylar canal	/	/	/	/	/	/	/	/	/	/	/	/	0	0	/	/	/	/
10 Precondylar tubercle	/	/	/	/	/	/	/	/	/	/	0	0	1	1	/	/	0	0
11 Paracondylar process	/	/	/	/	/	/	/	/	/	/	0	0	0	0	/	/	0	0
12 Hypoglossal canal bridging	/	/	/	/	/	/	/	/	/	/	0	0	0	0	/	/	0	0
13 Foramen of Huschke	0	/	/	/	0	1	/	0	/	0	0	0	0	0	/	/	0	0
14 Ovale-spinosum open	/	0	/	/	0	/	0	0	/	/	1	0	1	0	0	0	0	0
15 Foramen of Vesalius	/	0	/	/	0	/	/	/	/	/	0	0	/	/	/	/	/	/
16 Pterygospinous foramen	0	0	/	/	0	0	0	0	/	/	0	0	0	0	0	0	0	0
17 Medial palatine canal	0	/	0	/	0	0	1	0	/	/	0	0	0	0	0	0	0	0
18 Transv. zygomatic suture (5mm-)	0	/	/	/	0	0	0	/	/	/	/	/	0	/	0	0	/	0
19 Jugular foramen bridging	/	/	/	/	/	/	/	/	/	/	0	0	0	0	/	/	/	/
20 Sagittal sinus groove left*	1		0		0		/		/		/		0		/		/	
21 Clinoid bridging	/	/	/	/	/	/	/	/	/	/	/	/	/	/	/	/	/	/
22 Mylohyoid bridging	0	0	0	0	0	0	0	0	/	/	0	0	0	0	0	1	0	0

Appendix 4.1 (continued 3).

No.	MB07H2M19		MB07H2M22		MB07H2M24		MB07H2M27		MB07H2M30		MB07H2M32	
Age/Sex	adult/male		adult/female		adult/female		adult/male		adult/male		adult/male	
	R	L	R	L	R	L	R	L	R	L	R	L
1 Metopism*	/		0		0		0		0		0	
2 Supraorbital nerve groove	/	/	1	1	0	0	0	0	0	0	0	0
3 Supraorbital foramen	/	/	0	0	0	0	1	1	0	1	1	1
4 Ossicle at lambda*	/		0		0		0		0		0	
5 Biasterionic suture (10mm-)	/	/	0	0	0	0	0	0	0	0	0	/
6 Asterionic bone	/	/	0	0	/	0	0	0	/	/	0	0
7 Occipitomastoid bone	/	/	0	0	/	0	0	0	/	/	/	0
8 Parietal notch bone	/	/	0	0	0	0	0	0	0	0	0	0
9 Condylar canal	/	/	/	/	/	/	/	/	1	/	1	/
10 Precondylar tubercle	/	/	0	0	/	/	0	0	/	0	0	0
11 Paracondylar process	/	/	0	/	/	/	/	/	/	/	/	/
12 Hypoglossal canal bridging	/	/	0	1	/	/	0	0	0	0	1	0
13 Foramen of Huschke	/	/	0	0	0	0	0	0	0	0	1	/
14 Ovale-spinosum open	/	/	0	0	/	/	1	0	/	0	0	/
15 Foramen of Vesalius	/	/	/	/	/	/	/	/	/	/	/	/
16 Pterygospinous foramen	/	/	0	0	/	/	0	0	/	0	0	/
17 Medial palatine canal	/	/	0	0	0	/	0	0	0	0	/	0
18 Transv. zygomatic suture (5mm-)	/	/	0	0	/	/	0	0	0	/	0	0
19 Jugular foramen bridging	/	/	0	0	/	/	/	0	/	/	/	/
20 Sagittal sinus groove left*	/		/		0		/		0		0	
21 Clinoid bridging	/	/	/	/	/	/	/	/	/	/	/	/
22 Mylohyoid bridging	0	0	0	0	/	/	0	/	1	0	0	0

5

Quantitative and Qualitative Dental-Morphology at Man Bac

Hirofumi Matsumura[1]

Department of Anatomy, Sapporo Medical University, Japan.

The aim of this chapter is to explore the local population history of northern Vietnam, specifically the relationship between the Man Bac sample and mid-Holocene Da But (represented by the cemetery site Con Co Ngua) and late Pleistocene/early Holocene Bac Son/Hoabinhian communities. Additionally, any potential relationship with Metal period Dong Son and present-day Vietnamese is explored. Moreover, this study will also provide a test of the "Two-layer" hypothesis. For nearly a century it has been argued that Southeast Asia was initially settled by people akin to present-day Australo-Melanesians that, in the later neolithic, underwent substantial genetic modification due to the influx of immigrants associated with the spread of agriculture from southern China (Callenfels, 1936; Mijsberg, 1940; Barth, 1952; von Koenigswald, 1952; Coon, 1962; Thoma, 1964; Jacob, 1967, 1975; Brace, 1976; Howells, 1976; Brace et al., 1991). A number of recent archaeological reviews conclude that food producing communities spread south from the Yangtze Basin into mainland and island Southeast Asia (Bellwood, 1987, 1997; Spriggs, 1989; Glover and Higham, 1996; Bellwood et al., 1992). In order to test this scenario, the biological relationships between neolithic and pre-neolithic communities throughout Southeast Asia need to be examined in more detail. The Man Bac sample provides a crucial data set in such an examination (see also Chapter 3). Along with the above-mentioned aims, this chapter compares Man Bac metric and non-metric dental data with early and modern samples from East/Southeast Asia and the west Pacific.

MATERIALS AND METHODS

Man Bac Specimens and Comparative Samples

The dental sample derives from all four seasons of excavation at Man Bac. This is the only study in this monograph to estimate the sex of subadult individuals, in this instance using Schutkowski's (1993) protocols based on mandibular morphology.

Some 41 adult and subadult individuals contributed to the permanent tooth sample, while 17 subadults contributed to the deciduous tooth sample. Multivariable statistical procedures were undertaken to assess the population affinities between Man Bac and the comparative samples (see Table 5.1) including:

Vietnam, Laos, Thailand, Malaysia, Indonesia, China and Japan, as well as modern samples from East/Southeast Asia and the Pacific. All comparative data are from males for the odontometric analysis and from a sex-combined sample for the nonmetric trait comparisons.

Crown measurements and observations of non-metric traits were undertaken for teeth on the right side, or antimere substitutions, where necessary. Odontometric data and the presence of nonmetric traits recorded for the Man Bac individuals are given in Appendix 5.1 and 5.2 (this chapter), combining all four season's datasets.

Recording System of Quantitative and Qualitative Dental Morphology

Quantitative dental morphological data was represented by tooth crown diameters, which were recorded as maximum diameters according to the Fujita (1949) system. Measurements of the permanent dentition were only undertaken for sex-identified individuals. Due to difficulties in estimating the sex of subadults, crown diameters of deciduous teeth were recorded for all available specimens regardless of sex.

Qualitative dental morphology was recorded for 21 nonmetric dental traits of the permanent dentition, which were scored using protocols and criteria given in Matsumura (1995, see also Table 5.2). All traits were scored for both sexes on the basis of presence/absence to facilitate statistical comparisons, although males and females were combined given the low to minimal sexual dimorphism expected for these traits (Turner et al., 1991).

Statistical Procedures

Dental metric comparisons were made using mesiodistal and buccolingual crown diameters. In the first step, both the metric and nonmetric data recorded for the Man Bac specimens were compared with those of present-day Vietnamese. In univariate comparisons, Student's t-test and chi-square tests were employed to assess any significant differences in tooth dimensions and the frequency of the presence of nonmetric traits, respectively. In order to compare the magnitude of intra-group odontometric variation, coefficients of variation (CV = SD / M X 100: SD = standard deviation, M = mean value) were calculated for each measurement.

The next step included multivariate comparative analyses between Man Bac and other samples using male permanent tooth data. Similarities in odontometric proportions were estimated by Q-mode correlation coefficients based on odontometric data sets. Following this, measurement data were standardised using grand mean values of all comparative samples and standard deviations of the modern Vietnamese sample.

Population affinities, based on odontometric proportions, were estimated by calculating Q-mode correlation coefficients on the basis of full sets of 28 crown diameters. To aid in the interpretation of the matrix of inter-population phonetic distances, the Neighbour Joining method of Saitou and Nei (1987) was applied to the distance matrix (1-r) transformed from Q-mode correlation coefficients (r), using the software package "Splits Tree Version 4.0" provided by Huson and Bryant (2006).

Table 5.1 Comparative population samples, providing permanent dental data, from mainland East/Southeast Asia and the west Pacific.

Sample	Locality	Sample Period	Remarks	Metric Data	Non-metric Data
Early Holocene Vietnam & Laos	Northern Vietnam and Laos	Early Holocene (Hoabinhian-Neolithic)	Bac Son and Da But Cultural sites in Vietnam, Tam Hang and Tam Pong sites in Laos (Mansuy and Colani, 1925; Huard and Saurin, 1938)	-	Matsumura and Hudson, 2005
Hoabinhian	Vietnam	Hoabinhian Culture (c. 11,000 - 8,000 BP)	Sites of Mai Da Nuoc, Mai Da Dieu (Cuong, 1986), Dong Truong, Du Sang and Lan Bon	Matsumura and Hudson, 2005	-
Bac Son	Northern Vietnam	Bac Son Culture (c.8,000 BP)	Sites of Pho Binh Gia, Lang Cuom, and Cua Gi	Matsumura and Hudson, 2005	-
Con Co Ngua	Site in Thanh Hoa Prov., Nrt. Vietnam	Da But Culture (c.5,000 BP)	Patte, 1965; Duy, 1967; Bui, 1991; Thuy, 1990	Matsumura et al., 2001	-
An Son	Site in Long An Prov., Sth Vietnam	Late Neolithic (c.3,800 BP)	samples in Long An Museum, Vietnam (Cuong, 2006)	unpublished	-
Dong Son	Northern Vietnam	Early Iron age (c. 3,000-1,700 BP)	Sites of Vinh Quang, Chau Son, Doi Son, Nui Nap, Minh Duc, Dong Xa (Thuy, 1993 Cuong, 1996)	Matsumura et al., 2001	Matsumura et al., 2001
Hoa Diem	Site in Khanh Hoa Prov., Sth Vietnam	Early Metal age (c.2,000 BP)		unpublished	-
Gua Cha	Kelantan, Malaysia	Late Pleistocene - Early Holocene (Hoabinhian and Neolithic Culture)	Sieveking, 1954; Bulbeck 2000	Matsumura and Hudson, 2005	-
Mesolithic Flores	Flores Island	Early Holocene (c.7,000-4,000 BP)	Sites of Liang Momer, Linag Toge, Liang X, Gua Alo, Aimere, Sampung and Gua Nempong (Verhoeven, 1958; Jacob, 1967)	Matsumura and Pookajorn, 2005	-
Early Flores and Malay	Malay and Flores Island	Hoabinhian-Neolithic	Gua Cha, Guar Kepah Sites and Mesolithic sites in Flores (Jacob, 1967)	-	Matsumura and Hudson, 2005
Khok Phanom Di	Chonburi Prov., Thai	Late Neolithic (c. 4,000-3,500 BP)		Tayles, 1999	-
Weidun & Songze	Sites in Jiangsu Prov., Sth China	Neolithic Majiabang Culture (c.5,000 BP)	Chang, 1986; Nakahashi and Li, 2002	Matsumura 2002	-
Anyang (Yin-Shang)	Henan Province, China	Bronze age (c. 3,300BP)	IHIA, and CASS, 1982, samples in Academia Sinica, Taipei	unpublished	unpublished
Jiangnan	Yangtze River region	Zhou and Western Han periods (c. 2,770-1,992 BP)	Nakahashi and Li, 2002	Matsumura 2002	-
Jomon	Japan	Late Jomon (c.5,000-2,300 BP)	Akazawa and Aikens, 1986	Matsumura 1989	Matsumura1995
Yayoi	Western Japan	Early Metal age (c.2,000 BP)	Kanaseki et al., 1960; Nakahashi, 1989; Hudson, 1990	Matsumura and 1994	Matsumura1995
Australia	Australian aborigines	Modern		Matsumura and Hudson, 2005	Matsumura and Hudson, 2005
New Britain Islanders	New Britain Island	Modern		Matsumura 1995	Matsumura1995
Loyalty	Loyalty Islands	Modern		Matsumura and Hudson, 2005	Matsumura and Hudson, 2005
Andaman	Andaman Islands	Modern		Matsumura and Hudson, 2005	Matsumura and Hudson, 2005
Malay	Mainland Malay	Modern		Matsumura and Hudson, 2005	-
Dayak	Sarawak, Malaysia	Modern		Matsumura and Hudson, 2005	Matsumura and Hudson, 2005
Lesser Sunda	Sulawesi, Timor and Java	Modern		Matsumura and Hudson, 2005	Matsumura and Hudson, 2005
Vietnam	Vietnam	Modern		Matsumura et al., 2010	Matsumura et al., 2010
Laos	Laos	Modern		Matsumura et al, 2010	-
Thai	Bangkok, Thailand	Modern		Matsumura 1994	Matsumura1995
Myanmar	Myanmar	Modern		Matsumura et al., 2010	Matsumura et al., 2010
Atayal	Taiwan	Modern	samples in National Taiwan Univ.	unpublished	unpublished
Hainan	Hainan Island in Sth. China	Modern	samples in National Taiwan Univ.	unpublished	unpublished

Table 5.2 Criteria for scoring presence of the 21 non-metric dental traits.

Trait	Tooth	Description	Criteria	Presence
shoveling	UI1, UI2	Hanihara et al., 1970	Depth of Lingual Fossa (DFL)	DLF >= 0.5mm
double shoveling	UI, UI2	Suzuki and Sakai, 1973	3=+++(strong), 2=++(moderate), 1=+(weak)	2-3
dental tubercle	UI, UI2	Turner II et al., 1991	0=none, 1=faint, 2=trace, 3(strong ridging) - 6(strong cusp)	3-6
spine	UI1	0:none 1:present	1=single, 2=double, 3=triple	1
interruption groove	UI2	Turner II et al., 1991	0=none, 1=M(mesial), 2=Med(central), 3=d(distal)	1-3
winging (bilateral)	UI1	Enoki and Dahlberg, 1958	0=straight, 1=counter wing, 2=bilateral wing, 3=uni-counter wing, 4=uni-lateral wing	1
De Terra's tubercle	UP1	Saheki, 1958	0=none, 1=+(faint ridging), 2=++(small cusp), 3=+++(large cusp)	+,++,+++
double roots	UP1, UP2	Turner II et al., 1991	1=single, 2=double, 3=triple	2-3
Carabelli's trait	UM1	Dahlberg's P-plaque	0=a(none), 1=b, 2=c, 3=d, 4=e, 5=f, 6=g	d - g
hypocone reduction	UM2	Dahlberg's P-plaque	0=3(none), 1=3+(faint hyp cusp), 2=4-(small hyp cusp) , 3=4-(large hyp cusp) 4=4(full size hyp cusp)	3+
sixth cusp	LM1	Turner II et al., 1991	0=none, 1(much small cusp) - 5(much larger cusp)	1 - 5
seventh cusp	LM1	Turner II et al., 1991	0=none, 1(faint)-4(large)	2-4
protostylid	LM1	Dahlberg's P-plaque	0=none, 1=pit, 2=curved groove, 3(secondary groove) - 5(free apex)	3-5
deflecting wrinkle	LM1	Turner II et al., 1991	0=none, 1=faint, 2=moderately deflect, 3=L-shape	2-3
groove pattern Y	LM1	Jørgensen, 1955	1=Y, 2=+, 3=X	Y
groove pattern X	LM2	Jørgensen, 1955	1=Y, 2=+, 3=X	X
number of cusps (hypoconulid reduction)	LM2	Turner II et al., 1991	4=0(no hyld), 5=1(small hyld) - 5(very large hyld), 6=with cusp 6	4 (4 cusps molar)

U:Upper, L:Lower, I:Incisor, C:Canine, P:Premolar, M:Molar.

In the odontometric analyses, mean values obtained from small sample sizes were utilised for statistical procedures, but for the nonmetric trait comparisons small sample sizes skew the frequency data. Statistically, population comparisons using frequency data require larger sample sizes in each sub-sample than the odontometric comparisons. For this reason some population samples with small sample size were combined or excluded from the non-metric comparisons, as summarised in Table 5.1. C.A.B. Smith's distances (Berry and Berry, 1967), often referred to as "mean measure of divergence" values, were calculated to evaluate population affinities based on the presence/absence frequencies of the 21 non-metric traits. Finally, the Neighbour Joining analysis was applied to the Smith's distance matrix in order to provide a summarised pattern of population affinities in the non-metric trait battery.

RESULTS

Quantitative Data Comparisons

Summary Man Bac and modern Vietnamese permanent and deciduous odontometric statistics are presented in Tables 5.3 and 5.4. Buccolingual diameters of deciduous teeth were not measured for the modern Vietnamese specimens as the material used was plaster casts taken from living residents in which the maximum diameter point at the crown was covered by gingiva. Significant differences were found only in a few measurements for the deciduous teeth between the Man Bac and modern Vietnamese. The former sample possesses larger anterior teeth as compared to modern Vietnamese.

In comparing the permanent dentition, only four male crown diameters are statistically significantly different to those of the modern Vietnamese sample, in each case the Man Bac diameter is smaller. Regarding females, only a single statistically significant difference was found between Man Bac and modern Vietnamese crown diameters, in this case Man Bac UM1 BL diameters are larger.

When comparing the coefficients of variation (CV) between the Man Bac and modern Vietnamese series, no specific pattern of variability in either the permanent or deciduous dentition was observed.

Table 5.5 gives the distance matrix (1-r) transformed from the Q-mode correlation coefficients (r). Figure 5.1 displays the results of the Neighbour Joining analysis applied to the distance matrix of Table 5.5. The Man Bac sample is quite close to present-day populations from Laos, Thailand and Malaysia. The modern samples from Myanmar, Vietnam and the Metal Period Dong Son Vietnamese, as well as the contemporaneous series Khok Phanom Di from Thailand and the Neolithic southern Chinese from Weidun and Songze, are clustered in a second degree of proximity to Man Bac. In contrast, mid-Holocene samples, such as Con Co Ngua (representative of Da But communities) and late Pleistocene/early Holocene Bac Son and Hoabinhian series are grouped in the other major cluster, consisting of Australo-Melanesians, Andaman Islanders, and early Malay and Flores samples (including Gua Cha). Two late samples from Jomon (Japan) and one from neolithic An Son (southern Vietnam) are located intermediately within this schema.

Table 5.3 Summary statistics of mesiodistal and buccolingual crown diameters of Man Bac and present-day Vietnamese.

	Man Bac								Vietnamese									
	Males				Females				Males					Females				
	n	Mean	SD	CV	n	Mean	SD	CV	n	Mean	SD	CV	t-value	n	Mean	SD	CV	t-value
Mesiodistal diameters (mm)																		
UI1	17	8.51	0.52	6.1	11	8.53	0.66	7.8	19	8.73	0.44	5.1	1.37	8	8.35	0.33	3.9	0.71
UI2	6	7.24	0.44	6.0	4	7.11	1.29	18.1	18	7.14	0.50	7.0	0.44	8	6.64	0.41	6.2	0.98
UC	17	7.96	0.46	5.8	10	7.70	0.33	4.3	21	8.08	0.39	4.8	0.87	8	7.90	0.24	3.1	1.43
UP1	15	7.24	0.49	6.7	7	7.35	0.78	10.6	32	7.54	0.52	6.9	1.88	8	7.42	0.32	4.3	0.23
UP2	18	6.80	0.67	9.8	8	6.95	0.49	7.1	31	7.19	0.74	10.3	1.84	8	6.98	0.28	4.0	0.15
UM1	17	10.42	0.49	4.7	8	10.44	0.68	6.5	48	10.53	0.46	4.4	0.83	9	10.05	0.34	3.4	1.52
UM2	18	9.27	0.68	7.3	9	9.44	0.68	7.2	40	9.58	0.51	5.3	1.93	6	9.29	0.31	3.3	0.50
LI1	12	5.50	0.33	6.0	7	5.34	0.63	11.9	19	5.49	0.32	5.8	0.08	8	5.33	0.16	3.0	0.04
LI2	12	6.12	0.29	4.7	7	5.90	0.64	10.8	23	6.07	0.34	5.6	0.43	8	5.92	0.20	3.5	0.08
LC	17	6.95	0.49	7.0	9	6.79	0.62	9.1	25	7.14	0.38	5.4	1.41	8	6.76	0.20	3.0	0.13
LP1	16	7.06	0.55	7.8	7	6.75	0.50	7.5	26	7.43	0.47	6.3	2.32 *	8	7.08	0.34	4.7	1.51
LP2	16	7.02	0.58	8.2	10	6.98	0.48	6.9	24	7.61	0.43	5.7	3.70 ***	8	7.12	0.24	3.4	0.75
LM1	17	11.72	0.46	3.9	8	11.37	0.75	6.6	31	11.63	0.45	3.9	0.66	8	11.04	0.27	2.5	1.17
LM2	18	10.61	0.73	6.9	8	10.54	0.96	9.1	25	10.98	0.84	7.6	1.50	6	10.02	0.44	4.4	1.22
Buccolingual diameters (mm)																		
UI1	17	7.20	0.35	4.8	11	7.02	0.45	6.4	4	7.86	0.21	2.7	3.58 ***	2	7.75	0.36	4.7	2.15
UI2	6	6.73	0.32	4.7	4	6.34	0.82	12.9	4	7.07	0.42	6.0	1.46	2	6.49	0.33	5.1	0.24
UC	18	8.21	0.65	8.0	10	7.73	0.65	8.4	11	8.55	0.58	6.8	1.42	3	7.69	0.07	0.9	0.10
UP1	15	9.58	0.50	5.2	12	9.20	0.56	6.1	18	9.62	0.61	6.4	0.20	8	9.35	0.45	4.8	0.63
UP2	18	9.38	0.47	5.1	13	9.10	0.57	6.2	17	9.40	0.63	6.7	0.11	8	9.15	0.44	4.8	0.21
UM1	20	11.80	0.77	6.5	9	12.06	0.78	6.5	35	11.71	0.59	5.1	0.49	9	11.02	0.50	4.5	3.37 ***
UM2	17	11.30	0.71	6.3	10	11.08	1.00	9.1	30	11.64	0.69	5.9	1.61	7	10.79	0.39	3.6	0.72
LI1	12	5.82	0.42	7.2	7	5.67	0.56	9.8	9	5.93	0.51	8.5	0.54	5	5.94	0.25	4.2	1.00
LI2	12	6.16	0.33	5.4	7	5.97	0.58	9.7	12	6.30	0.49	7.8	0.82	5	6.30	0.32	5.1	1.14
LC	18	7.56	0.55	7.3	12	7.24	0.61	8.4	12	8.35	0.44	5.2	4.16 ***	3	6.97	0.28	4.1	0.73
LP1	18	8.08	0.63	7.8	12	7.63	0.44	5.8	25	8.25	0.55	6.7	0.94	8	7.88	0.52	6.6	1.16
LP2	18	8.36	0.53	6.4	12	7.97	0.56	7.0	24	8.66	0.48	5.6	1.92	8	8.22	0.53	6.4	1.00
LM1	19	10.93	0.70	6.4	10	10.57	0.61	5.8	33	10.87	0.57	5.2	0.34	8	10.32	0.47	4.6	0.95
LM2	18	10.20	0.69	6.8	9	9.89	0.81	8.2	29	10.38	0.63	6.1	0.92	5	10.20	0.43	4.2	0.79

CV=Coefficient of variation, ***: significance level at 0.5%, **: 1%, and *: 5%by t-test.

Table 5.4 Summary statistics of mesiodistal and buccolingual crown diameters of Man Bac and present-day Vietnamese.

	Man Bac				Vietnamese				
	n	Mean	SD	CV	n	Mean	SD	CV	t-value
Mesiodistal diameters (mm)									
udi1	11	7.04	0.37	5.2	15	6.62	0.29	4.4	3.207 ***
udi2	14	5.92	0.32	5.4	15	5.58	0.38	6.8	2.652 *
udc	15	7.04	0.21	3.0	15	6.68	0.40	6.1	3.021 **
udm1	13	7.30	0.37	5.0	15	7.48	0.36	4.9	1.275
udm2	13	9.14	0.67	7.4	15	9.15	0.56	6.2	0.052
ldi1	9	4.51	0.29	6.3	15	4.26	0.34	7.9	1.860
ldi2	10	5.04	0.22	4.3	15	4.78	0.38	7.9	1.914
ldc	12	5.99	0.19	3.1	15	5.90	0.30	5.1	0.897
ldm1	14	8.47	0.45	5.3	15	8.28	0.60	7.2	0.960
ldm2	15	10.50	0.35	3.3	15	10.44	0.50	4.8	0.365
Buccolingual diameter (mm)									
udi1	9	4.96	0.32	6.4					
udi2	13	4.87	0.34	7.0					
udc	14	5.96	0.30	5.1					
udm1	12	8.97	0.45	5.1		no data			
udm2	11	10.16	0.35	3.4					
ldi1	8	3.74	0.11	2.9					
ldi2	10	4.34	0.35	8.2					
ldc	11	5.53	0.45	8.1					
ldm1	13	7.06	0.34	4.8					
ldm2	14	8.77	0.43	4.9					

u: upper, l:lower, d:deciduous, i:incisor, c:canine, m:moler, CV=Coefficient of variation, significance level at ***: 0.5%, **: 1%, and *: 5% by t-test.

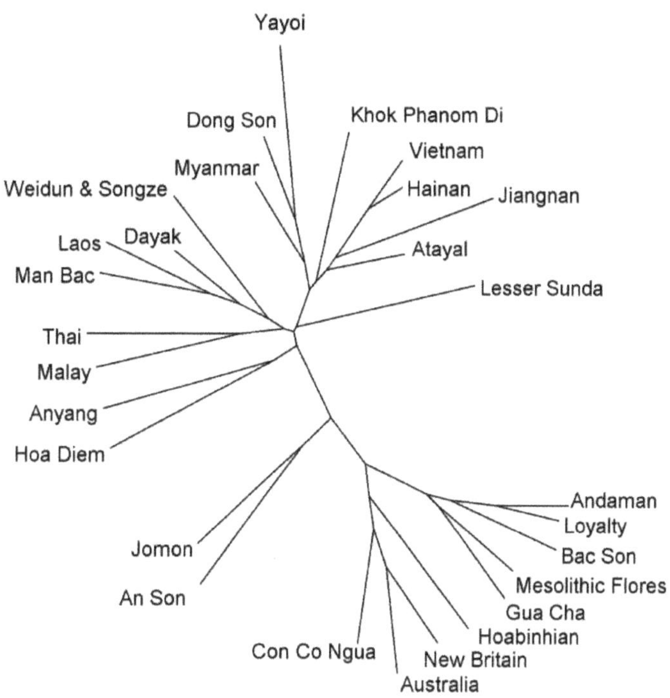

Figure 5.1 An un-rooted tree of neighbour joining analysis applied to the distance matrix of Q-mode correlation coefficients in Table 5.5, using 28 crown diameters of the male permanent dentition.

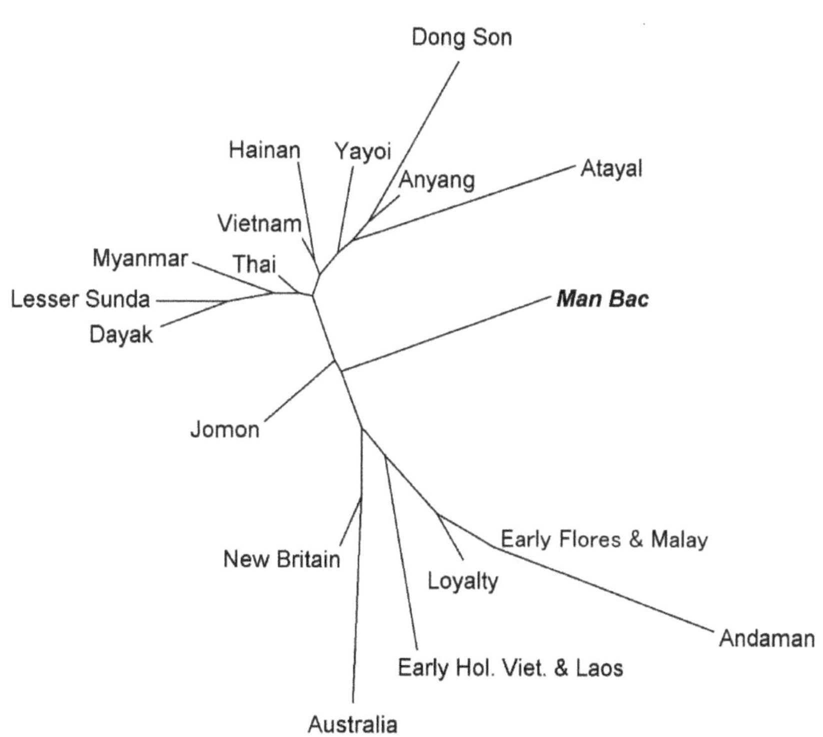

Figure 5.2 An un-rooted tree of neighbour joining analysis applied to Smith's distance matrix of Table 5.7, using frequency data of 21 nonmetric dental traits in the permanent dentition (sexes combined).

Qualitative Data Comparisons

Table 5.6 provides the frequency of the 21 non-metric dental traits recorded for the Man Bac and present-day Vietnamese assemblages. Statistically significant differences were detected in four of the 21 traits. The Man Bac sample shows higher occurrences of UI1 dental tubercle and LM1 seventh cusp, with lower frequencies of UI1 shoveling and UP1 De Terra's tubercle, as compared with modern Vietnamese.

Smith's distances computed using the 21 trait frequencies are presented in Table 5.7. An un-rooted tree of the Neighbour Joining analysis applied to the Smith's distance matrix is depicted in Figure 5.2. The samples compared are clearly divided into two major clusters. The first consists of early and modern East Asian samples and a sub-cluster of modern mainland and island Southeast Asians. The Metal Period Dong Son branches off from this assemblage. The other major cluster encompasses the remaining Hoabinhian-Neolithic Malay, Australian Aborigines, Melanesians and Andaman Islanders. The early Vietnamese series, consisting of the Bac Son and Da But (Con Co Ngua) series, also branch off from this assemblage. Man Bac, as well as the Jomon series, are positioned intermediately between the two major clusters.

DISCUSSION AND CONCLUSIONS

The 'Two Layer' model, or 'Immigration' hypothesis, supported by a wide array of archaeological, historical linguistic and genetic studies, is important for our understanding of the complexities of the population history of Southeast Asia. The prehistoric expansion of language families, specifically the Austronesian and Austroasiatic, can be correlated with the Neolithic dispersal of food producing populations (Renfrew, 1987, 1989, 1992; Bellwood, 1991, 1993, 1997; Hudson, 1994, 1999, 2003; Higham 1998, 2001; Hill, 2001; Bellwood and Renfrew, 2003; Diamond and Bellwood, 2003). In regards to the contributions of studies of human skeletal remains, there have been long term debates on this issue (for review see also Oxenham and Tayles, 2006). In contrast to the traditional "Two Layer" model, some recent cranial and dental studies (eg. Turner 1989, 1990, 1992, Hanihara, 1993, 1994, Pietrusewsky 1992, 1994, 1999) propose that the evolution of many present-day Southeast Asians was by local adaptation, and not by significant admixture with new food producing communities expanding from a source somewhere in mainland East Asia. In terms of craniodental morphology a difficulty arises in distinguishing between *in situ* local modernisation and gene flow mediated change (so-called "Mongoloidisation" by Bulbeck, 1982). Although regional population groupings such as those sometimes termed "Mongoloid" and "Australo-Melanesian" cannot be seen as modernisation in the sense of monophyletic groups, such a conundrum nevertheless remains central to the interpretation of the data analysed, including the Man Bac series. A dental morphological approach may help shed light on the debate given the generally accepted better heritability of dental traits in comparison to cranial morphology.

Previous dental analyses (Matsumura et al. 2001) demonstrated a large morphological gap between later Dong Son and early Holocene samples represented by Bac Son and Da But (Con Co Ngua) people. It was argued that the discontinuity

Table 5.5 Distance (1-r) transformed from q-mode correlation coefficients, on the bases of 28 crown diameters of the male permanent dentition.

		[1]	[2]	[3]	[4]	[5]	[6]	[7]	[8]	[9]	[10]	[11]	[12]	[13]	[14]	[15]	[16]	[17]	[18]	[19]	[20]	[21]	[22]	[23]	[24]	[25]	[26]	[27]
[1]	Andaman																											
[2]	Yayoi	1.02																										
[3]	Khok Phanom Di	1.52	0.79																									
[4]	Malay	1.22	1.19	0.77																								
[5]	An Son	0.85	1.41	1.23	1.06																							
[6]	Anyang	1.21	0.89	0.90	1.09	1.55																						
[7]	Atayal	1.53	1.12	0.52	0.72	1.08	0.99																					
[8]	Australia	0.71	0.92	1.18	1.26	0.85	1.29	1.54																				
[9]	Bac Son	0.55	0.96	1.50	1.12	1.17	0.66	1.38	0.81																			
[10]	Con Co Ngua	0.78	0.96	0.93	1.22	0.79	1.19	1.35	0.50	1.19																		
[11]	Dayak	1.56	1.31	0.64	0.73	0.75	1.05	0.40	1.31	1.60	1.11																	
[12]	Dong Son	1.41	0.59	0.68	1.28	1.25	0.72	0.61	1.40	1.15	1.42	0.96																
[13]	Gua Cha	0.56	1.04	1.49	1.41	0.83	1.32	1.23	0.72	0.56	1.10	1.27	1.08															
[14]	Hainan	1.37	1.16	0.63	0.62	1.39	0.85	0.33	1.36	1.26	1.37	0.60	0.71	1.28														
[15]	Hoabinhian Vietnam	0.86	1.02	1.03	1.17	1.18	0.80	1.62	0.52	0.72	0.84	1.45	1.22	1.02	1.33													
[16]	Lesser Sunda	1.11	1.23	0.86	0.85	1.10	0.99	0.77	1.33	1.43	1.02	0.78	0.92	1.37	0.68	1.07												
[17]	Jiangnan	1.20	1.05	0.90	1.02	1.13	1.04	0.62	1.05	1.05	1.34	0.78	0.69	0.61	0.64	1.11	1.22											
[18]	Jomon	0.78	0.74	0.81	1.21	0.71	1.31	1.16	0.85	1.42	0.60	0.88	1.01	1.11	1.33	1.05	1.07	1.10										
[19]	Laos	1.13	1.22	0.97	0.98	0.75	1.27	0.66	1.31	1.53	0.80	0.50	1.12	1.15	0.90	1.43	0.68	1.04	0.76									
[20]	Loyalty	0.32	1.04	1.34	1.40	1.07	1.06	1.48	0.40	0.52	0.71	1.65	1.23	0.59	1.20	0.63	1.06	1.12	1.14	1.40								
[21]	Mesolithic Flores	0.66	0.86	1.37	1.34	1.02	1.25	1.05	0.86	0.64	0.93	1.53	0.87	0.48	1.18	1.16	1.33	0.93	1.18	1.18	0.57							
[22]	Myanmar	1.66	0.75	0.64	0.96	1.29	0.75	0.44	1.48	1.21	1.20	0.75	0.45	1.17	0.71	1.33	0.62	0.88	1.34	0.86	1.42	0.96						
[23]	New Britain	0.65	1.21	1.28	1.30	0.90	1.16	1.24	0.44	0.87	0.63	1.36	1.36	0.93	1.27	0.72	1.02	1.14	0.95	1.00	0.47	0.68	1.33					
[24]	Thai	1.17	0.94	1.11	0.69	1.01	1.00	0.93	1.35	1.07	1.50	0.71	0.92	1.09	0.73	1.18	0.89	1.10	1.28	0.81	1.29	1.35	0.91	1.58				
[25]	Vietnam	1.26	0.68	0.60	0.85	1.61	0.91	0.54	1.27	1.18	1.35	0.96	0.53	1.14	0.22	1.17	0.78	0.62	1.21	1.07	1.07	0.91	0.59	1.20	0.85			
[26]	Weidun & Songze	1.43	1.07	0.96	0.93	0.86	0.84	0.49	1.41	1.12	1.41	0.53	0.68	1.04	0.81	1.28	0.95	0.73	1.09	0.81	1.6	1.13	0.59	1.24	0.8	0.99		
[27]	Hoa Diem	1.07	1.01	1.11	0.77	1.05	0.82	0.94	1.14	1.06	1.01	1.02	0.96	1.36	0.93	0.99	0.90	1.14	1.17	1.05	0.96	0.95	0.93	0.85	0.84	0.98	0.93	
[28]	Man Bac	1.30	1.09	0.82	0.84	0.85	0.92	0.78	1.58	1.51	1.07	0.53	1.01	1.58	1.00	1.34	0.82	1.28	0.65	0.53	1.64	1.45	0.94	1.23	0.77	1.15	0.83	0.85

Table 5.6 Frequencies of 21 non-metric dental traits: Man Bac and modern Vietnamese permanent teeth.

		Man Bac			Vietnamese				
		O	A	Freq.	O	A	Freq.	chi-value	
shoveling	UI1	29	18	62.1	41	34	82.9	3.869	*
shoveling	UI2	12	4	33.3	39	19	48.7	0.877	
double shoveling	UI1	33	5	15.2	41	6	14.6	0.004	
double shoveling	UI2	15	0	0.0	40	1	2.5	0.382	
dental tubercle	UI1	33	8	24.2	38	1	2.6	7.452	**
dental tubercle	UI2	14	1	7.1	36	2	5.6	0.045	
spine	UI1	30	8	26.7	40	10	25.0	0.025	
interruption groove	UI2	14	4	28.6	29	5	17.2	0.732	
winging (bilateral)	UI1	29	0	0.0	41	4	9.8	3.001	
De Terra's tubercle	UP1	12	0	0.0	60	16	26.7	4.114	*
double rooted	UP1	11	4	36.4	50	25	50.0	0.672	
double rooted	UP2	11	0	0.0	49	3	6.1	0.709	
Carabelli's trait	UM1	23	10	43.5	92	26	28.3	1.981	
hypocone reduction	UM2	31	6	19.4	75	15	20.0	0.006	
sixth cusp	LM1	20	6	30.0	63	15	23.8	0.308	
seventh cusp	LM1	25	4	16.0	66	2	3.0	4.952	*
protostylid	LM1	19	0	0.0	65	1	1.5	0.296	
deflecting wrinkle	LM1	18	3	16.7	53	19	35.8	2.312	
groove pattern Y	LM1	23	16	69.6	57	43	75.4	0.292	
groove pattern X	LM2	29	6	20.7	65	21	32.3	1.322	
hypoconulid reduction	LM2	25	15	60.0	66	25	37.9	3.602	

O: observed numbers of dentitions, A: affected numbers of dentitions, Freq: frequency (%).
significance level at 5%, and **: 1% by chi-square test.

between early and later Holocene populations was due to considerable levels of gene flow into what is now northern Vietnam from migrants moving in from the northern or eastern peripheral areas. Subsequent studies, using the early Hoabinhian Hang Cho specimen, and Man Bac remains excavated from 1999-2005, reconfirmed the large morphological discontinuity between these two sequences (Matsumura et al., 2008a and 2008b). The Hang Cho specimen was posited as representing an ancestral population of subsequent early to mid Holocene people, whereas the majority of the Man Bac assemblage (1999-2005 series), as well as the later Dong Son and modern Vietnamese, were thought to have closer genetic ties with immigrants from the northern peripheral area of what is now Vietnam and southern China.

In order to develop a more comprehensive interpretation of Man Bac population affinities this study has utilised a wide range of dental data and employed a geographically large and temporally deep comparative data set. This analysis has demonstrated that: (1) Man Bac odontometric variance is greater than that of modern Vietnamese; (2) in terms of odontometric proportions, the Man Bac sample has a closer affinity with modern Vietnamese than the earlier late Pleistocene/early to mid Holocene series (e.g. Con Co Ngua (Da But), Bac Son and Hoabinhian), who are in turn phenotypically akin to Australo-Melanesian populations; (3) with respect to non-metric dental traits, the Man Bac series is situated midway between earlier groups (e.g. Bac Son and Con Co Ngua (Da But) and modern Vietnamese. Inconsistencies in the results exhibited by the metric and non-metric data might

Table 5.7 Smith's distances based on 21 non-metric dental traits of the permanent dentition (sexes combined).

		[1]	[2]	[3]	[4]	[5]	[6]	[7]	[8]	[9]	[10]	[11]	[12]	[13]	[14]	[15]	[16]	[17]
[1]	Andaman																	
[2]	Dayak	0.33																
[3]	Early Flores & Malay	0.13	0.18															
[4]	Lesser Sunda	0.33	0.08	0.19														
[5]	E-Hol. Viet. & Laos	0.20	0.20	0.20	0.17													
[6]	Loyalty	0.14	0.21	0.13	0.21	0.23												
[7]	Jomon	0.30	0.15	0.17	0.14	0.23	0.16											
[8]	Yayoi	0.50	0.20	0.30	0.17	0.33	0.31	0.12										
[9]	Thai	0.35	0.08	0.21	0.08	0.21	0.20	0.10	0.07									
[10]	Australia	0.38	0.27	0.20	0.29	0.26	0.35	0.27	0.25	0.23								
[11]	New Britain	0.20	0.15	0.10	0.18	0.23	0.15	0.17	0.21	0.11	0.14							
[12]	Vietnam	0.37	0.10	0.20	0.12	0.21	0.19	0.15	0.10	0.07	0.23	0.18						
[13]	Dong Son	0.48	0.28	0.25	0.26	0.29	0.29	0.20	0.20	0.20	0.26	0.31	0.13					
[14]	Myanmar	0.42	0.10	0.22	0.13	0.29	0.25	0.22	0.15	0.09	0.29	0.19	0.08	0.24				
[15]	Atayal	0.44	0.28	0.39	0.32	0.41	0.28	0.26	0.18	0.18	0.41	0.29	0.18	0.26	0.29			
[16]	Hainan	0.49	0.17	0.34	0.16	0.33	0.22	0.17	0.14	0.10	0.40	0.29	0.07	0.21	0.13	0.20		
[17]	Anyang	0.55	0.20	0.34	0.21	0.38	0.28	0.19	0.09	0.11	0.31	0.28	0.07	0.12	0.11	0.15	0.10	
[18]	Man Bac	0.32	0.24	0.23	0.21	0.25	0.16	0.18	0.30	0.24	0.39	0.25	0.18	0.24	0.30	0.32	0.18	0.30

reflect sample bias or perhaps different patterns of genetic inheritance. Nevertheless, it can be concluded that the dental affinities of the Man Bac people indicate a considerable level of gene flow from Neolithic East Asia. Similarities between late Pleistocene and early to mid Holocene series and Man Bac, in terms of nonmetric dental morphology, suggest Man Bac was a population in genetic transition.

SUMMARY

This chapter provides quantitative (metric) and qualitative (non-metric) dental data recorded for the Man Bac series. It presents results of analyses using batteries of tooth traits useful for assessing biological affinities. Multivariate comparisons using odontometric data sets revealed the closer affinity of the Man Bac people to later Metal Age Dong Son and present-day Southeast Asians than to earlier populations such as the Con Co Ngua (Da But) and Bac Son/Hoabinhian series, who are in turn phenotypically more akin to Australo-Melanesian populations. Although the analysis of the non-metric trait battery suggests that Man Bac people still partially preserved genetic features of earlier indigenous populations, this study concludes that the population structure of Man Bac was affected by major gene flow that can likely be sourced to new immigrants from peripheral northern or eastern areas of East Asia, including southern China.

ACKNOWLEDGMENTS

The author is grateful to Former Director Dr. Ha Van Phung, and Vice Director Dr. Nguyen Giang Hai, the Vietnamese Institute of archaeology, for their permission and corroboration to excavate the Man Bac site. Thanks are due to Professor Dr. Hoang Tu Hung, Dean of Faculty of Odonto-Stomatology, the University of Medicine and Pharmacy at Ho Chi Minh City, Dr. Nguyen Kim Thuy and Dr. Nguyen Lan Cuong, the Vietnamese Institute of Archaeology, Hanoi, Dr. Bui Chi Hoang, Southern Institute of Sustainable Development, Mr. Nguyen Tam, Khanh Hoa Provincial Museum, Dr. Bui Phat Diem and Dr. Vuong Thu Hong, the Long An Provincial Museum, Dr. Tsai Hsi-Kue, National Taiwan University, College of Medicine and Dr. Wang Daw-Hwan, IHP, Academia Sinica, Taipei, for the permission to study the comparative dental specimens.

This study was supported in part by a Grant-in-Aid in 2003-2005 (No. 15405018) and 2008-2009 (No. 20370096) from the Japan Society for the Promotion of Science, and by the Toyota Foundation in 2006-2007 (No. D06-R-0035).

LITERATURE CITED

Akazawa T, Aikens CM. 1986. Introduction. In: Akazawa T, Aikens CM, editors. Prehistoric Hunter-Gatherers in Japan: New research Methods. Bulletin of University Museum of the University of Tokyo 27: 9-10.

Barth F. 1952. The southern Mongoloid migration. Man 52:5-8.

Bellwood P. 1987. The prehistory of island Southeast Asia: a multidisciplinary review of recent research. J World Prehistory 1:171-224.

Bellwood P. 1991. The Austronesian dispersal and the origin of languages. Scientific American 265:88-93.

Bellwood P. 1993. An archaeologist's view of language macrofamily relationships. Bull Indo-Pacific Prehist Assoc 13:46-60.

Bellwood P. 1997. Prehistory of the Indo-Malaysian archipelago, revised edition. Honolulu: Univ Hawai'i Press.

Bellwood P, Gillespie R, Thompson GB, Vogel JS, Ardika IW, Datan I. 1992. New dates for prehistoric Asian rice. Asian Perspectives 31:161-170.

Bellwood P, Renfrew C. eds. 2002. Examining the farming/language dispersal hypothesis. Cambridge: McDonald Institute for Archaeological Research.

Berry AC, Berry RJ. 1967. Epigenetic variation in the human cranium. J Anat 101:361-379.

Brace CL. 1976. Tooth reduction in the Orient. Asian Perspectives 19:203-219.

Brace CL, Tracer DP, Hunt KD. 1991. Human craniofacial form and the evidence for the peopling of the Pacific. Bull Indo-Pacific Prehist Assoc 12:247-269.

Bui V. 1991. The Da But culture in the stone age of Vietnam. Bull Indo-Pacific Prehist Assoc 10:127-131.

Bulbeck D. 1982. A re-evaluation of possible evolutionary processes in Southeast Asia since the late Pleistocene. Bull Indo-Pacific Prehist Assoc 3:1-21.

Bulbeck D. 2000. Dental morphology at Gua Cha, West Malaysia, and the implications for "Sundadonty". Bull Indo-Pacific Prehist Assoc 19:17-41.

Callenfels VS. 1936. The Melanesoid civilizations of Eastern Asia. Bulletin of the Raffles Museum Series B,1: 41-51.

Chang KC. 1986. The archaeology of ancient China, fourth edition. New Haven: Yale Univ Press.

Coon CS. 1962. The origin of races. New York: Alfred A Knoph.

Cuong NL. 1986. Two early Hoabinhian crania from Thanh Hoa province, Vietnam. Zeitschrift für Morphologie und Anthropologie 77: 11-17.

Cuong NL. 1996. Anthropological characteristics of Dong Son population in Vietnam. Hanoi: Social Sciences Publishing House (in Vietnamese with English title and summary).

Cuong N.L. (2006) About the ancient human bones at An Son (Long An) through the third excavation. Khao Co Hoc (Vietnamese Archaeology) 6-2006: 39-51 (in Vietnamese with

English title and summary).

Diamond J, Bellwood P. 2003. Farmers and their languages: the first expansions. Science 300:597-603.

Duy N. 1967. Etat actuel de l'etude raciale des cranes anciens decouverts au Vietnam. L'Anthropologie No. 3-4. Hanoi: Social Sciences Publishing House (in French).

Enoki K, Dahlberg AA. 1958. Rotated maxillary central incisors. Orthodontic J Japan 17:157.

Fujita T. 1949. On the standard for measurement of teeth. J Anthropol Soc Nippon 61:27-32 (in Japanese).

Glover IC, Higham CFW. 1996. New evidence for early rice cultivation in South, Southeast and East Asia. In: Harris DR, editor. The Origins and Spread of Agriculture and Pastoralism in Eurasia. London: UCL Press. p 413-441.

Hanihara K, Tanaka T, Tamada M. 1970. Quantitative analysis of the shovel-shaped character in the incisors. J Anthropol Soc Nippon 78:90-93.

Hanihara T. 1993. Population history of East Asia and the Pacific as viewed from craniofacial morphology: The basic populations in East Asia, IV. Am J Phys Anthropol 91:173-187.

Hanihara T. 1994. Craniofacial continuity and discontinuity of Far Easterners in the late Pleistocene and Holocene. J Hum Evol 27:417-441.

Higham CFW. 1998. Archaeology, linguistics and the expansion of the East and Southeast Asian Neolithic. In: Blench R, Spriggs M, editors. Archaeology and Language II: Archaeological Data and Linguistic Hypotheses. London: Routledge. p 103-114.

Higham CFW. 2001. Prehistory, language and human biology: is there a consensus in East and Southeast Asia? In: Jin L, Seielstad M, Xiao CJ, editors. Genetic, Linguistic and Archaeological Perspectives on Human Diversity in Southeast Asia. Singapore: World Scientific. p 3-16.

Hill J.H. 2001. Proto-Uto-Aztecan: a community of cultivators in central Mexico? Am Anthropologists 103:913-934.

Howells WW. 1976. Physical variation and history in Melanesia and Australia. Am J Phys Anthropol 45:641-650.

Huard P, Saurin E. 1938. État actuel de la craniologie Indochinoise. Bull Service Géologique de l'Indochine XXV (in French).

Hudson MJ. 1990. From Toro to Yoshinogari: changing perspectives on Yayoi period archeology. In: Barnes GL, editor. Hoabinhian, Jomon, Yayoi, Early Korean States: Bibliographic Reviews of Far Eastern Archaeology 1990. Oxford: Oxbow. p 63-111.

Hudson MJ. 1994. The linguistic prehistory of Japan: some archaeological speculations. Anthropol Sci 102:231-255.

Hudson MJ. 1999. Japanese and Austronesian: n archeological perspective on the proposed linguistic links. In: Omoto K, editor. Interdisciplinary Perspectives on the Origins of the Japanese. Kyoto: International Research Center for Japanese Studies. p 267-279.

Hudson MJ. 2003. Agriculture and language change in the Japanese islands. In: Bellwood P, Renfrew C, editors. Examining The Farming/Language Dispersal Hypothesis. Cambridge: McDonald Institute for Archaeological Research. p 311-318.

Huson DH, Bryant D. 2006. Application of phylogenetic networks in evolutionary studies. Molecular Biology and Evolution 23: 254-267.

Institute of History, Institute of Archaeology (IHIA), Chinese Academy of Social Science (CASS), editors. 1982. Contributions to the study on human skulls from the Shang sites at Anyang. Beijing: Cultural Relics Publishing House (in Chinese with English summary).

Jacob T. 1967. Some problems pertaining to the racial history of the Indonesian region.

Utrecht: Ph.D. dissertation, Univ Utrecht.

Jacob T. 1975. Morphology and paleontology of early man in Java. In: Tuttle RH, editor. Paleoanthropology, Morphology, and Paleoecology. Paris: Hague. p 311-324.

Jørgensen 1955. The Dryopithecus pattern in recent Danes and Dutchmen. J Dent Res 34: 195-208.

Kanaseki T, Nagai M, Sano H. 1960. Craniological studies of the Yayoi-period ancients excavated at the Doigahama site, Yamaguchi prefecture. Quarterly J Anthropol VII (3-4). Supplement:1-35 (in Japanese with English summary).

Mansuy H, Colani M. 1925. Contribution à l'etude de la préhistoire de l'Indochine VII. Néolithique inférieur (Bacsonien) et Néolithique supérieur dans le Haut-Tonkin. Bull Service Géologique de l'Indochine XII (in French).

Matsumura H. 1989. Geographical variation of dental measurements in the Jomon population. J Anthropol Soc Nippon 97:493-512.

Matsumura H. 1994. A microevolutional history of the Japanese people from a dental characteristics perspective. Anthropol Sci 102:93-118.

Matsumura H. 1995. Dental characteristics affinities of the prehistoric to the modern Japanese with the East Asians, American natives and Australo-Melanesians. Anthropol Sci 103:235-261.

Matsumura H. 2002. The possible origin of the Yayoi migrants based on the analysis of the dental characteristics. In: Nakahashi T, Li M, editors. Ancient people in the Jiangnan Region, China. Fukuoka: Kyushu University Press. p 61-72.

Matsumura H, Hudson MJ. 2005. Dental perspectives on the population history of Southeast Asia. Am J Phys Anthropol 127:182-209.

Matsumura H. and Pookajorn S. 2005. Morphometric analysis of the Late Pleistocene human remains from Moh Khew Cave in Thailand. Homo – Journal of Comparative Human Biology 56: 93-118.

Matsumura H, Cuong NL, Thuy NK, Anezaki T. 2001. Dental morphology of the early Hoabinhian, the Neolithic Da But and the Metal Age Dong Son Cultural people in Vietnam. Zeitschrift Morphol Anthropol 83:59-73.

Matsumura H, Oxenham MF, Dodo Y, Domett K, Cuong NL, Thuy NK, Dung K, Huffer D, Yamagata M. 2008a. Morphometric affinity of the late Neolithic human remains from Man Bac, Ninh Binh Province, Vietnam: key skeletons with which to debate the 'Two layer' hypothesis. Anthropol Sci 116:135-148.

Matsumura H, Yoneda M, Dodo Y, Oxenham MF, Dodo Y, Thuy NK, Cuong NL, Dung LM, Long VT, Yamagata M, Sawada J, Shinoda K, Takigawa W. 2008b. Terminal Pleistocene human skeleton from Hang Cho cave, northern Vietnam: implications for the biological affinities of Hoabinhian people. Anthropol Sci 116;201-217.

Matsumura H, Domett K, O'Reilly JWO. 2010. On the origin of pre-Angkorian peoples: perspectives from cranial and dental affinity of the human remains from Iron Age Phum Snay, Cambodia. Anthropological Science (in press).

Mijsberg WA. 1940. On a Neolithic Paleo-Melanesian Lower Jaw Found in Kitchen Midden at Guar Kepah, Province Wellesley, Straits Settlements. Singapore: Proceedings of 3rd Congress of Prehistorians of the Far East. p 100-118.

Nakahashi T. 1989. The Yayoi people. In: Nagai M, Nasu T, Kanaseki Y, Sahara M, editors. Yayoi bunka no kenkyu [Research on Yayoi Culture]. Tokyo: Yuzankaku. p 23-51 (in Japanese).

Nakahashi T, Li M, editors. 2002. Ancient People in the Jiangnan Region, China. Fukuoka: Kyushu Univ Press.

Oxenham MF, Tayles N, editors. 2006. Bioarchaeology of Southeast Asia. Cambridge: Cambridge University Press. p 335-349.

Patte E. 1965. Les ossements du kjokkenmodding de Da But. Hanoi: Bull Service Ethnologie de Indochine. Vol.XL, p 1-87 (in French).

Pietrusewsky M. 1992. Japan, Asia and the Pacific: a multivariate craniometric investigation. In: Hanihara K, editor. Japanese as A Member of the Asian and Pacific Populations. Kyoto: International Research Center for Japanese Studies. p 9-52.

Pietrusewsky M. 1994. Pacific-Asian relationships: a physical anthropological perspective. Oceanic Linguistics 33:407-429.

Pietrusewsky M. 1999. A multivariate craniometric study of the inhabitants of the Ryukyu Islands and comparison with cranial series from Japan, Asia and the Pacific. Anthropol Sci 107:255-281.

Renfrew C. 1987. Archaeology and Language: the Puzzle of Indo-European Origins. London: Jonathan Cape.

Renfrew C. 1989. Models of change in language and archaeology. Trans Philol Soc 87:103-155.

Renfrew C. 1992. World languages and human dispersals: a minimalist view. In: Hall JA, Jarvie IC, editors. Transition to Modernity: Essays on Power, Wealth and Belief. Cambridge: Cambridge Univ Press. p 11-68.

Saheki M.1958. On the heredity of the tooth crown configuration studied in twins. Acta Anat Nipponica 33:456-470 (in Japanese with English summary).

Saitou N, Nei M. 1987. The neighbor-joining method: a new method for reconstructing phylogenetic tree. Molecular Biology and Evolution 4: 406–425.

Schutkowski H. 1993. Sex determination of infant and juvenile skeletons: I. morphognostic features. Am J Phys Anthropol 90:199-205.

Sieveking GG. 1954. Excavations at Gua Cha, Kelantan 1954. Part 1. Federation Mus J 1:75-143.

Spriggs M. 1989. The dating of the island Southeast Asian Neolithic: an attempt at chronometric hygiene and linguistic correlation. Antiquity 63:587-613.

Suzuki M, Sakai T. 1973. The Japanese dentition. Matsumoto: Shinshu Univ Press.

Tayles, N. 1999 The excavation of Khok Phanom Di: A prehistoric site in Central Thailand. Vol. 5: The People. London: The Society of Antiquaries of London, Research Report L.

Thoma A. 1964. Die entstehung der Mongoliden. Homo 15:1-22 (in German).

Thuy NK. 1990. Ancient human skeletons at Con Co Ngua. Khao Co Hoc (Vietnamese Archaeology) 3-1990:37-48 (in Vietnamese with English title and summary).

Thuy NK. 1993. Ancient skulls at Minh Duc. Khao Co Hoc (Vietnamese Archaeology) 3-1993:1-8 (in Vietnamese with English title and summary).

Turner CG II 1989. Teeth and prehistory in Asia. Scientific American 260:70-77.

Turner CG II. 1990. Major features of Sundadonty and Sinodonty, including suggestions about East Asian microevolution, population history and late Pleistocene relationships with Australian Aborigines. Am J Phys Anthropol 82:295-317.

Turner CG II. 1992. Microevolution of East Asian and European populations: a dental perspective. In: Akazawa T, Aoki K, Kimura T, editors. The Evolution and Dispersal of Modern Humans in Asia. Tokyo: Hokusensha. p 415-438.

Turner CG II, Nichol CR, Scott GR.1991. Scoring procedures for key morphological traits of the permanent dentition: The Arizona State University dental anthropology system. In: Kelly MA, Larsen CS, editors. Advances in Dental Anthropology. New York: Wiley-Liss. p 13-31.

Verhoeven TH. 1958. Pleistozane Funde in Flores. Anthropos 53:264–265.

Von Koenigswald GHR. 1952. Evidence of a prehistoric Australo-Melanesoid population in Malaya and Indonesia. Southwestern J Anthropol 8:92-96.

Appendix 5.1 Crown measurements (mm) and the presence of non-metric dental traits of the permanent dentition of the Man Bac series.

Sample Number		99M2	99M3	01M1	01M5	01M9	01M10	05M9	05M10	05M11
Sex		Female	Male	Unknown	Male	Male	Male	Female	Male	Male
Mesiodistal diameters (mm)	UI1	7.95	8.18		8.22	8.35	8.48	8.34	9.15	
	UI2	6.78				6.75			7.87	
	UC	7.30	7.35		7.82	7.61	7.50	7.44		8.83
	UP1	6.81	7.27			7.11				7.41
	UP2	6.85	7.02		6.53	5.60	6.42			7.11
	UM1	9.73	9.93		10.23	10.75			10.68	10.93
	UM2	8.78	8.35		9.57	9.11	9.15	9.98	9.81	10.20
	LI1	4.94	5.40		5.65		4.98	4.93	5.92	5.62
	LI2	5.45	5.96		5.83		6.08	5.33	6.33	6.42
	LC	6.32	6.56		6.51	6.90	6.94	6.87		7.84
	LP1	6.42	7.53		6.23		7.02			7.64
	LP2	6.33	7.53		6.52			7.22		7.53
	LM1	11.11	11.82		11.24			11.23	12.18	12.11
	LM2	9.53	10.29		10.12		10.18	11.04		11.61
Buccolingual diameters (mm)	UI1	6.56	6.53		6.98	7.53	7.42	7.01	7.13	
	UI2	5.74				6.82			6.80	
	UC	6.88	7.51		7.52	7.72	6.62	8.01		8.49
	UP1	8.30	9.13			9.60		9.43		9.62
	UP2	8.38	9.37		8.86	8.91	9.18	8.85		9.39
	UM1	11.06	11.46		11.40	11.51	11.37	11.77	11.37	12.47
	UM2	10.04	10.81		10.94	11.64	11.39	11.73		12.24
	LI1	4.81	5.42		5.58		5.20	5.65	6.01	5.73
	LI2	5.31	5.78		6.29		5.82	6.10	6.07	6.34
	LC	6.68	6.87		7.20	8.04	7.17	7.25		7.80
	LP1	6.85	7.82		7.28		8.24	7.97		8.47
	LP2	6.78	8.00		7.81		8.06	8.50		8.56
	LM1	9.67	10.04		10.65		10.19	10.32	10.67	11.75
	LM2	8.67	9.28		9.64		9.41	9.98		11.02
shoveling	UI1	-	+	-	+	-	+		+	
shoveling	UI2	-				-			+	
double shoveling	UI1	-	-	-	+	-	+	-	-	
double shoveling	UI2	-		-		-			-	
dental tubercle	UI1	-	-	-	-	-	-	-	-	
dental tubercle	UI2	-				-			-	
spine	UI1	-	-	-	-	-	-		-	
interruption groove	UI2	-				-			+	
winging (bilateral)	UI1	-	-	-	-	-		-	-	
De Terra's tubercle	UP1	-	+			+				
double rooted	UP1	-			-	+				-
double rooted	UP2	-	-		-	-	-			-
Carabelli's trait	UM1	-	+	-	+	-			-	
hypocone reduction	UM2	+	-	-	-	-	-	-	-	+
sixth cusp	LM1	-	+	-					-	+
seventh cusp	LM1	-	+	-					-	-
protostylid	LM1	-	-	-					-	-
deflecting wrinkle	LM1	-	-	+					-	-
groove pattern Y	LM1	+	+	+	+				+	-
groove pattern X	LM2	-	-	-	-		-	-	-	-
number of cusps	LM2	+	+				-	-	-	-

U:Upper, L:Lower, I:Incisor, C:Canine, P:Premolar, M:Moler, +: present, -: absent

Appendix 5.1 (Continued 1).

Sample Number		05M13	05M15	05M20	05M24	05M25	05M28	05M29	05M31	05M32
Sex		Unknown	Female	Male	Unknown	Unknown	Female	Male	Male	Male
Mesiodistal	UI1	8.44		8.15	8.25		8.49	8.64		9.10
diameters (mm)	UI2	7.23						6.85		
	UC	7.79	8.36	7.86			7.48	8.19	8.70	
	UP1	7.32	7.98				7.29	7.39	7.28	8.27
	UP2	7.33	7.36	7.07			7.46	6.92	7.14	8.64
	UM1	10.64	10.73	10.43	11.53		10.53	10.59	9.66	11.32
	UM2	9.29	9.04	9.44			9.73	10.20	9.17	
	LI1	5.39	5.62	5.21	5.23			5.70		6.12
	LI2	5.98	6.54	6.34				6.27		
	LC	6.66	7.12	7.30				7.20	7.45	
	LP1	7.04	7.31	7.09				7.17	7.69	7.83
	LP2	7.26	7.10	7.14				7.13	7.47	7.71
	LM1	11.15	12.23	11.39	12.00			12.31	11.21	12.17
	LM2	11.03	9.82	9.38	6.06		10.63	11.16	10.66	12.02
Buccolingual	UI1	8.03		6.83	6.89		6.70	7.86		7.46
diameters (mm)	UI2	7.10						7.29		
	UC	9.09	8.48	7.64			7.77	8.84	8.74	
	UP1	9.51	9.64				9.33	10.36	9.78	9.34
	UP2	9.75	9.53	9.49			9.65	10.05	9.59	9.81
	UM1	11.41	12.91	11.55	11.87			13.24	11.69	11.87
	UM2	11.73	10.29	11.21			11.57	12.50	11.96	
	LI1	5.95	5.77	5.59				6.60		6.49
	LI2	6.82	5.88	6.12				6.80		
	LC	7.69	7.81	7.20				8.37	8.00	
	LP1	8.38	7.97	7.97				9.48	8.54	8.60
	LP2	9.04	8.03	7.91				9.17	8.41	8.24
	LM1	10.77	11.21	10.46	10.79			12.11	10.99	10.91
	LM2	9.99	10.08	9.94			10.03	11.13	10.54	10.80
shoveling	UI1	+		-	+					+
shoveling	UI2	-			+					
double shoveling	UI1	-		+	+		-	-		-
double shoveling	UI2	-			-			-		
dental tubercle	UI1	-		-	-		-	-		+
dental tubercle	UI2	-			-			-		
spine	UI1	-		-	-		-			+
interruption groove	UI2	-			-			+		
winging (bilateral)	UI1	-		-	-			-		
De Terra's tubercle	UP1	-	-							
double rooted	UP1		+	-						-
double rooted	UP2		-						-	
Carabelli's trait	UM1	+	-	-	+					+
hypocone reduction	UM2	-	-	-			-	-	-	
sixth cusp	LM1	-	-	-	-	-				-
seventh cusp	LM1	+	-	-	+	-			-	-
protostylid	LM1	-	-	-	-					-
deflecting wrinkle	LM1	-	+		-	-				
groove pattern Y	LM1	-	+	-	+	-				+
groove pattern X	LM2	+	-	+			-		-	-
number of cusps	LM2	+	-	+			+		+	

U:Upper, L:Lower, I:Incisor, C:Canine, P:Premolar, M:Moler, +: present, -: absent

Appendix 5.1 (Continued 2).

Sample Number		05M34	07H1M1	07H1M3	07H1M4	07H1M5	07H1M8	07H1M9	07H1M10	07H1M11
Sex		Female	Unknown	Female	Female	Male	Male	Male	Female	Female
Mesiodistal	UI1	8.52		7.83	8.24	9.10	7.80	8.19	9.58	8.48
diameters (mm)	UI2			6.01				7.52		
	UC	7.73		7.36	7.68	8.23	8.00	8.47	7.94	
	UP1	7.05		6.67		7.36	7.29	7.48		
	UP2	6.68		6.56		6.09	6.76	7.17		
	UM1	10.18		9.75		10.97	10.11	10.46	10.75	
	UM2			9.21		9.42	7.59	9.70	9.64	8.67
	LI1			4.69			5.23		5.98	
	LI2			5.31			5.75	6.57	5.92	
	LC	6.86		6.24			6.69	7.41	6.77	6.45
	LP1	6.51		6.21			6.80	7.15		6.59
	LP2	7.56		6.46			6.53	7.11	7.48	6.85
	LM1			10.68			11.04	11.49		10.94
	LM2	10.58		9.82		11.04	9.83	10.63		
Buccolingual	UI1	7.14		6.85	6.73	7.22	7.34	6.91	6.97	7.49
diameters (mm)	UI2			5.82				6.47		
	UC	7.54		7.13	7.67	8.53	8.05	8.79	7.67	
	UP1	9.21		8.90	9.64	9.62	9.70	9.56	9.20	9.21
	UP2	9.05		8.63	9.37	9.62	9.37	9.09	9.16	9.52
	UM1	12.35		11.27		11.93	12.21	11.36	12.17	12.20
	UM2			10.72		11.63	10.50	11.73	11.28	10.78
	LI1			5.33			5.81		6.00	
	LI2			5.63			5.80	6.39	6.12	
	LC	7.52		6.82	6.98	8.05	6.26	7.64	7.30	7.47
	LP1	7.80		7.18	7.75	8.78	7.50	7.99	7.28	8.10
	LP2	8.00		7.49	8.47	9.28	8.06	7.77	7.39	8.19
	LM1			9.84	10.10	11.48	10.81	11.08		10.80
	LM2	10.91		8.82		10.79	10.76	10.36		
shoveling	UI1	+	+	+	-	-	-	+	+	+
shoveling	UI2		-	-				-		
double shoveling	UI1	-	-	-	-	-	-	-	-	-
double shoveling	UI2		-	-				-		
dental tubercle	UI1	+	+	-	-	+	-	-	-	+
dental tubercle	UI2		+	-				-		
spine	UI1	+	-	-	-	-	-	-	-	-
interruption groove	UI2		-	-				+		
winging (bilateral)	UI1	-	-	-	-	-	-	-	-	-
De Terra's tubercle	UP1		+	-				+		
double rooted	UP1									+
double rooted	UP2				-					
Carabelli's trait	UM1	+	-	-				-		
hypocone reduction	UM2		-	-		-	-	+	-	-
sixth cusp	LM1		-	-						
seventh cusp	LM1		-	-			-	-		
protostylid	LM1		-	-				-		
deflecting wrinkle	LM1		-	-				-		
groove pattern Y	LM1		+	+				-		
groove pattern X	LM2	-	-	+			+	+		
number of cusps	LM2	+	-	-		-		-		

U:Upper, L:Lower, I:Incisor, C:Canine, P:Premolar, M:Moler, +: present, -: absent

Appendix 5.1 (Continued 3).

Sample Number		07H2M1	07H2M2	07H2M10	07H2M12	07H2M13	07H2M15	07H2M17	07H2M18	07H2M19
Sex		Male	Female	Male	Female	Unknown	Unknown	Unknown	Male	Male
Mesiodistal	UI1	7.76	9.98						8.67	7.92
diameters (mm)	UI2		8.97	7.01					7.43	
	UC	7.25	8.04	7.51					7.74	
	UP1	5.86	8.80	7.26					7.16	
	UP2	5.63	7.68	6.69	6.32				7.13	
	UM1	10.06	11.79	10.04					9.66	
	UM2	8.53	10.81	9.75					9.03	8.67
	LI1		6.32	5.43					5.57	
	LI2		6.97	6.26					6.00	
	LC	6.17	8.22	6.98					6.41	6.20
	LP1	5.80	7.59	7.54					6.64	
	LP2	5.60	7.61	7.77	6.46				6.55	6.58
	LM1	11.29	12.81	11.55	10.78				11.13	11.80
	LM2	9.94	12.56	10.90					10.05	9.62
Buccolingual	UI1	7.02	8.15						6.86	6.99
diameters (mm)	UI2		7.51	6.60					6.42	
	UC	8.25	9.00	8.39					7.66	8.35
	UP1	8.44	10.27	9.38	8.26				9.64	
	UP2	8.11	10.18	9.73	8.07				9.39	
	UM1	11.98	13.44	11.65					11.38	10.69
	UM2	11.13	13.47	11.79					10.36	10.23
	LI1		6.60	5.74					5.54	
	LI2		7.12	6.11					5.83	
	LC	7.39	8.72	7.75	7.07				7.06	7.89
	LP1	6.68	8.27	8.21	7.31				7.67	8.26
	LP2	7.53	8.83	8.93	7.69				7.99	8.50
	LM1	10.47	11.67	10.63	10.59				10.30	10.43
	LM2	9.17	11.08	10.38					8.99	10.00
shoveling	UI1	+	+					+	-	-
shoveling	UI2		+	+				-	-	
double shoveling	UI1	-	+					-	-	-
double shoveling	UI2		-	-				-	-	
dental tubercle	UI1	-	+					-	-	-
dental tubercle	UI2		-	-				-	-	
spine	UI1	-	+					-	-	-
interruption groove	UI2		+	-				-	-	
winging (bilateral)	UI1	-	-					-	-	-
De Terra's tubercle	UP1		-					-	-	
double rooted	UP1				-				+	
double rooted	UP2				-					
Carabelli's trait	UM1		+	-			+	+	-	-
hypocone reduction	UM2	-	-	-				-	+	-
sixth cusp	LM1		+		+	-		+	-	
seventh cusp	LM1		-		-	-		-	-	-
protostylid	LM1		-		-	-		-	-	
deflecting wrinkle	LM1		-		-	-		+	-	
groove pattern Y	LM1	+	+		-	+		+	+	
groove pattern X	LM2		-	-				-	-	+
number of cusps	LM2	+	-	+				+	+	+

U:Upper, L:Lower, I:Incisor, C:Canine, P:Premolar, M:Moler, +: present, -: absent

Appendix 5.1 (Continued 4).

Sample Number		07H2M22	07H2M24	07H2M27	07H2M30	07H2M32
Sex		Female	Female	Male	Male	Male
Mesiodistal	UI1	8.35	8.08	9.47	8.48	9.09
diameters (mm)	UI2	6.68				
	UC	7.64		7.88	8.01	8.39
	UP1	6.83		6.84	7.15	7.42
	UP2	6.65		6.60	6.89	7.01
	UM1	10.04		10.28		11.05
	UM2	9.07		9.13		10.02
	LI1	4.88			5.15	
	LI2	5.76			5.66	
	LC	6.27		6.92	7.60	7.11
	LP1	6.60		6.56	7.05	7.19
	LP2	6.74		6.61	7.33	7.14
	LM1	11.20		11.90	12.25	12.43
	LM2	10.35		10.82	11.16	11.51
Buccolingual	UI1	6.76	6.86	7.11	7.55	7.67
diameters (mm)	UI2	6.30				
	UC	7.11		8.75	9.31	8.58
	UP1	8.97		9.03	10.17	10.34
	UP2	9.07	8.81	9.06	10.12	9.64
	UM1	11.38		10.87	14.06	11.84
	UM2	10.57	10.39	10.15		11.82
	LI1	5.53			6.15	
	LI2	5.65			6.58	
	LC	6.49	6.76	7.19	8.04	8.18
	LP1	7.24	7.83	7.53	8.38	8.02
	LP2	8.10	8.15	8.29	9.28	8.66
	LM1	10.66	10.79	10.63	12.78	11.37
	LM2	9.56	9.90	10.06	11.18	10.13
shoveling	UI1		+	-	-	+
shoveling	UI2					
double shoveling	UI1	-	-	-	-	-
double shoveling	UI2	-				
dental tubercle	UI1	-	+	-	+	-
dental tubercle	UI2	-				
spine	UI1	-	-	-		-
interruption groove	UI2	-				
winging (bilateral)	UI1	-		-	-	-
De Terra's tubercle	UP1					-
double rooted	UP1		-			
double rooted	UP2				-	
Carabelli's trait	UM1				+	-
hypocone reduction	UM2	+		+		-
sixth cusp	LM1			-		+
seventh cusp	LM1			-	+	-
protostylid	LM1					-
deflecting wrinkle	LM1					-
groove pattern Y	LM1			+		+
groove pattern X	LM2	-	-	-	-	-
number of cusps	LM2	+			+	+

U:Upper, L:Lower, I:Incisor, C:Canine, P:Premolar, M:Moler, +: present, -: absent

Appendix 5.2 Crown measurements (mm) of the deciduous dentition of the Man Bac series (sex unknown).

	05M1	05M3	05M5	05M10	05M12	05M14	05M18	05M24	05M25
Mesiodistal diameters (mm)									
udi1	7.32				6.64	7.33			
udi2	6.55	5.75			5.65	6.17	6.28		5.60
udc			6.93	6.99	7.12	6.68	6.88	7.12	7.14
udm1	7.31			7.30	6.54	6.75	7.36	7.04	7.87
udm2	9.06			9.12	8.22	8.96		9.25	9.01
ldi1			4.52		4.28	4.93	4.51		
ldi2	5.25		4.88		4.91	5.25	5.19		
ldc			5.72	5.94	5.82	6.07		5.81	6.35
ldm1	7.66		9.14	8.18	7.78	9.01	8.22	8.17	8.71
ldm2	10.03		10.28	10.48	9.88	10.75	10.41	10.46	10.71
Buccolingual diameter (mm)									
udi1					5.12	5.18			
udi2	4.70	5.53			5.04	4.90	4.86		4.96
udc			5.56	6.14	6.18	6.12	5.67	6.11	6.45
udm1	9.34			9.31	8.91	9.07	9.35	8.49	8.78
udm2	9.66			9.87	9.64	10.64		9.98	10.28
ldi1			3.89		3.58	3.76	3.65		
ldi2	4.64		4.10		4.20	4.54	5.06		
ldc			6.64	5.54	5.42	5.69		5.72	5.72
ldm1	7.09			7.56	6.92	7.34	6.88	7.63	7.24
ldm2	8.56			8.98	8.67	9.04	8.75	9.56	8.98

	05M30	07H2M6	07H2M7	07H2M13	07H2M15	07H2M16	07H2M26	07H2M31
Mesiodistal diameters (mm)								
udi1	7.40	6.53	6.81	7.27	6.46	7.47	7.08	7.08
udi2	6.07	6.16	5.90	5.68	5.39	6.16	5.80	5.75
udc	7.08	6.82	6.92	7.03	6.84	7.39	7.46	7.16
udm1	7.61		7.33	7.65	7.16	7.57		7.43
udm2	9.44		9.32	9.78	7.69	9.14	10.50	9.31
ldi1	4.95	4.19		4.52	4.15			4.51
ldi2			4.78	5.00	4.68	5.12		5.31
ldc		6.16	5.96	5.85	6.16	5.91		6.14
ldm1			8.17	8.60	8.56	8.79	8.93	8.71
ldm2		10.20	10.37	10.85	10.46	10.94	11.20	10.45
Buccolingual diameter (mm)								
udi1		5.10	4.63	4.94	4.38	5.42	4.78	5.12
udi2		4.90	4.42	4.55	4.49	5.38	4.50	5.10
udc		5.72	5.35	6.01	6.30	6.05	5.85	5.95
udm1			8.21	8.35	9.75	9.09		9.01
udm2				10.52	10.33	10.02	10.35	10.52
ldi1		3.83		3.70	3.65			3.84
ldi2			3.93	3.93	4.14	4.51		4.34
ldc			4.90	5.10	5.39	5.44		5.29
ldm1			6.64	6.64	6.82	6.83	7.40	6.81
ldm2		8.85	7.74	8.72	8.60	8.79	9.27	8.27

u: upper, l:lower, d:deciduous, I:incisor, c:canine, m:moler

6

Quantitative Limb Bone-Morphology at Man Bac

Hirofumi Matsumura[1], Wataru Takigawa[2], Nguyen Kim Thuy[3] and Nguyen Anh Tuan[3]

[1]*Department of Anatomy, Sapporo Medical University, Japan.*
[2]*Department of Physical Therapy, International University of Health and Welfare, Japan*
[3]*The Vietnamese Institute of Archaeology*

The aim of this chapter is to: (1) morphometrically describe the Man Bac adult infracranial remains, essentially the major long bones; (2) estimate sex specific stature in the series; and (3) compare the relative level of Man Bac infracranial robusticity with other samples in the region. It is hoped this analysis will contribute to a better understanding of generalised behaviours the Man Bac community may have been engaged in.

MATERIALS AND METHODS

The long bone sample for this study derives from the 2005 and 2007 excavation seasons. Measurements, left side only or right if missing, were taken for the humerus, radius, ulna, femur, tibia, and fibula, following Martin and Saller's (1957) methodology. For incomplete long bones, estimates of maximum length were based on Wright and Vasquez's (2003) methodology.

A range of indices were derived from the measurement suite. For instance, the cross sectional index of a long bone shaft expresses the relative roundness or flatness of the diaphysis in terms of minimum diameter to maximum diameter, or of transverse diameter to sagittal diameter. Upper and lower limb proportions were evaluated by way of the brachial and crural indices. The brachial index is the length ratio of the radius to humerus, while the crural index expresses the proportion of tibial to femoral length. Calculated indices for the Man Bac series were compared to Pietrusewsky and Douglas' (2002) study of the Ban Chiang, Thailand, (4,100BP-1,900 BP) skeletal series, which includes a global infracranial comparative data set.

Due to the lack of an ancient Vietnamese-specific set of stature regression functions, stature was estimated using a range of published methods: Stevenson (1929) derived from a northern Chinese sample; Trotter and Gleser (1958) derived from Asian Americans; Mo (1983) from a southern Chinese series; Fujii (1960) from a Japanese sample; Sangvichien et al. (1985) using a Thai/Chinese sample and Sjøvold's (1990) generic functions. Among these, only Fujii and Sangvichien et al. provided sex specific equations. In order to assess the most suitable stature estimation equations for the Man Bac series, this study compared estimated statures calculated using different kinds of long bones (humerus, femur and tibia), and assessed discrepancies among the results of the different regression equations. The regression equation set, which provided the smallest discrepancy between estimated stature in the Man Bac series, based on different long bones, was regarded as the most appropriate set.

RESULTS

A summary of limb measurements and indices for individual specimens from Man Bac are given in Tables 6.1 and 6.2. The specimen MB07H1M9 has been excluded from subsequent analyses, including mean limb dimension statistics, due to its pathologically abnormal limb dimensions and morphology (see Oxenham et al. 2009).

The overall size of the long bones will be evaluated in terms of stature comparisons with neighbouring prehistoric samples. In terms of the indices, specific comparisons with Ban Chiang are noted. Unless otherwise stated, all reported measurements are in millimetres. The mid-shaft cross-sectional index of the humerus is greater in the males (82.1) than in the females (70.9), indicating a more rounded humeral shaft for males. The Man Bac values are a little bit greater than the Ban Chiang averages (males 78.7, females 71.7), which were categorised as possessing moderate roundness, so-called 'eurybrachia'.

The mid-shaft indices of the radius are similar for males (71.3) and females (69.6). However, the cross-sectional index of the ulna is greater in the males (91.8) than in the females (82.0), suggesting that the male shaft is flatter than the female diaphysis. These mid-shaft indices were not compared with Ban Chiang due to differences in measurement landmarks.

The mid-shaft cross-sectional index of the femur, 'pilastric index', relates to the degree of development of the linear aspera. Man Bac males (109.5) show greater values than females (102.3), indicating that males have a more robust pilastric form associated with well developed linea aspera. Although the Man Bac values show slight differences when compared to the Ban Chiang samples (males 112, females 102.9), both samples are within the range of medium levels of development globally.

The proximal femoral shaft cross-sectional index, 'platymeric index', reflects the degree of flatness in the upper portion of the femoral diaphysis. There is very little sexual dimorphism difference in the Man Bac series using this index (males 78.9, females 75.0). Man Bac averages are slightly greater than the Ban Chiang males (77.7), which are classified as having moderate flatness in global terms.

The average cross-sectional index of the tibia, which evaluates transverse flatness or sagittal thickness of the tibial shaft at the level of the nutrient foramen, is similar in both the Man Bac (males 66.7, females 67.6) and Ban Chiang (males 68.7, females 67.9) series, suggesting shafts in the moderate range (mesocnemic).

The brachial and crural indices, which evaluate arm and leg length ratios, respectively, are given in Table 6.3. There is little sexual dimorphism in either index (brachial: males 81.6 females 78.1, crural: males 85.5, females 83.9). Figure 6.1 plots these two indices for Man Bac males (open circles) together with comparative population data. Several individuals show relatively very long forearms

Table 6.1 Humerus, radius and ulna measurements (mm) recorded for limb bones of the Man Bac series.

Sample No.	MB05 M9	MB05 M11	MB05 M15	MB05 M16	MB05 M28	MB05 M29	MB05 M31	MB05 M34	MB07 H1M4	MB07 H1M5	MB07 H1M8	MB07 H1M9	MB07 H1M10	MB07 H1M11
Sex	female	male	male	male	female	male	male	female	male	male	male	male	female	male
Humerus	right	right		right	left	right	left	right	right	right	right	right	right	right
1. Max. length	272	288	-	298	-	-	282	295	-	292	307	-	-	269
2. Total length	270	285	-	295	-	-	277	291	-	288	303	-	-	264
4. Bi-epicondylar width	57	63	-	59	-	70	63	57	-	64	62	-	61	60
5. Max. mid-shaft diameter	23	21	-	28	21	27	25	21	24	23	22	16	23	22
6. Min. mid-shaft diameter	15	15	-	17	15	21	21	16	16	17	19	14	18	16
9. Transv. head diameter	40	42	-	41	-	45	46	37	-	43	41	-	-	39
10. Max. sagittal head diameter	37	40	-	42	-	-	40	40	38	44	44	-	-	42
6/5. Mid-shaft cross-section index	64.4	69.0	-	60.7	69.0	77.8	84.0	76.2	66.7	73.9	86.4	87.5	78.3	72.7
9/10. Head cross-section index	109.6	106.3	-	97.6	-	-	115.2	92.5	0.0	97.7	93.2	-	-	92.9
Radius	right	right	-	left	right	right	left	right	right	right	right	right	left	left
1. Max. length	216	239	-	239	-	241	229	216	-	245	241	-	234	207
2. Physiological length	204	229	-	234	-	236	214	212	-	239	237	213	230	202
4. Max. Transv. shaft diameter	14	14	-	16	15	18	19	14	16	15	16	13	15	16
5. Sagittal shaft diameter	11	12	-	11	10	14	13	10	10	10	12	7	11	11
5/4. Mid-shaft cross-section index	77.8	85.2	-	68.8	63.3	77.8	65.8	71.4	62.5	66.7	75.0	53.8	73.3	68.8
Ulna	right	right	right	left	left	right	left	left	left	left	left	right	left	left
1. Max. length	235	264	-	253	-	257	249	234	-	263	262	245	261	227
2. Physiological length	206	234	-	222	-	221	217	205	-	231	232	213	228	199
6. Olecranon breadth	22	23	-	24	-	26	26	23	-	24	25	25	25	23
11. Dorso-ventral shaft diameter	12	12	12	13	11	20	15	14	-	14	14	9	11	12
12. Transv. shaft diameter	16	15	18	16	14	15	16	12	-	16	15	12	16	17
11/12. Shaft cross-section index	71.9	79.3	65.7	81.3	81.5	133.3	90.6	116.7	-	87.5	93.3	75.0	68.8	70.6

For a definition of the measurements see Martin R and Saller K. 1957

Table 6.1 (continued).

Sample No.	MB07 H1M13a	MB07 H2M1	MB07 H2M10	MB07 H2M12	MB07 H2M19	MB07 H2M24	MB07 H2M27	MB07 H2M30	MB07 H2M32	Man Bac Average					
Sex	unknown	male	male	female	male	female	male	male	male	males			females		
Humerus	left	left	left	left	right	right	left	left	left	n	Mean	SD	n	Mean	SD
1. Max. length	-	313	327	281	300	-	282	351	322	10	306.4	22.3	5	283.0	13.1
2. Total length	-	309	321	281	296	-	278	343	315	10	301.5	21.0	5	280.2	13.3
4. Bi-epicondylar width	-	66	74	58	65	59	62	65	66	11	65.4	3.7	7	58.7	1.6
5. Max. mid-shaft diameter	22	22	24	19	27	22	22	28	27	12	23.7	3.4	9	22.5	2.5
6. Min. mid-shaft diameter	15	18	19	14	20	17	21	25	23	12	19.4	3.3	9	15.8	1.3
9. Transv. head diameter	-	43	45	39	46	-	38	-	44	10	43.2	2.3	5	39.2	1.5
10. Max. sagittal head diameter	-	45	48	39	49	-	42	49	47	10	44.7	3.6	6	39.6	2.2
6/5. Mid-shaft cross-section index	68.2	81.8	79.2	73.7	72.6	76.1	95.3	89.5	87.6	12	82.1	7.8	9	70.9	6.0
9/10. Head cross-section index	-	95.6	93.8	100.0	92.8	-	92.2	0.0	92.5	10	87.9	31.8	6	82.1	40.7
Radius	right	left	right	-	left	right	right	right	left	n	Mean	SD	n	Mean	SD
1. Max. length	-	255	278	227	249	227	229	273	261	11	249.1	16.3	7	223.7	11.3
2. Physiological length	-	251	273	222	245	220	223	269	256	12	240.4	19.5	7	217.7	12.3
4. Max. Transv. shaft diameter	14	16	20	15	18	15	15	19	18	12	16.7	2.3	9	15.1	0.9
5. Sagittal shaft diameter	11	12	14	10	14	11	12	13	12	12	11.9	1.9	9	10.5	0.6
5/4. Mid-shaft cross-section index	78.6	75.0	70.0	66.7	75.2	73.5	76.2	68.1	66.3	12	71.3	7.9	9	69.6	5.0
Ulna	left	left	right	left	left	right	right	left	left	n	Mean	SD	n	Mean	SD
1. Max. length	-	279	299	246	265	245	250	-	283	11	265.1	16.2	7	243.0	11.8
2. Physiological length	-	244	268	213	230	215	215	-	249	11	232.2	16.5	7	212.6	10.1
6. Olecranon breadth	-	27	31	22	27	23	26	27	27	12	26.1	2.0	7	23.0	1.2
11. Dorso-ventral shaft diameter	12	15	17	12	14	13	14	14	14	12	14.3	2.6	9	12.1	1.1
12. Transv. shaft diameter	15	12	16	15	18	13	16	20	19	12	15.8	2.5	9	15.1	1.9
11/12. Shaft cross-section index	80.0	125.0	106.3	80.0	79.1	101.6	92.2	67.6	72.8	12	91.8	20.5	9	82.0	16.8

Table 6.2 Femur, tibia and fibula measurements (mm) recorded for limb bones of the Man Bac series.

Sample No.	MB05 M9	MB05 M11	MB05 M15	MB05 M16	MB05 M28	MB05 M29	MB05 M31	MB05 M34	MB07 H1M4	MB07 H1M5	MB07 H1M8	MB07 H1M9	MB07 H1M10	MB07 H1M11
Sex	female	male	female	female	female	male	male	female	male	male	male	male	female	female
Femur	right	right	right	left	left	right	right	right	left	left	left	left	right	left
1. Max. length	377	401	-	420	-	394	407	400	-	398	421	-	(406)	399
2. Physiological length	376	398	-	415	-	391	402	398	-	396	419	-	-	397
6. Sagittal mid-shaft diameter	24	26	28	25	24	27	27	25	24	25	27	14	27	27
7. Transv. mid-shaft diameter	25	21	24	27	24	26	25	24	26	27	25	16	25	29
9. Sagittal proximal shaft diameter	25	23	22	23	-	24	25	22	21	23	23	14	24	24
10. Transv. proximal shaft diameter	27	23	28	31	-	33	30	29	34	29	31	20	31	33
18. Vertical head diameter	42	43	44	46	-	46	49	41	45	46	46	-	45	46
19. Sagittal head diameter	41	42	43	45	-	46	48	41	-	47	46	-	45	46
21. Bicondylar width	69	74	-	77	-	81	80	71	-	82	77	-	-	78
6/7. Mid-shaft cross-section index (Pilastric)	94.0	124.4	119.1	92.6	102.1	103.8	108.0	104.2	92.3	92.6	108.0	87.5	108.0	93.1
9/10. Prox. shaft cross-section index (Platymeric)	108.0	100.0	78.6	134.8	-	137.5	120.0	131.8	161.9	126.1	134.8	142.9	129.2	137.5
19/18. Head cross-section index	98.8	97.7	97.7	97.8	-	100.0	99.0	100.0	0.0	102.2	100.0	-	100.0	100.0
Tibia	right	-	right	-	-	left	left	right	right	left	left	left	right	right
1. Total length	326	-	-	-	-	(334)	340	320	-	342	357	-	(337)	326
1a. Max. length	329	-	-	-	-	(337)	343	323	-	335	360	-	-	328
3. Epicondylar breadth	-	-	-	-	-	79	-	66	-	77	75	-	-	73
8. Max. sagittal mid-shaft diameter	28	-	29	-	-	31	33	26	27	29	30	19	28	30
9'. Transv. mid-shaft diameter	18	-	20	-	-	24	20	18	-	19	21	13	19	18
8a. Max. sagit. diam. at nutrient foramen	34	-	32	-	-	36	36	28	19	35	33	22	35	34
9a'. Transv. diameter at nutrient foramen	21	-	22	-	-	27	22	20	-	20	23	-	25	21
9'/8. Mid-shaft cross-section index	65.5	-	70.2	-	-	77.4	60.6	69.2	-	65.5	70.0	68.4	67.9	60.0
9a'/8a. Shaft cross-section index	62.7	-	67.2	-	-	75.0	61.1	71.4	-	57.1	69.7	-	71.4	61.8
Fibula	right	-	right	-	right	right	right	-	-	left	left	left	right	right
1. Max. length	323	-	-	-	-	341	334	-	-	339	-	-	-	-
2. Max. mid-shaft diameter	15	-	16	-	15	17	17	-	-	14	15	9	17	18
3. Min. mid-shaft diameter	11	-	10	-	10	12	12	-	-	11	10	8	10	8
3/2. Mid-shaft cross-section index	70	-	63	-	69	71	68	-	-	79	67	89	59	44

Parenthesis values estimated via formulae from Wright and Vasquez (2003), (05MB29: T1-T7=322mm; 07MB H1-10: F0-F5=375mm, T1-T6=305mm)

Table 6.2 (continued).

Sample No.	MB07 H1M13a	MB07 H2M1	MB07 H2M10	MB07 H2M12	MB07 H2M19	MB07 H2M24	MB07 H2M27	MB07 H2M30	MB07 H2M32	Man Bac Average					
Sex	female	male	male	female	male	female	male	male	male	males			females		
Femur	right	right	right	left	right	left	left	right	left	n	Mean	SD	n	Mean	SD
1. Max. length	420	445	470	407	400	416	390	482	452	11	423.6	32.9	8	405.6	14.2
2. Physiological length	418	443	467	403	398	412	388	481	450	11	421.2	33.3	7	402.7	14.4
6. Sagittal mid-shaft diameter	27	29	33	24	27	25	29	36	31	12	27.5	5.3	11	25.4	1.6
7. Transv. mid-shaft diameter	24	26	26	23	27	24	25	28	29	12	25.0	3.5	11	24.9	1.8
9. Sagittal proximal shaft diameter	21	24	25	21	25	23	24	29	24	12	23.5	3.4	10	22.6	1.4
10. Transv. proximal shaft diameter	32	30	34	29	31	29	30	33	35	12	29.9	4.4	10	30.3	2.3
18. Vertical head diameter	44	47	47	41	47	43	43	50	47	11	46.3	2.1	10	43.6	2.0
19. Sagittal head diameter	44	47	46	41	47	43	42	50	48	11	46.2	2.3	9	43.2	1.9
21. Bicondylar width	-	78	85	72	79	71	78	85	82	11	80.2	3.3	6	73.0	3.6
6/7. Mid-shaft cross-section index (Pilastric)	112.5	111.5	126.9	104.3	102.8	103.3	113.5	125.9	108.5	12	109.5	12.3	11	102.3	8.8
9/10. Prox. shaft cross-section index (Platymeric)	152.4	125.0	136.0	138.1	125.1	126.7	126.9	115.7	144.9	12	78.9	8.5	10	75.0	8.3
19/18. Head cross-section index	100.0	100.0	97.9	100.0	98.9	99.3	99.4	100.8	102.1	11	99.8	1.5	10	89.4	31.4
Tibia	right	left	right	left	right	right	right	left	right	n	Mean	SD	n	Mean	SD
1. Total length	350	384	409	337	346	333	332	412	381	10	363.7	30.5	7	332.7	9.9
1a. Max. length	357	386	415	342	350	-	330	409	378	10	364.3	31.1	5	335.8	13.8
3. Epicondylar breadth	72	-	81	67	77	-	-	78	75	7	77.5	2.1	4	69.5	3.5
8. Max. sagittal mid-shaft diameter	30	31	32	26	32	26	28	38	31	11	30.4	4.6	9	27.7	1.6
9'. Transv. mid-shaft diameter	21	20	22	18	19	18	18	23	22	11	20.2	3.0	8	18.8	1.2
8a. Max. sagit. diam. at nutrient foramen	32	35	37	32	38	34	32	41	35	11	34.6	4.7	9	31.1	5.0
9a'. Transv. diameter at nutrient foramen	21	22	34	20	21	24	21	24	24	10	23.7	4.1	8	21.7	1.9
9'/8. Mid-shaft cross-section index	70.0	64.5	68.8	69.2	60.7	69.0	66.0	60.6	71.0	11	66.7	5.2	8	67.6	3.4
9a'/8a. Shaft cross-section index (Tibial thickness)	65.6	62.9	91.9	62.5	55.6	70.6	64.2	57.8	67.5	10	66.3	10.8	8	66.7	4.1
Fibula	left	left	left	left	right	left	left	left	left	n	Mean	SD	n	Mean	SD
1. Max. length	-	-	-	-	357	323	324	-	373	6	344.7	17.6	2	323.0	0.0
2. Max. mid-shaft diameter	15	17	18	15	18	15	16	18	19	11	16.1	2.8	8	15.7	1.2
3. Min. mid-shaft diameter	11	11	10	9	13	10	13	13	11	11	11.2	1.5	8	9.9	0.9
3/2. Mid-shaft cross-section index	73	65	56	60	71	69	86	69	59	11	70.7	10.3	8	63.4	9.3

Table 6.3 Indices of limb bone proportions and estimated statures (cm) of the Man Bac series.

Sample No.	MB05 M9	MB05 M11	MB05 M16	MB05 M29	MB05 M31	MB05 M34	MB07 H1M5	MB07 H1M8	MB07 H1M10	MB07 H1M11	MB07 H1M13a
Sex	female	male	male	male	male	female	male	male	female	male	female
Limb length proportion											
Brachial index (Rad.1/Hum.1)	79.4	83.0	80.2		81.2	73.2	83.9	78.5		77.0	
Crural index (Tib.1a/Fem.1)	87.3			85.0	84.3	80.8	84.2	85.5		82.2	85.0
Stature estimation											
Stevenson (1929): northern Chinese (femur)	153.6	159.5	164.1	157.8	160.9	159.2	158.7	164.4	160.7	159.0	164.1
Stevenson (1929): northern Chinese (tibia)	157.9			160.3	162.1	156.1	162.7	167.3		157.9	165.1
Stevenson (1929): northern Chinese (humerus)	158.0	162.5	165.3		160.8	164.5	163.7	167.9		157.2	
Trotter & Gleser (1958): USA Asians (femur)	153.6	158.8	162.9	157.3	160.1	158.6	158.1	163.1	160.1	158.4	162.9
Trotter & Gleser (1958): USA Asians (tibia)	160.1			162.0	163.4	158.6	161.5	167.5		159.8	166.8
Mo (1983): southern Chinese (femur)	149.0	154.4	158.7	152.8	155.8	154.2	153.7	158.9	155.6	154.0	158.7
Mo (1983): southern Chinese (tibia)	153.2			155.6	157.4	151.4	155.0	162.5		152.9	161.6
Fujii (1960): Japanese (femur)	145.5	153.9	158.6	152.2	156.2	150.6	153.2	158.9	152.0	153.5	155.1
Fujii (1960): Japanese (tibia)	150.3			157.2	158.7	148.9	156.7	162.9		155.0	162.2
Fujii (1960): Japanese (humerus)	146.0	153.6	156.4		151.9	151.5	154.7	158.9		148.3	
Sangvichien et al. (1985): Thai/Chinese (femur)	**146.6**	**157.5**	**160.7**	**156.3**	**158.5**	**152.5**	**156.9**	**160.9**	**154.1**	**157.1**	**157.7**
Sangvichien et al. (1985): Thai/Chinese (tibia)	149.4			155.8	157.5	147.6	155.3	162.2		153.3	157.7
Sjovold (1990): nonspecific populations (femur)	148.0	154.5	159.7	152.6	156.2	154.3	153.7	160.0	155.9	154.0	159.7
Sjovold (1990): nonspecific populations (tibia)	155.6			158.2	160.2	153.6	157.6	165.8		155.3	164.8
Sjovold (1990): nonspecific populations (humerus)	144.7	152.1	156.7		149.3	155.3	153.9	160.8		143.3	
Difference of stature estimation based on comparisons between humeral and femoral lengths											
Stevenson (1929)	4.4	3.1	1.2		0.1	5.3	4.9	3.5		1.8	
Fujii (1960)	0.5	0.4	2.3		4.3	0.9	1.5	0.0		5.2	
Sjovold (1990)	3.4	2.5	3.0		6.9	1.0	0.2	0.9		10.7	
Difference of stature estimation based on comparisons between femoral and tibial lengths											
Stevenson (1929)	4.3			2.5	1.2	-3.1	4.0	2.9		-1.1	1.0
Trotter & Gleser (1958)	6.5			4.7	3.4	0.1	3.4	4.4		1.5	3.9
Mo (1983)	4.2			2.8	1.6	-2.8	1.2	3.5		-1.1	2.9
Fujii (1960)	4.8			5.0	2.5	-1.7	3.5	4.0		1.6	7.1
Sangvichien et al. (1985)	2.8			-0.4	-1.0	-4.9	-1.7	1.3		-3.8	0.0
Sjovold (1990)	7.6			5.6	4.0	-0.7	3.8	5.8		1.3	5.1

bold: stature estimation regarded as most suitable for the Man Bac series.

Table 6.3 (continued).

Sample No.	MB07 H2M10	MB07 H2M12	MB07 H2M19	MB07 H2M24	MB07 H2M27	MB07 H2M30	MB07 H2M32	Man Bac Average males n	Mean	SD	females n	Mean	SD
Sex	male	female	male	female	male	male	male	n	Mean	SD	n	Mean	SD
Limb length proportion													
Brachial index (Rad.1/Hum.1)	85.0	80.8	83.0		81.2	77.8	81.1	10	81.6	2.3	5	78.1	3.1
Crural index (Tib.1a/Fem.1)	88.3	84.0	87.5		84.6	84.9	83.6	10	85.5	1.5	5	83.9	2.5
Stature estimation													
Stevenson (1929): northern Chinese (femur)	176.3	160.9	159.2	163.1	156.8	179.2	171.9	11	165.0	8.0	8	160.6	3.5
Stevenson (1929): northern Chinese (tibia)	183.0	161.2	163.9	160.0	159.7	183.9	174.5	10	169.3	9.2	6	159.7	3.2
Stevenson (1929): northern Chinese (humerus)	173.5	160.6	165.9		160.8	180.3	172.1	10	167.7	6.3	5	161.1	3.7
Trotter & Gleser (1958): USA Asians (femur)	173.6	160.1	158.6	162.0	156.4	176.2	169.8	11	163.7	7.1	8	159.8	3.1
Trotter & Gleser (1958): USA Asians (tibia)	180.6	163.2	165.1		160.3	179.2	171.8	10	168.5	7.4	5	161.7	3.3
Mo (1983): southern Chinese (femur)	170.0	155.8	154.2	157.8	151.9	172.7	166.0	11	159.5	7.4	8	155.5	3.2
Mo (1983): southern Chinese (tibia)	179.0	157.1	159.5		153.5	177.2	167.9	10	163.8	9.3	5	155.2	4.1
Fujii (1960): Japanese (femur)	171.0	152.2	153.7	154.2	151.2	174.0	166.5	11	159.6	8.1	8	152.0	3.3
Fujii (1960): Japanese (tibia)	176.5	153.1	160.4		155.5	175.0	167.4	10	164.0	7.7	5	152.9	5.4
Fujii (1960): Japanese (humerus)	164.5	148.2	156.9		151.9	171.2	163.1	10	158.7	6.2	5	148.7	3.0
Sangvichien et al. (1985): Thai/Chinese (femur)	**169.4**	**154.3**	**157.3**	**156.6**	**155.6**	**171.5**	**166.3**	**11**	**161.4**	**5.7**	**8**	**154.0**	**3.7**
Sangvichien et al. (1985): Thai/Chinese (tibia)	177.4	153.2	159.4		153.9	175.7	167.2	10	163.4	8.6	5	151.3	4.2
Sjovold (1990): nonspecific populations (femur)	173.2	156.2	154.3	158.6	151.6	176.5	168.4	11	160.7	8.9	8	155.8	3.9
Sjovold (1990): nonspecific populations (tibia)	183.9	159.9	162.5		155.9	181.9	171.7	10	167.2	10.2	5	157.8	4.5
Sjovold (1990): nonspecific populations (humerus)	170.1	148.8	157.6		149.3	181.2	167.8	10	160.6	10.3	5	149.7	6.1
Difference of stature estimation based on comparisons between humeral and femoral lengths								Mean stature difference (humerus - femur)					
Stevenson (1929)	2.8	0.4	6.7		4.0	1.0	0.2			2.7			
Fujii (1960)	6.5	4.0	3.2		0.7	2.8	3.5			2.7			
Sjovold (1990)	3.2	7.3	3.3		2.3	4.7	0.6			3.5			
Difference of stature estimation based oncomparisons between femoral and tibial lengths								Mean stature difference (humerus - femur)					
Stevenson (1929)	6.7	0.3	4.7		2.9	4.7	2.6			2.6			
Trotter & Gleser (1958)	7.0	3.1	6.5		3.9	3.0	2.0			3.9			
Mo (1983)	9.0	1.3	5.3		1.5	4.5	2.0			2.8			
Fujii (1960)	5.5	0.9	6.7		4.3	1.1	0.8			3.2			
Sangvichien et al. (1985)	8.0	-1.1	2.1		-1.7	4.3	0.9			0.6			
Sjovold (1990)	10.6	3.7	8.2		4.4	5.4	3.3			5.1			

and lower legs, while the Man Bac mean is close to the Ban Chiang mean (Pietrusewsky and Douglas, 2002). In general, Man Bac individuals are dispersed around a range of population means including: Dayak (Yokoh, 1940), Jomon (Takigawa, 2005) and Tasmanians (Roth, 1899). A single specimen shows a close affinity with Chinese (Olivier, 1969) and Iron Age Yayoi Japanese (Wakebe, 2002). Australian aborigines, Hawaiians (Olivier, 1969), Japanese (Takigawa, 2005), Han Chinese and Weidun Neolithic southern Chinese (Wakebe, 2002) are somewhat distant from the Man Bac sample. The Man Bac series, as well as the Ban Chiang sample, are characterised as possessing proportionally longer forearms and lower limbs in comparison with modern Chinese and Japanese.

The results of stature estimation, using several sets of regression equations, are given in Table 6.3. As expected, estimated stature varied by formulae and the specific lone bone employed, with some of this variation caused by sample-specific limb ratio differences (see above). In order to assess the most suitable set of equations for Man Bac, the consistency of estimated statures were compared between the sets of regression formulae. Table 6.3 gives the difference in stature estimation based on comparisons between humeral and femoral lengths, and that based on comparisons between femoral and tibial lengths. In terms of humeral-femoral comparisons Stevenson (1929) and Fijii's (1960) formulae show the smallest differences. Examining estimates based on femoral and tibial lengths, Sangvichien et al.'s (1985) equations provide the smallest differences, more so even than the humeral-femoral comparisons, and are deemed the most appropriate functions for estimating Man Bac stature from the tibia and femur.

Accordingly, average estimated stature for the Man Bac series is 161.4cm (males) and 154.0cm (females). These values can be compared to prehistoric samples from Thailand (Domett, 2001; Pietrusewsky and Douglas, 2002) also based on

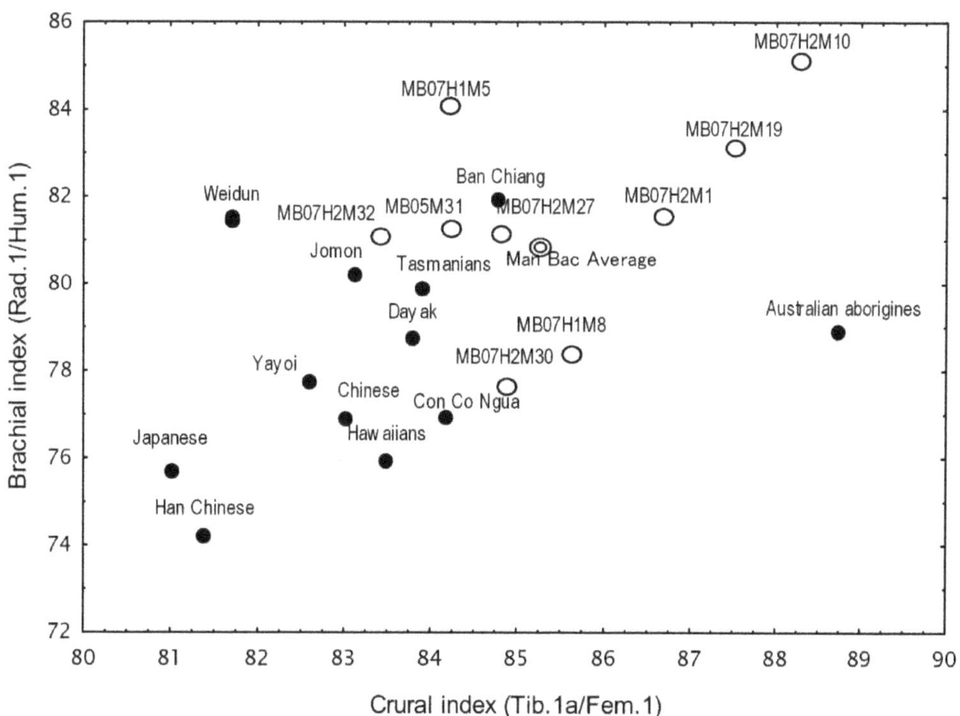

Figure 6.1 Crural and brachial indices for the Man Bac and comparative samples.

Sangvichien et al.'s (1985) equations: Ban Chiang males 166.2 cm, females 154.4 cm; Khok Phanom Di males 162.2, females, 154.3 cm; Bronze Age Nong Nor males 167.2 cm, females 156.1 cm; Bronze Age Ban Lum Khao males 164.7 cm, females 154.7 cm; Bronze-Iron Age Ban Na Di males 168.0 cm, females 155.9 cm. In terms of stature, at least, Mac Bac is closest to the near contemporaneous series from Khok Phanom Di.

DISCUSSION

Man Bac cross-sectional indices, expressing relative roundness of limb diaphyses, are, for the most part, consistent with neighbouring neolithic, Bronze and Iron Age samples and indicate an intermediate position between gracility and robusticity.

The brachial and crural indices are considered useful in evaluating ancestral features with respect to body proportions. Low values, reflecting shorter forearms or lower legs, are associated with cold climate adaptation. The Man Bac series, along with Ban Chiang, is characterised by relatively long forearms and lower limbs.

Stature estimation was carried out using several different sets of regression equations, with Sangvichien et al.'s (1985) functions derived from modern Thai and Chinese cadavers determined to be the most appropriate for the Man Bac series. Male and female Man Bac stature falls within the range of neighbouring prehistoric Thai samples, falling closest to the contemporaneous Khok Phanom Di series.

While the analysis of long bone morphometrics in this chapter is preliminary, the findings and reported data will contribute to more extensive studies addressing questions of physical activity, health, nutrition conditions and genetic relationships in comparison with other populations.

SUMMARY

This chapter has examined limb bone morphometrics of the Man Bac series in comparison with neighbouring assemblages. Relative to neolithic, Bronze and Iron Age samples from Thailand, Man Bac limb bones are neither particularly robust nor gracile. The limb length proportions represented by radial-humeral and tibial-femoral indices also fall in the global intermediate range. Regarding stature estimations, Sangvichien et al.'s (1985) formulae based on the lower limbs was determined to be the most appropriate for the Man Bac assemblage, providing mean stature estimates of 161.4 cm for males and 154.0 cm for females. These values are consistent with those seen in the near contemporaneous Khok Phanom Di series from Thailand.

ACKNOWLEDGMENTS

The authors are grateful to Former Director Dr. Ha Van Phung, and Vice Director Dr. Nguyen Giang Hai, the Vietnamese Institute of archaeology, for their permission to excavate the Man Bac site.

This study was supported by a Grant-in-Aid in 2003-2005 (No. 15405018) and 2008-2009 (No. 20370096) from the Japan Society for the Promotion of Science, and by support from the Toyota Foundation in 2006-2007 (Grant No. D06-R-0035).

LITERATURE CITED

Domett K. 2001. Health in late prehistoric Thailand. BAR International Series 946. Oxford: Archaeopress.

Fujii A. 1960. On the relation of long bone lengths of limb to stature. Bulletin of the School of Physical Education, Juntendo Univ 3: 49-61 (in Japanese with English summary).

Martin R, Saller K. 1957. Lehrbuch der Anthropologie, Band 1. Stuttgart: G Fischer.

Mo S. 1983. Estimation of stature by long bones of Chinese male adults in South China. Acta Anthropologica Sinica 2:80-85 (in Chinese with English abstract).

Olivier G. 1969. Practical Anthropology. Springfield: Charles C. Thomas.

Oxenham M, Tilley L, Matsumura H, Cuong NL, Thuy NK, Dung NK, Domett K, Huffer D. 2009. Paralysis and severe disability requiring intensive care in Neolithic Asia. Anthropol Sci 117: 107-112.

Pietrusewsky M, Douglas MT. 2002. Ban Chiang: a prehistoric village site in northeast Thailand. I: The human skeletal remains. Philadelphia: Museum of Archaeology and Anthropology, University of Pennsylvania.

Roth HL. 1899. Aborigines of Tasmania. Halifax: F. King & Sons Co.

Sangvichien SJ, Srisurin V, Watthanayingsakul. 1985. Estimation of stature of Thai and Chinese from the length of femur, tibia and fibula. Siriraj Medical Journal 37: 215-218.

Sjøvold T. 1990. Estimation of stature from long bones utilizing the line of organic correlation. J Hum Evol 5:431-447.

Stevenson PH. 1929. On racial differences in stature long bone regression formulae, with special reference to stature reconstruction formulae for the Chinese. Biometrika 21: 3030-321.

Takigawa W. 2005. Metric comparison of limb bone characteristics between the Jomon and Hokkaido Ainu. Anthropol Sci (Japanese series) 113: 43-61 (in Japanese with English title and summary).

Trotter M, Gleser GCG. 1958 A re-evaluation of stature based on measurements of stature during life and of long bones after death. Am J Phys Anthropol 16: 79-123.

Wright LE and Vasquez MA 2003. Estimating the length of incomplete long bones: forensic standards from Guatemala. Am J Phys Anthropol 120: 233-251.

Yokoh Y. 1940. Beiträge zur Kraniologie der Dajak Beiträge zur Osteologie der Dajak. Japanese Journal of Medical Sciences, Part I Anatomy Vol. VIII No.3. Tokyo: The National Research Council of Japan (in Germany).

Wakebe T. 2002. Morphological characteristics of the limb bones of human skeletal remains excavated from Jiangnan area in China. In: Nakahashi T, Li M, editors. Ancient People in the Jiangnan Region, China. Fukuoka: Kyushu University Press. p 51-60.

7

Palaeohealth at Man Bac

Marc F. Oxenham[1] and Kate M. Domett[2]

[1]School of Archaeology and Anthropology, Australian National University
[2]School of Medicine and Dentistry, James Cook University, Australia

The purpose of this chapter is to review the evidence of adult and subadult health for individuals recovered from the Man Bac site during the 2005 and 2007 excavation seasons. A fuller appreciation of the inhabitants of Man Bac can only be realised through an examination of the nature and patterning of health markers in the context of other bio-variables such as preservation, demographic profile, stature, diet and genetic relationships with contemporaneous, previous and later populations in the region. To this end, the health profile of the Man Bac inhabitants has been developed towards the end of this monograph.

The palaeohealth of the ancient inhabitants of what is now Vietnam has been extensively examined and discussed in a number of studies (Oxenham et al., 2005; Oxenham, 2006; Oxenham et al., 2006). With respect to Man Bac specifically, limited examinations of childhood health, using remains from the 2005 season only, have been carried out in the context of broader mortuary archaeological questions (Oxenham, 2006). In this chapter, health variables are limited to two non-specific signatures of physiological impairment, cribra orbitalia and linear enamel hypoplasia, as well as a range of oral health indicators, including dental caries, alveolar defects (often termed abscesses) and antemortem tooth loss. Subsequent publications will review the evidence for other health variables including trauma and infectious disease.

MATERIALS AND METHODS

Only individuals excavated by the authors in the 2005 and 2007 seasons were included in this study and operational sample size varied according to the variable of interest. For the oral health assessment 29 adults and 11 subadults had assessable teeth, while 28 adults and 18 subadults possessed assessable alveoli. The sample for assessment of cribra orbitalia included 26 adults and 32 subadults. For the purposes of this study a subadult was any individual aged 15 years or younger, while an adult was any individual aged 16 years or older. Adults were further divided into two age groups (younger 16-29 years; older 30+ years) for assessing any possible age-dependent or correlated affects on the manifestation of health variables.

Oral health variables included caries, antemortem tooth loss (AMTL) and alveolar defects (AD). Carious lesion recording was based on Hillson (2001) and effectively included 10 categories of lesion (limited by the types of lesions occurring in the Man

Bac assemblage): (A) lesion initiated on aproximal (interproximal) attrition facet; (AG) gross lesion with unclear initiation site (includes aproximal facet); (BL) buccal or lingual lesions of crown (not CEJ, occlusal, aproximal etc); (BLG) BL lesion that includes other sites (initiation site unclear); (GG) massive crown/root destruction (initiation point unclear); (OG) occlusal gross (fissure system/occlusal facet initiation unclear); (OWFD) occlusal wear facet initiation (dentine exposed); (P) buccal molar or upper lingual incisor pit initiation; (R) lesion [groove] following cement-enamel junction or just on root; (RG) gross root lesion but also includes other sites (initiation site unclear). In addition to reporting by age and sex, carious lesions were recorded by maxillary and mandibular location as well as position (anterior compared to posterior dentition).

Antemortem tooth loss can be identified in a relatively straightforward manner, and Lukacs' (1989:271) definition was followed: "progressive resorptive destruction of the alveolus". Alveolar defects of pathological origin (AD), often erroneously referred to as abscesses in the literature, have been recorded following the same method, and reasoning, as Oxenham et al. (2005), where they were referred to as alveolar defects of pulpal origin. The further revision of this term avoids assumptions regarding the ultimate origin of the infection. In practical terms, AD were not recorded with reference to their precise location. Moreover, AD includes alveolar defects that are both isolated or circumscribed lesions in the alveolar bone and defects that are continuous with the margins of the alveoli.

Cribra orbitalia (CO) and linear enamel hypoplasia (LEH) were the two signatures of physiological disruption selected for analysis. Linear enamel hypoplastic (LEH) events expressed labially on assessable canines and incisors that met DDE index type 4 (Federation Dentaire International 1982) criteria were recorded. LEH is reported by position (maxillary, mandibular and combined) and tooth class (canine, incisor and combined) using the tooth count and individual count reporting protocols. Moreover, tooth count and individual count frequencies are also reported by age class and sex. An assessable tooth is defined as one where less than 50% of approximated crown height has been removed through wear. Only LEH severity categories 1 and 2, following Duray (1996), are presented in this analysis.

When recording cribra orbitalia (CO), the minimal requirement for inclusion in the analysis was the preservation of the anterolateral and anteromedial aspects of at least one orbital roof. CO was scored using the following categories: (1) absent; (2) presence of faint remodelling scars; (3) presence of clear remodelling scars; (4) presence of light to active lesions and remodelling scars; (5) presence of pronounced active lesions and remodelling scars; (6) presence of light to mild active lesions but an absence of remodelling scars; (7) presence of pronounced active lesions but an absence of remodelling scars. Faint remodelling scars are analogous to Webb's (1995: plate 5-1a) porotic form but without evidence of un-remodelled perforating lesions. Clear remodelling scars are analogous to Webb's (1995: Plate 5-2) recovery scars from remodelling. The use of the term 'remodelling scars', on its own, refers to any manifestation of faint to clear evidence of remodelling. Light to mild active (open) lesions are analogous to Webb's (1995: Plate 5-1a,b) porotic and cribrotic forms of CO, but exclude any reference to remodelling. Pronounced active lesions are analogous to Webb's (1995: Plate 5-1c) trabecular form of CO. For the purposes of reporting, frequencies are provided for these various categories but are collapsed into two broader categories for subsequent analysis and comparison: remodelled CO (includes only cases with remodelled CO and excludes any case that also displays active lesions); and active CO (includes all cases with active CO whether or not they also manifest remodelling).

RESULTS

Oral Health

Caries

Table 7.1 summarises oral structure sample preservation for adults and subadults. A total of 28 individuals (727 alveoli) with entirely adult dentitions and 44 individuals (537 alveoli) with mixed dentitions possessed assessable alveoli, or 29 individuals (581 teeth) with completely adult dentitions and 38 individuals (433 teeth) with mixed dentitions having assessable teeth. Note that Table 7.1 separates individuals with mixed dentitions into a group where only the permanent teeth are assessed and one where only the deciduous dentition is examined. Given good preservation in general there is a relatively high proportion of teeth relative to alveoli, both adult and subadult, in the sample. Comparing anterior relative to posterior teeth present by alveoli demonstrates the lower proportion of preserved anterior teeth. Contributing factors to tooth loss include antemortem loss (including tooth ablation) and postmortem loss. The anterior teeth are more susceptible to postmortem loss and were targeted for tooth ablation. There is slightly better retention of maxillary compared to mandibular teeth in the sample but not to a marked degree. Table 7.2 summarises the frequency and patterning of carious lesions, by tooth count reporting method, for all permanent (males, females and indeterminate sex aged 16+ years; subadults aged 6-15 years) and deciduous (subadults aged 9 months+) teeth. The overall frequency of carious lesions for the adult sample is 11.0%, which declines to 8.6% when the permanent dentition of subadults are included. A statistically significantly (see Table 7.7 for a summary of oral health statistical comparisons) higher frequency of lesions occur in the female dentition (15.5%) as compared to male teeth (7.6%). However, in terms of individuals, 58.3% of female individuals display carious lesions compared to 60.0% of males. Looking at carious lesions by age category shows the expected outcome of a statistically significantly (Table 7.7) higher level of lesions among older females (23.4% of all older female teeth compared to 3.6% of younger female teeth) and older males (10.8% of older male teeth compared to 3.8% of younger male teeth). Further, no subadult permanent teeth displayed carious lesions. In terms of tooth position, statistically significantly more posterior (14.7% of all lesions) permanent adult teeth were affected by lesions than anterior adult teeth (4.0% of all lesions). Finally, the distribution of lesions by upper and lower jaw was very similar with 11.1% of maxillary and 10.9% of mandibular teeth being carious.

When examining the various manifestations (including location) of carious lesions on permanent teeth, the most common type was R (48.4% of all lesions), or lesions following the cement-enamel junction or just on the root. Following R lesions come AG (gross lesion with unclear initiation site) at 18.8% of all lesions and then GG (massive crown/root destruction with unclear initiation point) at 17.2% of all

Table 7.1 Dental sample summary: Man Bac 2004/5-7 seasons.

	females[3]	males[3]	unsexed[3]	adult subtotal	unsexed SA[4]	adult & SA[4] Total	subadult[5]	TOTAL
N[1]	11	15	2	28	18	46	26	72
alveoli	270	429	28	727	208	935	329	1264
alveoli/N	24.5	28.6	14.0	26.0	11.6	20.3	12.7	17.6
N[2]	12	15	2	29	11	40	27	67
teeth	207	354	20	581	163	744	270	1014
teeth/N	17.3	23.6	10.0	20.0	14.8	18.6	10.0	15.1
teeth/alveoli %	76.7	82.5	71.4	79.9	78.4	79.6	82.1	80.2
preserved ant. alveoli	109	160	12	281	84	365	188	553
preserved ant. teeth	77	117	6	200	79	279	151	430
ant. teeth/alveoli %	70.6	73.1	50.0	71.2	94.0	76.4	80.3	77.8
preserved post. alveoli	161	269	16	446	124	570	141	711
preserved post. teeth	130	237	14	381	84	465	119	584
post. teeth/alveoli %	80.7	88.1	87.5	85.4	67.7	81.6	84.4	82.1
preserved max. alveoli	132	209	0	341	99	440	158	598
preserved max. teeth	99	180	0	279	77	356	139	495
max. teeth/alveoli %	75.0	86.1	0.0	81.8	77.8	80.9	88.0	82.8
preserved man. alveoli	138	220	28	386	109	495	171	666
preserved man. teeth	108	174	20	302	86	388	131	519
man. teeth/alveoli %	78.3	79.1	71.4	78.2	78.9	78.4	76.6	77.9

[1]any individual with an assessable alveolus
[2]any individual with an assessable tooth
[3]adult dentition
[4]subadults with partial adult dentition (only permanent teeth assessed)
[5]includes any individual with partial or complete deciduous dentition (only deciduous teeth assessed)
Note: positions/N is low for unsexed due to number of SAs with only partial permanent dentition

lesions. At 7.8% of all lesions, A type (lesion initiated on an interproximal attrition facet) is the fourth most common, followed by relatively rare forms at Man Bac including BL (buccal or lingual lesions of crown) at 3.1%, and several other forms at 1.6%: OG (occlusal gross), OWFD (occlusal wear facet initiation) and RG (gross root lesion but with unclear ignition point). This pattern is broadly maintained when looked at by sex, although GG is the second most common lesion among females (28.1% of all female lesions) but relatively uncommon among males (7.4% of all male lesions).

Table 7.2 also presents a summary of carious lesions affecting the deciduous dentition. Of the 270 deciduous teeth available for inspection, 3.7% (27.8% of subadult individuals with deciduous teeth) displayed carious lesions. Unlike the pattern seen in the permanent dentition, 5.3% of anterior teeth were carious, compared to 1.7% of posterior teeth. Further, again unlike the relatively even distribution of maxillary and mandibular lesions seen in the adult dentition, 5.0% of maxillary as compared to 2.3% of mandibular deciduous teeth were carious. In a pattern different to that seen in the permanent dentition, 60.0% of all lesions were evenly distributed among type A and BL forms with the remaining scattered among BLG, OG, P and R. Neither BLG (BL lesion that includes other sites) nor P (buccal molar or upper lingual incisor pit initiation) are forms seen in the permanent dentition.

Table 7.2 Caries profile (tooth count): Man Bac 2005-7 seasons.

		N[1]	n[2] ant/post	obs caries ant/post	proportion of caries ant/post	n[3] max/man	obs caries max/man	proportion of caries max/man	A	AG	BL	BLG	GG	OG	OWFD	P	R	RG	Total caries	%[4] caries	caries individ.[5]
Permanent Teeth																					
female	16-29yrs	83	31/52	0/3	0.0/5.8	41/42	2/1	4.9/2.4		1							2		3	3.6	1/4 (25.0)
	30+yrs	124	46/78	4/25	8.7/32.1	58/66	11/18	19.0/27.3	1	7			9				12		29	23.4	6/8 (75.0)
	subtotal	207	77/130	4/28	5.2/21.5	99/108	13/19	13.1/17.6	1	8			9				14		32	**15.5**	**7/12 (58.3)**
	%								3.1	25.0			28				44		100.0		
male	16-29yrs	159	50/109	2/4	4.0/3.7	80/79	5/1	6.3/1.3	1		2						3		6	3.8	2/7 (28.6)
	30+yrs	195	67/128	2/19	3.0/14.8	100/95	13/8	13.0/8.4	1	4			2	1	1		11	1	21	10.8	7/8 (87.5)
	subtotal	354	117/237	4/23	3.4/9.7	180/174	18/9	10.0/5.2	2	4	2		2	1	1		14	1	27	**7.6**	**9/15 (60.0)**
	%								7.4	15	7.4		7.4	3.7	3.7		52	3.7	100.0		
indet.	16-29yrs	0	0/0	0/0	0.0/0.0	0/0	0/0	0.0/0.0											0	0.0	0/0 (0.0)
	30+yrs	20	6/14	0/5	0.0/35.7	0/20	0/5	0.0/25.0	2								3		5	25.0	2/2 (100)
	subtotal	20	6/14	0/5	0.0/35.7	0/20	0/5	0.0/25.0	2								3		5	**25.0**	**2/2 (100)**
	%								40.0								60.0		100.0		
subtot.		**581**	**200/381**	**8/56**	**4.0/14.7**	**279/302**	**31/33**	**11.1/10.9**	5	12	2		11	1	1		31	1	64	**11.0**	**18/29 (62.1)**
	%								7.8	18.8	3.1		17.2	1.6	1.6		48.4	1.6	100.0		
SA	6-15yrs	163	79/84	0/0	0.0/0.0	77/86	0/0	0.0/0.0												**0.0**	**0/11 (0.0)**
Total		**744**	**279/465**	**8/56**	**2.9/12.0**	**356/388**	**31/33**	**8.7/8.5**	5	12	2		11	1	1		31	1	64	**8.6**	**18/40 (45.0)**
									7.8	18.8	3.1		17.2	1.6	1.6		48.4	1.6	100.0		
Deciduous Teeth																					
SA	9mths+	**270**	**151/119**	**8/2**	**5.3/1.7**	**139/131**	**7/3**	**5.0/2.3**	3		3	1		1		1	1		10	**3.7**	**5/18 (27.8)**
									30.0		30.0	10.0		10.0		10.0	10.0		100.0		
TOTAL		**1014**	**430/584**	**16/58**	**3.7/9.9**	**495/519**	**38/36**	**7.7/6.9**	8	12	5	1	11	2	1	1	32	1	74	**7.3**	**23/58 (39.7)**
%									11	16.2	6.8	1.4	14.9	2.7	1.4	1.4	43.2	1.4	100.0		

A lesion initiated on aproximal (interproximal) attrition facet
AG gross lesion with unclear initiation site (includes aproximal facet)
BL buccal or lingual lesions of crown (not CEJ, occlusal, aproximal etc)
BLG BL lesion that includes other sites (initiation site unclear)
GG massive crown/root destruction (initiation point unclear)
OG occlusal gross (fissure sys/occlusal facet initiation unclear)
OWFD occlusal wear facet initiation (dentine exposed)
P buccal molar or upper lingual incisor pit initiation
R lesion (groove) following cement-enamel junction or just on root
RG gross root lesion but also includes other sites (initiation site unclear)

[1] total preserved teeth
[2] preserved anterior teeth (incisors/canines)/ preserved posterior teeth (premolars/molars)
[3] preserved maxillary teeth/preserved mandibular teeth
[4] carious teeth/total assessable teeth for this category x 100
[5] obs/individual (%)

Antemortem tooth loss (AMTL)

Table 7.3 summarises the frequency and patterning of AMTL in all individuals with permanent alveoli. The data presented here exclude all cases of deliberate tooth ablation, a behaviour common amongst the adult Man Bac sample and including several combinations of maxillary and/or mandibular incisor extraction. The patterning and significance of tooth ablation is dealt with in a forthcoming publication. The overall frequency of AMTL for the adult sample is 2.6%, which declines to 2.0% when the permanent alveoli of subadults are included. A statistically significantly (Table 7.7) higher frequency of AMTL occurs in the alveoli of females (4.8%) compared to males (1.4%).

By age group no younger female alveoli show AMTL, compared to 7.5% of older female alveoli, and only 0.5% of younger male alveoli compared to 2.1% of older male alveoli display AMTL. Further, no subadult permanent or deciduous alveoli displayed AMTL. In terms of position, statistically significantly more posterior (3.8%) permanent adult alveoli were affected by AMTL than anterior adult alveoli (0.7%). The distribution of lesions by upper and lower jaw was somewhat similar although more AMTL affected mandibular adult alveoli (2.8%) than maxillary alveoli (2.3%).

Alveolar defects of pathological origin (AD)

The frequency and distribution of AD is summarised in Table 7.3. A slightly higher frequency of AD, by alveoli count, occurs in males (1.9% of all alveoli) than females (1.5%), albeit not to a statistically significant degree. By age, older females and older males both display more AD than their younger age counterparts, however, this is only a statistically significant difference among males. No cases of subadults with either permanent or deciduous dentitions displayed AD. The majority of cases of AD occurred in the posterior alveolar bone (2.7% of all adult alveoli) to a statistically significant degree. The distribution of AD was the same, at 1.8%, for both adult mandibular and maxillary alveoli.

Physiological Health

Cribra orbitalia

Table 7.4 provides information on the frequency and type of CO by age and sex for Man Bac adults and subadults. While the raw frequencies of each of the various forms of CO are presented, small sample sizes mean that meaningful comparisons can only be made in terms of remodelled CO (individuals with remodelling scars but without active lesions) and active CO (individuals with active lesions, regardless of whether or not they also have evidence for remodelling). Males display a higher frequency of CO (92.3%, χ^2 3.128, p 0.077, Yates corrected), active and remodelled combined, than females (53.8%). CO does not appear to vary with age as a similarfrequency of CO occurs among younger (50.0%) and older females (55.6%) and younger males (100%) and older males (87.5%). However, age does appear to be an important factor in the type of CO seen in adults, with active CO absent in females and accounting for only 15.4% of CO among males. While the overall frequency of adult CO is 73.1%, 89.5% of all adult individuals with CO (17/19

Table 7.3 Antemortem Tooth Loss (AMTL) and Alveolar Defect (AD) Profile: Man Bac 2004/5-7 Seasons.

			n[2]	n[3]	obs AMTL[4]	% AMT	obs AMTL[4]	% AMTL	**Total %**	obs AD	% AD	obs AD	% AD	**Total %**
		N[1]	ant/post	max/man	ant/post	ant/post	max/man	max/man	**AMTL**	ant/post	ant/post	max/man	max/man	**AD**
Permanent Alveoli														
female	16-29yrs	97	37/60	51/46	0/0	0.0/0.0	0/0	0.0/0.0	0.0	0/0	0.0/0.0	0/0	0.0/0.0	0.0
	30+yrs	173	72/101	81/92	0/13	0.0/12.9	4/9	4.9/9.8	7.5	0/4	0.0/4.0	2/2	2.5/2.2	2.3
	subtotal	270	109/161	132/138	0/13	0.0/8.1	4/9	3.0/6.5	4.8	0/4	0.0/2.5	2/2	1.5/1.4	1.5
male	16-29yrs	193	72/121	95/98	1/0	4.4/0.0	0/1	0.0/1.0	0.5	0/0	0.0/0.0	0/0	0.0/0.0	0.0
	30+yrs	236	88/148	114/122	1/4	1.1/2.7	4/1	3.5/0.8	2.1	1/7	1.1/4.7	4/4	3.5/3.3	3.4
	sub total	429	160/269	209/220	2/4	1.3/1.5	4/2	1.9/0.9	1.4	1/7	0.6/2.6	4/4	1.9/1.8	1.9
indet.	16-29yrs	0	0/0	0/0	0/0	0.0/0.0	0/0	0.0/0.0	0.0	0/0	0.0/0.0	0/0	0.0/0.0	0.0
	30+yrs	28	12/16	0/28	0/0	0.0/0.0	0/0	0.0/0.0	0.0	0/1	0.0/6.3	0/1	0.0/3.6	3.6
	subtotal	28	12/16	0/28	0/0	0.0/0.0	0/0	0.0/0.0	0.0	0/1	0.0/6.3	0/1	0.0/3.6	3.6
subtot.		**727**	**281/446**	**341/386**	**2/17**	**0.7/3.8**	**8/11**	**2.3/2.8**	**2.6**	**1/12**	**0.4/2.7**	**6/7**	**1.8/1.8**	**1.8**
SA	6-15yrs	208	84/124	99/109	0/0	0.0/0.0	0/0	0.0/0.0	0.0	0/0	0.0/0.0	0/0	0.0/0.0	0.0
Total		**935**	**365/570**	**440/495**	**2/17**	**0.5/3.0**	**8/11**	**1.8/2.3**	**2.0**	**1/12**	**0.3/2.1**	**6/7**	**1.4/1.4**	**1.4**
Deciduous Alveoli														
SA[5]	9mths+	329	188/141	158/171	0/0	0.0/0.0	0/0	0.0/0.0	0.0	0/0	0.0/0.0	0/0	0.0/0.0	0.0
TOTAL %		**1264**	**553/711**	**598/666**	**2/17**	**0.4/2.4**	**8/11**	**1.3/1.7**	**1.5**	**1/12**	**0.3/1.7**	**6/7**	**1.0/1.1**	**1.0**

[1]total preserved alveoli

[2]preserved anterior alveoli (incisors/canines)/ preserved posterior alveoli (premolars/molars)

[3]preserved maxillary alveoli/ preserved mandibular alveoli

[4]AMTL antemortem tooth loss, note excludes cases of deliberate tooth ablation

[5]excludes cases of natural deciduous tooth exfoliation

Table 7.4 Cribra Orbitalia Profile: Man Bac 2004/5-7 Seasons.

	Adult orbits		CO	faint	clear	LAL & RS	PAL & RS			total[2] CO remod.	total[3] CO active	TOTAL[4] CO
		N[1]	absent	RS	RS	WT	WT	LAL	PAL	obs (%)	obs (%)	obs (%)
female	16-29yrs	4	2	1	1					2 (50.0)	0 (0.0)	2 (50.0)
	30+yrs	9	4	3	2					5 (55.6)	0 (0.0)	5 (55.6)
	subtotal	13	6	4	3					7 (53.8)	0 (0.0)	7 (53.8)
male	16-29yrs	5	0	2	2	1				4 (80.0)	1 (20.0)	5 (100)
	30+yrs	8	1	3	3	1				6 (75.0)	1 (12.5)	7 (87.5)
	sub total	13	1	5	5	2				10 (76.9)	2 (15.4)	12 (92.3)
subtot.		**26**	**7**	**9**	**8**	**2**				**17 (65.4)**	**2 (7.7)**	**19 (73.1)**
subadults	6-15yrs	6	1	1		2		2		1 (16.7)	4 (66.7)	5 (83.3)
	2.5-5yrs	5	0					5		0 (0.0)	5 (100)	5 (100)
	12-29mths	10	1	2	1	1		4	1	3 (30.0)	6 (60.0)	9 (90.0)
	6-11mths	6	1					5		0 (0.0)	5 (83.3)	5 (83.3)
	0-5mths	5	0	1				3	1	1 (20.0)	4 (80.0)	5 (100)
subtot.		**32**	**3**	**4**	**1**	**3**		**19**	**2**	**5 (15.6)**	**24 (75.0)**	**29 (90.6)**
Total		**58**	**10**	**13**	**9**	**5**		**19**	**2**	**22 (37.9)**	**26 (44.8)**	**48 (82.8)**

L & R orbits assessed whenever possible; most severe lesion form in any give orbit used to score the individual
CO absent: clear of lesions
faint RS: faint remodelling scars
clear RS: clear remodelling scars
LAL & RS: light to mild active lesions and remodelling scars
PAL & RS: pronounced active lesions and remodelling scars
LAL: light to mild active lesions only
PAL: pronounced active lesions only

[1]sample of individuals with at least one assessable orbit
[2]only individuals with 'remodelled only' CO (active cases excluded)
[3]individuals with active CO including those with or without remodelling
[4]all individuals displaying signs of CO (remodelled, active, remodelled & active lesions)

cases) have a remodelled only form, with only 10.5% (2/19 cases) showing evidence of active or un-remodelled lesions.

A higher frequency of subadult CO (90.6%) is seen compared to adults, but not to a statistically significant degree (χ^2 1.988, p 0.159, Yates corrected). The chief difference between subadult and adult CO is that the vast majority of subadults display active or un-remodelled lesions (75.0% of all subadult CO is active). The majority of subadult lesions are characterised as light to mild active lesions (19/29 of all forms of subadult CO, or 65.5%). There does not appear to be any correlation between increasing subadult age and the proportion of active lesions, but this may be due to the limited sample sizes within each subadult age category.

Linear enamel hypoplasia (LEH)

Tooth count

Only permanent incisor and canine LEH is examined in this study. Deciduous tooth hypoplasia, particularly localised hypoplasia of primary canines (LHPC), is examined in detail in a forthcoming publication. Of the total assessable sample of permanent incisors and canines, 181/279 (64.9%) display evidence for LEH (Table 7.5). A higher frequency of female combined canines and incisors displayed LEH (80.5%) than males (67.5%) to a statistically significant degree (χ^2 3.951, p 0.047). The frequency of combined incisor and canine LEH is higher in younger females (87.1%) than their older counterparts (76.1%; χ^2 0.815, p 0.367, Yates corrected) and this is also seen among males (younger 80.0%, older 58.2%; χ^2 6.200, p 0.013), although differences by age class are only statistically significant among males. While a higher degree of LEH is seen among younger adult females and males, the frequency of LEH in the permanent combined canines and incisors of subadults is comparatively low at 45.6%. Further, the frequency of subadult combined canine and incisor LEH (36/79, 45.6%) is statistically significantly lower than the frequency of combined male and female younger adult LEH (67/81, 82.7%: χ^2 24.063, p 0.000) and combined male and female older adult LEH (74/113, 65.5%: χ^2 7.538, p 0.006).

The frequency of female incisor LEH is a little higher (82.5%) than canine LEH (78.4%). This pattern is reversed for males, where the frequency of canine LEH (79.2%) is statistically significantly greater than incisor LEH (57.8%, χ^2 6.073, p 0.014). For subadults with permanent teeth, the frequency of canine LEH (75.0%) is also statistically significantly greater than incisor LEH (35.6%, χ^2 9.351, p 0.002).

The frequency of female incisor LEH is somewhat higher among the maxillary teeth (90.9%) than mandibular teeth (72.2%) (χ^2 1.275, p 0.259, Yates corrected), with the distribution of canine LEH being similar in maxillary (77.8%) and mandibular (78.9%) canines (χ^2 0.00, p 1.0, Yates corrected). For males a similar pattern is seen with a higher frequency of maxillary incisor LEH (68.8%) than mandibular incisor LEH (46.9%, χ^2 3.139, p 0.076). Further, the frequency of male maxillary canine LEH (77.8%) is quite similar to the frequency of mandibular canine

LEH (80.8%) ($\chi^2$0.072, p 1.0). Regarding subadults with permanent teeth, a statistically significantly higher frequency of maxillary incisor LEH (57.1%) compared to mandibular incisor LEH (16.1%, $\chi^2$10.795, p 0.001) occurs. In terms of subadult permanent canines, a similar frequency of maxillary (77.8%) and mandibular (72.7%) LEH is seen ($\chi^2$0.00, p 1.0, Yates corrected).

Table 7.5 Incisor & canine linear enamel hypoplasia profile: Man Bac 2004/5-7 seasons.

Permanent Teeth

		N[1]	I n[2] max/man	C n[3] max/man	obs I LEH[4] max/man	% I LEH[5] max/man	% I LEH total	obs C LEH max/man	% C LEH max/man	% C LEH total	**% Total LEH combin.[6]**
female	16-29yrs	31	9/10	6/6	9/8	100/80.0	89.5	6/4	100/66.7	83.3	**87.1**
	30+yrs	46	13/8	12/13	11/5	84.6/62.5	76.2	8/11	66.7/84.6	76.0	**76.1**
	subtotal	77	22/18	18/19	20/13	90.9/72.2	82.5	14/15	77.8/78.9	78.4	**80.5**
male	16-29yrs	50	14/13	12/11	12/8	85.7/61.5	74.1	10/10	83.3/90.9	87.0	**80.0**
	30+yrs	67	18/19	15/15	10/7	55.6/36.8	45.9	11/11	73.3/73.3	73.3	**58.2**
	sub total	117	32/32	27/26	22/15	68.8/46.9	57.8	21/21	77.8/80.8	79.2	**67.5**
indet.	16-29yrs	0	0/0	0/0	0/0	0.0/0.0	0.0	0/0	0.0/0.0	0.0	**0.0**
	30+yrs	6	0/4	0/2	0/2	0.0/50.0	50.0	0/2	0.0/100	100.0	**66.7**
	subtotal	6	0/4	0/2	0/2	0.0/50.0	50.0	0/2	0.0/100	100.0	
											66.7
subtot.		**200**	**54/54**	**45/47**	**42/30**	**77.8/55.6**	**66.7**	**35/38**	77.8/80.9	**79.3**	72.5
SA	6-15yrs	79	28/31	9/11	16/5	57.1/16.1	35.6	7/8	77.8/72.7	75.0	**45.6**
Total		**279**	**82/85**	**54/58**	**58/35**	**70.7/41.2**	**55.7**	**42/46**	**77.8/79.3**	**78.6**	64.9

[1]total preserved maxillary and mandibular incisors and canines
[2]preserved maxillary/ mandibular incisors
[3]preserved maxillary/ mandibular canines
[4]observed LEH count for tooth class
[5]% LEH for tooth class
[6]% LEH for combined incisors and canines

Individual count

Table 7.6 presents the frequency of individuals, by age class and sex, with LEH in either their canines or incisors. In lieu of matching LEH events across the teeth in any given individual, Table 7.6 presents a way of determining the minimum number of LEH formation events affecting any individual by reporting the maximum number of observable LEH events as determined by the tooth with the most hypoplastic signatures.

While the vast majority of adults displayed LEH, the majority of adults had either a minimum of two LEH events (9/28, 32.1%) or 3+ LEH events (17/28, 60.7%). Low sample sizes mean it is difficult to assess any potential differences in the number of LEH events by age category. What can be said is that a higher proportion of young females have 3+ LEH events (75%) than older females (62.5%). The same general pattern is seen for males with a much higher proportion of younger males have 3+ LEH events (85.7%) and older males (37.5%). For subadult individuals with permanent teeth, 72.7% (8/11) of individuals displayed LEH, with 18.2% of those with LEH having at least one or a minimum of two events and 36.4% having 3+ LEH events.

Table 7.6 Incisor & canine linear enamel hypoplasia by individual: Man Bac 2004/5-7 seasons.

Permanent Teeth		N[1]	I/C n[2]	I/C[3] LEH obs	I/C[4] LEH%	I C combin.[5] LEH obs (%)	I C combin.[6] 1 LEH obs (%)	I C combin.[6] 2 LEH obs (%)	I C combin.[6] 3+ LEH obs (%)
female	16-29yrs	4	4/4	4/4	100/100	4 (100)	0 (0.0)	1 (25.0)	3 (75.0)
	30+yrs	8	6/8	5/7	83.3/87.5	8 (100)	1 (12.5)	2 (25.0)	5 (62.5)
	subtotal	12	10/12	9/11	90.0/91.7	12 (100)	1 (8.3)	3 (25.0)	8 (66.7)
male	16-29yrs	7	7/6	7/6	100/100	7 (100)	0 (0.0)	1 (14.3)	6 (85.7)
	30+yrs	8	8/8	6/8	75.0/100	8 (100)	1 (12.5)	4 (50.0)	3 (37.5)
	sub total	15	15/14	13/14	86.7/100	15 (100)	1 (6.7)	5 (33.3)	9 (60.0)
indet.	16-29yrs	0							
	30+yrs	1	1/1	1/1	100/100	1 (100)	0 (0.0)	1 (100)	0 (0.0)
	subtotal	1	1/1	1/1	100/100	1 (100)	0 (0.0)	1 (100)	0 (0.0)
subtot.		**28**	**26/27**	**23/26**	**88.5/96.3**	**28 (100)**	**2 (7.1)**	**9 (32.1)**	**17 (60.7)**
SA	6-15yrs	11	10/7	6/7	60.0/100	8 (72.7)	2 (18.2)	2 (18.2)	4 (36.4)
Total		**39**	**36/34**	**29/33**	**80.6/97.1**	**36 (92.3)**	**4 (10.3)**	**11 (28.2)**	**21 (58.3)**
Deciduous Teeth									
SA	9mths+	24	21/20	0/4	0.0/20.0	4 (16.7)	4 (16.7)	0 (0.0)	0 (0.0)
TOTAL %		**63**	**57/54**	**29/37**	**50.9/68.5**	**40 (63.5)**	**8 (12.7)**	**11 (17.5)**	**21 (33.3)**

[1]total individuals with canines and/or incisors
[2]total individuals with incisors/canines
[3]number of individuals with at least one LEH event affecting incisors/canines
[4]% of individuals with at least one LEH event affecting incisors/canines
[5]% of individuals with at least one LEH event affecting incisors and/or canines combined
[6]number and proportion of individuals with 1 only, 2, and 3 or more LEH events affecting any given incisor and/or canine

DISCUSSION

For the purposes of this discussion, oral and physiological signature comparisons will be limited to other northern Vietnamese assemblages: Mid Holocene Da But and early Metal Period materials. Comparative Vietnamese data referred to is from Oxenham (2006) unless otherwise stated.

Oral Health

Caries

The overall frequency of caries by the tooth count reporting method is 11.0%, or 8.6% if subadult permanent teeth are included. This rate is considerably higher than that seen for either the temporally earlier Da But period (1.5%) or later Metal Period (2.3%). Incidentally, this is the highest (Khok Phanom Di just lower at 10.9%) rate of caries reported for an ancient Southeast Asian site to date (Tayles, 1999; Oxenham et al., 2006). Caries by individual in the Man Bac sample is 62.0% of adults, in comparison to 13.8% of Da But (Con Co Ngua) and 20.8% of metal period adults. The trend toward higher caries rates in females (Da But females 2.1% of teeth, 21.4% of individuals; males 1.6% of teeth, 13.9% of individuals: metal period females 3.7% of teeth, 37.0% of individuals; males 1.4% of teeth, 19.2% of individuals) is also evident in the Man Bac assemblage, with 15.5% of female teeth carious compared to 7.6% of male teeth, although 60% of male individuals compared to 58.3% of female individuals suffered from caries at Man Bac.

In terms of the frequency of caries by age-at-death, the expected pattern of more carious teeth and a higher proportion of older people displaying caries occurred at Man Bac. This pattern is also seen in the Da But assemblage where 28.4% (2.8% of teeth) of 40+ years, 4.8% (0.4% of teeth) of 30-39 years and only 4.0% (0.7% of teeth) of <30 years individuals displayed lesions. The older Man Bac and even earlier Da But series differ from the Metal Period where the reverse pattern was seen; 12.5% (1.6% of teeth) of 40+ years, 18.8% (1.9% of teeth) of 30-39 years and only 27.5% (2.8% of teeth) of <30 years individuals displaying lesions.

Table 7.7 Summary of oral health statistical comparisons.

AMTL	X^2	p	GV[2]
female young/old	**6.11***	**0.013**	**old**
male young/old	0.98*	0.322	old
female/male	**7.31**	**0.014**	**fem**
anterior/posterior[1]	**5.35***	**0.021**	**post**
maxillary/mandibular[1]	0.18	0.848	man
AD			
female young/old	0.97*	0.325	old
male young/old	**5.78***	**0.016**	**old**
female/male	0.09*	0.764	male
anterior/posterior[1]	2.60*	0.107	post
maxillary/mandibular[1]	0.06*	0.815	max
Caries			
female young/old	**13.40***	**0.000**	**old**
male young/old	**6.08**	**0.014**	**old**
female/male	**8.51**	**0.004**	**fem**
anterior/posterior[1]	**15.31**	**0.000**	**post**
maxillary/mandibular[1]	0.01	0.944	max

Bold = statistically significantly different
* Yates Corrected
[1]Adults only
[2]GV = Greatest Value

Regarding lesion location, the higher proportion of posterior relative to anterior lesions is to be expected with the frequency of lesions seen at Man Bac (see discussion of differential susceptibility of teeth to caries in Hillson, 2001). Regarding the type of lesions seen here, 48.4% of Man Bac carious lesions occurred on the root or followed the cement-enamel junction. This is in contrast to 35.7% of Da But and 26.9% of Metal Period lesions manifesting in this manner (Oxenham, 2000). While it has been suggested that an increase in the grain component of a diet can increase the risk of such lesions (e.g. Molnar and Molnar, 1985; Moore, 1993), increases in agricultural intensification in southeast Asian assemblages is not associated with an increase in root/CEJ caries (Oxenham et al., 2006).

In the context of ancient Southeast Asia, the very high rate of caries at Man Bac is intriguing. Given the complete lack of physical evidence for rice agriculture or rice consumption it is unlikely that this particular food stuff contributed to poor oral health. Moreover, rice consumption in the region is not believed to be associated with caries anyway (Oxenham et al., 2006; Tayles et al., 2009). Preliminary stable isotopic work on a small Man Bac sample suggests more than 50% of the protein component of the diet derived from marine and/or freshwater sources and later Metal Period populations consumed more C3 plants (including rice) than was the case at Man Bac (Yoneda, 2008; Bower et al., 2006). The only other Southeast Asian assemblage with a comparably high rate of caries is Khok Phanom Di, where it has been suggested that the consumption of cariogenic foodstuffs such as taro, yam and banana may have had a contributory role (Tayles, 1999). The presence of such crops has not been identified at Man Bac.

While a similar proportion of adult male and female individuals displayed carious lesions, females with caries had a much greater number of affected teeth. Interestingly the lesion rate per tooth (males 7.6%, females 15.5%) is very similar to that seen at Khok Phanom Di (males 6.9%, females 14.6%). In both cases females have a rate more than twice that of males, and it is the female rate that contributes significantly to the overall high frequency of carious lesions in these two early, and essentially contemporaneous, assemblages. While the nature of the Man Bac diet is still unclear (apart from the significant aquatic food component and relatively low C3 component), Yoneda's (2008) preliminary isotopic results suggest females and males had slightly different diets (females show more negative d13C values and lower d15N values), which may have been a contributing factor to higher female rates of caries. Without knowing the Man Bac diet specifically, a range of possible reasons may be contributing to an elevated risk of caries in females including: rate and composition of female saliva; differential diet, genetic factors as well as the deleterious affects of pregnancy (see Ferraro and Vieira, 2010; Lukacs and Largaespada, 2006 for recent reviews). Regarding pregnancy, Lukacs (2008) has argued for a direct link between high fertility and greater rates of caries in females. The poor level of oral health in Man Bac females, particularly, is consistent with the elevated level of fertility suggested for the site (see Chapter 2). Clearly, oral disease is aetiologically multifactorial, however, whatever the range of proximate causes for poor oral health at Man Bac, the effects of pregnancy in the female sample is likely to have been a significant contributor.

Before moving on to consider other oral health variables some mention of subadult caries is required. Using a much smaller sample of Man Bac subadults, 50% of individuals (n=6) and 8.5% of deciduous teeth were previously reported as carious by Oxenham et al. (2008). With a much larger sample assessed (18 individuals and 290 teeth) the risk of subadult caries is somewhat lower (27.8% of individuals, 3.7% of teeth) making the rate of subadult caries similar to that seen at contemporaneous Khok Phanom Di (33.3% of individuals, 4.8% of teeth). It is somewhat intriguing that the Khok Phanom Di series displays very similar rates and patterns of carious lesions in the adult and subadult portions of their respective assemblages. Risk factors for subadult caries in ancient Southeast Asian assemblages are discussed in Oxenham et al. (2008) and include issues surrounding fluoride levels, oral hygiene, predisposing risks from hypocalcifications and LEH, and breast feeding practices.

Antemortem tooth loss (AMTL)

The overall adult level of AMTL in the Man Bac sample (2.6% of alveoli) is lower than that seen in both the Da But (4.8% of alveoli) and Metal Period samples (3.0% of alveoli). However, it is interesting to note that the frequency of Man Bac female AMTL by alveoli is nearly 3.5 times greater than that seen for males. Not only is the elevated rate of female AMTL consistent with the high rate of caries seen in Man Bac females, but given the correlation between AMTL and caries, is further evidence for the link between elevated fertility and poor oral health in females. It is worth noting that at least two papers have argued in support of the oral health and fertility hypothesis based on the use of AMTL alone (Watson et al., 2010; Fields et al., 2009). The distribution of AMTL by both age-at-death and location (posterior vs anterior) is consistent with expectations.

Alveolar defects of pathological origin (AD)

The overall adult level of AD in the Man Bac sample (1.8% of alveoli) is slightly higher than seen in the Da But sample (1.5% of alveoli) and somewhat lower than the Metal Period sample (2.6% alveoli). Given the much higher rates of female caries and AMTL at Man Bac it is a little surprising that there is a lower, albeit very slight, level of female AD compared to males. Part of the reason is likely the very low level of AD in general seen at Man Bac and the possibility that AD operates under different aetiological constraints than either caries or AMTL. The elevated level of AD seen in the Metal Period sample was attributed to the effects of often using the anterior teeth as tools (Oxenham et al., 2006).

Physiological Health

Cribra orbitalia (CO)

A remarkably high frequency of cribra orbitalia (CO) is seen in the Man Bac sample (73.1% of adults) relative to the Da But (28%) and Metal Periods (30%) of northern Vietnam. The even higher proportion of subadults exhibiting cribra orbitalia (the majority of which manifests as un-remodelled lesions in subadults) is indicative of a ubiquity of responsible stressors. The sedentary nature of the population, in addition to the elevated parasite loads of a tropical environment and riverine/marine focus in resource gathering, are likely the chief contributors to such physiological stressors (see discussion in Oxenham, 2006:228-9; Oxenham and Cavill, 2010). There is a clear, and expected, separation between adults and subadults in terms of the type of CO, with subadults having unremodelled lesions and adults displaying remodelled CO. There can be no doubt that whatever the stressors were, children were more vulnerable.

Linear enamel hypoplasia (LEH)

Compared to the Da But (72% of individuals; 63% male, 81% female) and Metal Period samples (67% of individuals; 65% males, 67% females), 100% (see Table 7.6) of Man Bac adult individuals displayed at least one canine and/or incisor LEH event. Clearly, whatever the aetiology of LEH at Man Bac, in terms of LEH this population appears to have been physiologically compromised relative to earlier and later populations in the region.

With every adult individual displaying at least one LEH event we need to turn to the distribution of LEH by sex, age and frequency of stressors in order to explore the implication of LEH at Man Bac. In terms of sex differences, females display a higher frequency of LEH by tooth count even though all adults have LEH. It would appear that greater male vulnerability was possibly offset at Man Bac by cultural and/or behavioural factors that elevated the risk of the development of LEH in female children.

In terms of LEH and age-at-death, the lower frequency of LEH affected teeth in the older male and female age cohorts might suggest a link between an increased level of LEH and lower age-at-death. Such a relationship has been suggested for both the Da But and Metal Period samples from Vietnam, as well as other studies globally (e.g. Duray, 1996; Goodman and Armelagos, 1988; Saunders and Keenleyside, 1999). Further support for this correlation can be observed when LEH is looked at by the minimum number of LEH events per individual. For instance, a greater proportion of individuals with 3 or more LEH events were found in the younger age cohort, particularly with regard to males.

Compromised Health Experience

This chapter set out to examine the evidence for adult and subadult health at Man Bac through the lens of oral (caries, antemortem tooth loss, and alveolar defects) and physiological health (cribra orbitalia and LEH). No matter what health indicator is looked at, the inhabitants of Man Bac appear to have been more compromised in their health than either the earlier Da But period or subsequent Metal Period assemblages. A number of factors, to a greater or lesser degree, may have contributed to the sub-optimal level of health seen at Man Bac: (1) colonising population; (2) migration; (3) sedentism; (4) adoption of new subsistence strategies; and (5) elevated fertility. Moreover, it is probable that whichever factors were responsible, they were probably not acting in isolation.

Evidence presented in Chapter 3 indicates that Man Bac is phenotypically and genetically heterogeneous, suggesting a population in biological transition. If indeed Man Bac is one of those archaeologically rare occurrences of a population undergoing rapid genetic change due to the effects of an influx of new migrants from the north, new genotypic expressions in an equally new environment may have resulted in a net health cost. Such a scenario is also consistent with some level of colonisation of the region as well as adoption of new subsistence strategies, all contributing to a greater impact on the health of the Man Bac community.

Intensification of agricultural practices in the later Metal Period have been argued to have been associated with a marked increase in the frequency of infectious disease in northern Vietnam (Oxenham et al., 2005). While infectious disease levels have not been examined for Man Bac yet, the base line health data discussed in this chapter are certainly consistent with the significant changes to the relationship between humans and the land. Sedentism in and of itself has been argued to be associated with compromised physiological health in earlier and later Vietnamese populations (Oxenham et al., 2006). Moreover, sedentism, the adoption of new subsistence strategies and increasing fertility are also familiar bedfellows and entirely consistent with the scenario being sketched for Man Bac some 3,500 years ago.

To conclude, while still early days in terms of analytical intensity, at face value it appears that the relatively poor base line health data for this population are consistent with a raft of biological and archaeological findings suggesting that Man Bac was a population in major transition.

LITERATURE CITED

Bower NW, Yasutomo Y, Oxenham MF, Nguyen LC, Nguyen KT. 2006. Preliminary reconstruction of diet at a neolithic site in Vietnam using stable isotope and Ba/Sr analyses. Bulletin of the Indo-Pacific Prehistory Association 26: 79-85.

Duray SM. 1996. Dental indicators of stress and reduced age at death in prehistoric native Americans. Am J Phys Anthropol 99:275-286.

Federation Dentaire International 1982. An epidemiological index of developmental defects of dental enamel (DDE Index). International Dental Journal 32:159-167.

Fields M, Herschaft EE, Martin DL, Watson JT. 2009. Sex and the agricultural transition: dental health of early farming females. Journal of Dentistry and Oral Hygiene 1:42-51.

Ferraro M, Vieira AR. 2010. Explaining gender differences in caries: a multifactorial approach to a multifactorial disease. International Journal of Dentistry, doi:10.1155/2010/649643.

Goodman, AH, Armelagos GJ. 1988. Childhood stress and decreased longevity in a prehistoric population. American Anthropologist 90:936-944.

Hillson S. 2001. Recording dental caries in archaeological remains. International Journal of Osteoarchaeology 11:249-289.

Lukacs JR. 1989. Dental paleopathology: methods for reconstructing dietary patterns. In: Iscan MY, Kennedy KAR, editors. Reconstruction of Life from the Skeleton. New York: Wiley-Liss. p 261-286.

Lukacs JR. 2008. Fertility and agriculture accentuate sex differences in dental caries rates. Current Anthropology 49: 901-914.

Lukacs JR, Largaespada LL. 2006. Explaining sex differences in dental caries prevalence: saliva, hormones, and "life-history" etiologies. Am J Hum Biol 18:540–555

Molnar S, Molnar I. 1985. Observations of dental diseases among prehistoric populations of Hungary. Am J Phys Anthropol 67:51-63.

Moore WF. 1993. Dental caries in Britain from Roman times to the nineteenth century. In: Geissler CG, Oddy DF, editors. Food, Diet and Economic Change Past and Present. Leicester: Leicester University Press. p 50-61.

Oxenham MF. 2000. Health and Behaviour During the Mid-Holocene and Metal Period of Northern Viet Nam. Unpublished PhD Thesis, Northern territory University, Darwin, NT, Australia.

Oxenham MF. 2006. Biological responses to change in prehistoric Vietnam. Asian Perspectives 45:212-239.

Oxenham MF, Cavill I. 2010. Porotic hyperostosis and cribra orbitalia: the erythropoietic response to Iron-deficiency anaemia. Anthropological Science DOI: 10.1537/ase.100302.

Oxenham MF, Nguyen KT, Nguyen LC. 2005. Skeletal evidence for the emergence of infectious disease in bronze and iron age northern Vietnam. Am J Phys Anthropol 126:359-376.

Oxenham MF, Nguyen LC, Nguyen KT. 2006. The oral health consequences of the adoption and intensification of agriculture in Southeast Asia. In: Oxenham M, Tayles N, editors. Bioarchaeology of Southeast Asia. Cambridge: Cambridge University Press. p 263-289.

Oxenham MF, Matsumura H, Domett K, Nguyen KT, Nguyen KD, Nguyen LC, Huffer D, Muller S. 2008. Health and the experience of childhood in late Neolithic Vietnam. Asian Perspectives 47:190-209.

Saunders SR, Keenleyside A. 1999. Enamel hypoplasia in a Canadian historic sample. Am J Hum Biol 11:513-524.

Tayles N. 1999. The People. Vol. 5 of The Excavation of Khok Phanom Di: A Prehistoric Site in Central Thailand. London; Society of Antiquities of London.

Tayles N, Domett K, Halcrow S. 2009. Can dental caries be interpreted as evidence of farming? The Asian experience. In: Kope T, Meyer G, Alt KW, editors. Comparative Dental Morphology. Basel: Karger. p 162-166.

Watson JT, Fields M, Martin D. 2010. Introduction of agriculture and its effects on women's oral health. Am J Hum Biol 22: 92-102.

Webb S. 1995. Palaeopathology of Aboriginal Australians: Health and Disease Across a Hunter-Gatherer Continent. Cambridge: Cambridge University Press.

Yoneda M. 2008. Dietary reconstruction of ancient Vietnamese based on carbon and nitrogen isotopes. Paper presented at the Man Bac Symposium, Institute of Archaeology, Hanoi, Vietnam, 19 July 2008

8
Mitochondrial DNA of Human Remains at Man Bac

Ken-ichi Shinoda

Department of Anthropology, National Museum of Nature and Science, Japan

Up until as recently as twenty years ago the genetic affinities, including intra and inter-sample comparisons, of skeletal remains in archaeological contexts was the domain of morphologists, using a suite of metric and non-metric skeletal characteristics believed to have an underlying genetic basis. Studies of the genes themselves were restricted to explorations of genetic variation, amongst contemporary, or living, human populations with subsequent inferences about their past evolutionary history (Ingman et al., 2000; Forster, 2004). While the latter approach has provided a panoramic, broad-stroke picture of our evolutionary past, such straightforward retrospective projections of the modern genetic composition and distribution on to the past have a number of inherent limitations. Relatively short-term and local biological and cultural processes such as epidemics, conquests, and forced relocations are likely to have contributed to and complicated the modern-day genetic landscape. A logical complement to studies of living DNA is clearly an examination of ancient DNA in archaeologically recovered regional populations. Recent advances in molecular biological techniques have facilitated the recovery and analysis of DNA from ancient material, thus providing a direct means of studying the genetic composition of past populations.

Because of its special characteristics, including small size, matrilineal inheritance, high copy number, and fast mutation rate, the majority of ancient DNA analysis is on mitochondrial DNA (mtDNA) (Alzualde et al., 2006; Maca-Mayer et al., 2005). This recently acquired ability to analyse mtDNA from archaeological remains yields more accurate genetic information than can be obtained through the morphological study of bones. Such information, combined with the results of archaeological investigation, should allow us to put forth and test new theories concerning the formation and subsequent history of past populations (Casas et al., 2006; Adachi et al., 2008). Notwithstanding this, there is clearly still a role for traditional bioarchaeological approaches when investigating such issues as such as age-at-death, pathology, and nutritional status of archaeological skeletal material. However, it is desirable that molecular biological analyses are used in conjunction with conventional bioarchaeological and morphological techniques in all studies of human skeletons.

In the present study, DNA analysis was performed on human skeletal remains excavated from Man Bac, Vietnam from 1999 to 2007. The significance of this study is increased, because of the lack of any previous DNA work on ancient human remains from Vietnam. The origin of the genetic diversity of human populations in Southeast Asia is still very controversial, despite the multidisciplinary approach of the research being used to address this question. Ancient DNA analysis can contribute to this debate by providing at least a piece of the genetic landscape at a precise time in the past, and so assists in shedding light on the origins of the genetic composition of present Southeast Asian populations.

MATERIALS AND METHODS

Of the skeletal sample excavated from 1999 to 2007, 35 well-preserved individuals were selected for DNA analysis. As tooth enamel forms a natural barrier to exogenous DNA contamination; and because DNA recovered from teeth appears to lack most of the inhibitors to the enzymatic amplification of ancient DNA (Woodward et al., 1994; Thomas et al., 2003) the majority of DNA samples were collected from teeth for this analysis. When teeth were not available, bone was used instead. A list of all the samples used in this study is presented in Table 8.1.

Authentication Methods

During analysis of ancient DNA samples it is necessary to exclude false positive results, that can arise because of postmortem damage and contamination with more recent DNA samples (Cooper and Poinar, 2000; Bandelt, 2005). In order to ensure the accuracy and reliability of results, standard contamination precautions, such as separation of pre- and post-PCR experimental areas, use of disposable laboratory ware and filter-plugged pipette tips, treatment with DNA contamination removal solution (DNA-OFF$_{TM}$; TaKaRa, Otsu, Japan), UV irradiation of equipment and benches, negative extraction controls and negative PCR controls, were employed in the present study. Other rigorous authentication methods were employed throughout the DNA-based analyses as described elsewhere (Shinoda et al., 2006). Bone or tooth preparation, DNA extraction, and PCR amplification were carried out in a physically separated room of a laboratory dedicated to the study of ancient DNA.

DNA Extraction and Purification

Bone and tooth samples were dipped in a DNA-OFF solution for 10 min to eliminate contamination, rinsed several times with DNase/RNase-free distilled water and air dried. When the samples were completely dry they were pulverised in a mill (Multi-beads shocker MB400U; Yasui Kikai, Osaka, Japan).

DNA was extracted in 2 steps using a DNA extraction kit (Mo Bio Co.). The pulverised tooth or bone powder (0.3 g) was placed in a 15-ml conical tube and demineralised in 5 ml of 0.5 M ethylene diamine tetra-acetic acid (EDTA). The samples were rotated and incubated at 37°C for 12–15 h. After digestion with proteinase K (0.5 mg/ml), the resultant pellet was used for DNA extraction. The eluted DNA (approximately 50 µl) was amplified by PCR, without prior processing. DNA extraction was performed only once; if the subsequent PCR amplification was not successful, no further extraction was carried out.

Table 8.1 Changes in the nucleotides of mtDNA and the haplogroups observed in the samples from Man Bac.

Co No.	Sex	Age class	Sample	Mutations in the segments HVR1 Sequence 121-238 (+16000)	HVR1 Sequence 2 09-402 (+16000)	HVR2 sequence 1 28-267	Haplogroup
99MB-3	M	18-20 years	Isolated tooth	N.E	N.D.	N.E	N.D.
01MB-9	F?	Young Adult	Isolated tooth	N.E	N.D.	N.E	N.D.
05MB-6		1-2 years	Fibula	N.E	223, 355, 362	N.E	D/G
05MB-9	F	Adult	Maxilla Right M3	N.E	256, 270	N.E	F
05MB-10		9 years	Costa fragment	N.E	N.D.	N.E	N.D.
05MB-11	M	Young Adult	Maxilla Left M3	N.E	N.D.	N.E	N.D.
05MB-12		3-4 years	Costa fragment	N.E	294, 296, 304	N.E	F
05MB-13		14-16 years	Costa fragment	N.E	223, 362	N.E	D/G
05MB-14		3-4 years	Mandible Right DM	N.E	N.D.	N.E	N.D.
05MB-15	M?	15-19 years	Costa fragment	N.E	209, 311	N.E	F
05MB-18		1-2 years	Isolated tooth	N.E	209, 311	N.E	F
05MB-19		Young Adult	Fibula	N.E	N.D.	N.E	N.D.
05MB-20	M	Adult	Mandible Right M3	N.E	223	N.E	N.D.
05MB-24		8 years	Costa fragment	N.E	223, 362	N.E	D/G
05MB-28	F	Adult	Fibula	N.E	N.D.	N.E	N.D.
05MB-31	M	Adult	Mandible Right M3	N.E	N.D.	N.E	N.D.
05MB-29		Adult	Mandible Right M3	N.D.	N.D.	N.D.	N.D.
05MB-34	F		Maxilla Right M2	C.R.S	311	N.D.	F
07MB H1-1		Child	Maxilla Left M3	184, 223	223, 318, 362	N.D.	D/G
07MB H1-4			Maxilla Left M3	N.D.	223, 284	150, 195	M*
07MB H1-5		Adult	Mandible Left M3	192, 223	223, 291, 362	150, 199	D/G
07MB H1-8		Adult	Mandible Left M3	N.D.	N.D.	N.D.	N.D.
07MB H2-1	M	Adult	Maxilla Right M3	223	223, 278, 362	150	D/G
07MB H2-5	F	Adult	Mandible Left M2	N.D.	N.D.	N.D.	N.D.
07MB H2-10	M	Adult	Mandible Left M3	N.D.	N.D.	N.D.	N.D.
07MB H2-11		Adult	Maxilla Right M3	N.D.	N.D.	N.D.	N.D.
07MB H2-12	F	Adult	Mandible Left M2	C.R.S	266G	N.D.	B5
07MB H2-18	M	Adolescent	Mandible Left M2	C.R.S	304, 335, 368	N.D.	F
07MB H2-19	M	Adult	Mandible Right M3	N.D.	N.D.	N.D.	N.D.
07MB H2-22	F	Adult	Maxilla Right M2	N.D.	N.D.	N.D.	N.D.
07MB H2-24	F	Adult	Maxilla Right M3	N.D.	217, 304, 311	N.D.	B4
07MB H2-27	M	Adult	Mandible Left M3	129, 162, 172	304	150, 152, 249d	F
07MB H2-30	M	Adult	Mandible Right M3	N.D.	232A, 249, 260, 304, 311	N.D.	F1b
07MB H2-32	M	Adult	Mandible Right M3	183, 189	311	N.D.	F

All polymorphic sites are numbered according to the revised Cambridge reference sequence (Andrews et al. 1999). C.R.S indicates that the sequence of the segment is identical to the revised Cambridge reference sequence, and diagnostic polymorphisms are emphasized by bold italic type. The suffix A and G indicates a transversion. N.D: Not Determined, N.E: Not Examined, Young adult: aged between 16 and 25 years. *denotes that the haplogroup status cannot be identified further.

Amplification and Sequencing of HVR1 and HVR2

Figure 8.1 shows the structure of the mitochondrial genome and the analytical portion used for this study. Segments of hyper variable region (HVR) 1 (nucleotide positions 16121–16238 and 16209–16402, as per the revised Cambridge reference sequence; Andrews et al., 1999) and HVR 2 (nucleotide positions 128–267) of the D-loop region were sequenced. Because ancient DNA is usually degraded to fragments that are typically hundreds of base pairs in length, the PCR was designed to amplify specific segments of mtDNA less than 250 bps long. The distribution of mutations in the D-loop region is significantly nonrandom. The primer set was designed to include the most variable region.

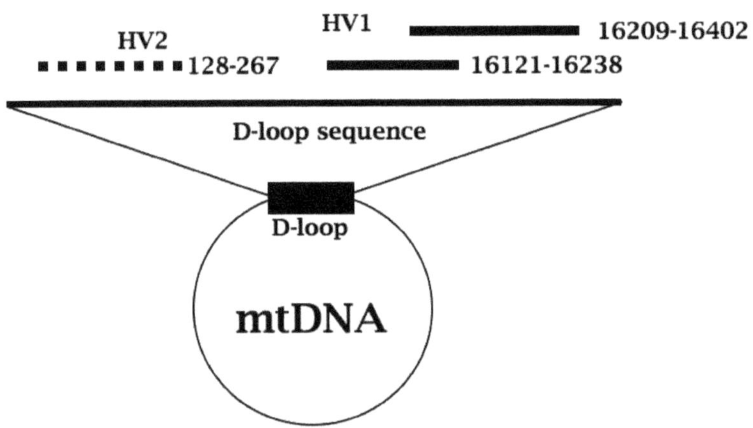

Figure 8.1 Map of human mitochondrion showing location of the D-loop region and analytical portion in this study.

Aliquots (2 μl) of the extracts were used as templates for PCR. Amplifications were carried out in a reaction mixture (total volume, 25 μl) containing 1 unit of Taq DNA polymerase (HotStarTaq™ DNA polymerase; Qiagen, Germany), 0.1 μM of each primer, and 100 μM of deoxyribo nucleoside triphosphates (dNTPs) in 1 × PCR buffer provided by the manufacturer. The PCR conditions were as follows: incubation at 95°C for 15 min; followed by 40 cycles of heat treatment at 94°C for 20 s; 50°C–56°C for 20 s; and 72°C for 15 s; and final extension at 72°C for 1 min.

The following primers were used to amplify HVR1 and HVR2.

L16120 5′-TTACTGCCAGCCACCATGAA-3′
H16239 5′-TGGCTTTGGAGTTGCAGTTG-3′
L16208 5′- CCCCATGCTTACAAGCAAG-3′
H16403 5′-TTGATTTCACGGAGGATGGTG-3′
L127 5′-AGCACCCTATGTCGCAGTAT-3′
H268 5′-GTTATGATGTCTGTGTGG-3′

The PCR products were subjected to agarose gel electrophoresis on a 1.5% gel and were recovered using a QIAEX II agarose gel extraction kit (Qiagen, Germany). Aliquots of the samples were prepared for sequencing on a BigDye cycle sequencing kit (Applied Biosystems, Foster City, CA, USA). The primers that were used in the PCR amplification were also used in the sequencing reaction. Sequencing was performed in both directions so as to enable identification of polymorphisms or ambiguous bases using a single primer. The sequencing reactions were performed on a DNA Sequencer (ABI model no, 3130) equipped with SeqEd software.

Bases at CRS positions 16209 to 16402 contain the majority of phylogenetically variable sites and were consequently subjected to more amplification than were the outer regions. The HVR 1 portion of the D-loop region of our experiments overlaps 30 bases. This allows analysis of whether the DNA sources are different. Moreover, if the separate fragments of the D-loop region are well in line with modern mtDNA lineages from different branches of the mtDNA phylogeny, we can determine whether they may have been derived from some artificial recombinant or different DNA source.

Data Analysis

The nucleotide diversity and the mean number of pairwise differences between the mitochondrial D-loop sequences were computed using the Arlequin software package version 3.0 (Excoffier et al., 2005), considering Tamura and Nei distances and a gamma parameter value of 0.26 (Mayer et al., 1999). The differences between the Man Bac sample and other populations were also computed using the Arlequin software (Raymond and Rousset, 1995). Neighbour-joining (NJ) trees were constructed on the basis of the pairwise Fst values by using the Mega 3.0 program (Kumar et al., 2004) in order to study the relationships between the populations.

The haplogroup status of mtDNA was tentatively assigned on the basis of a search for HVR1 motif specific to a haplogroup and by matching or almost matching the values with the mtDNA haplotypes in the global database. The haplogroup status was further characterised on the basis of other specific mutations in the HVR2 motif.

RESULTS AND DISCUSSION

Table 8.1 shows the list of materials that we used for this study. Bone and teeth samples belonging to 35 individuals were collected. Because of the generally poor quality of the mtDNA extracted from ancient materials, it was not possible to amplify all of the samples. Table 8.1 also shows the results of PCR amplification. Suspected false positive results stemming from contamination with contemporary DNA and other questionable data were omitted from this study and resulted in 34 out of 70 (approximately 49% success rate) PCR amplifications being successfully analysed. It is known from past studies that the success rate of DNA analysis of ancient human remains is between 60%–80% at best, even when well-preserved samples are used. Our results may suggest that the preservation conditions of DNA in the Man Bac samples are poor. In general, hot and humid conditions are unfavourable for the preservation of DNA in human skeletal remains; the possibility of finding well-preserved DNA in a tropical region such as Vietnam is low. However, the present experiment proved that sufficient amounts of DNA are retained in some human samples, even though the efficiency of analysis may be poor. For this reason it was decided that there was value in continuing the experiments to obtain more detailed data on the human skeletal remains from Man Bac.

Comparison with the revised Cambridge Reference Sequence for this region enabled the identification of 16 mitochondrial haplotypes that were defined on the basis of 21 segregating sites (Table 8.1). One of the main purposes of studying specimens from ancient burial sites is to clarify whether the human remains belong to unrelated individuals or to members of a single family or limited number of families. Since mtDNA is maternally inherited, the observation that the studied individuals shared the same haplotype suggests the possibility of a maternal relationship. Our mtDNA analysis has revealed some biological links. Kinship ties were defined among 4 out of 19 individuals. Among the samples analysed here, there may be related individuals from several generations. However, most individuals did not share the same haplotype, which could be due to the absence of close matrilineal relationships at this site.

Table 8.2 summarises the results of sequence analysis calculated from the HVR 1 region. The value of gene diversities, mean number of pairwise differences, and nucleotide diversity are presented, and can sometimes reflect relationships between populations. To clarify the genetic characteristics of the Man Bac sample, the mtDNA data were compared with the previous work on ancient Japanese and contemporary aboriginal Formosan populations (Table 8.2). The Man Bac series shows higher values of these parameters compared with ancient Japanese and is practically identical with aboriginal Formosan. It is known that the contemporary northern Vietnamese population possesses high genetic diversity and a large number of unique haplotypes (Irwin et al., 2008). It is possible that this tendency goes back to ancient times. However, in ancient DNA analysis it is necessary to take into account the possibility that the original sequences have changed due to the ageing of the DNA. Therefore, it is risky to regard as authentic all the base sequences determined in the investigation under discussion. It should be appreciated, in advance, that it is inevitable that such a limitation will occur in analyses of the scarce DNA that remains in ancient samples.

Table 8.2 mtDNA HVR I haplotype diversity indices for the Man Bac site and other populations.

Site	n	Gene diversity	Nucleotide diversity	Mean number of pairwise differences
Man Bac	19	1.00 +/- 0.017	0.020 +/- 0.011	3.673 +/- 1.943
Jomon (Kanto)	67	0.91 +/- 0.024	0.018 +/- 0.010	3.429 +/- 1.775
Yayoi (Kuma-Nishioda)	31	0.85 +/- 0.051	0.012 +/- 0.007	2.241 +/- 1.268
Aboriginal Formosan	28	1.00 +/- 0.010	0.022 +/- 0.012	4.161 +/- 2.133

Population references: Kanto Jomon (Shinoda and Kanai, 1999; Shinoda, 2003), Yayoi (Shinoda, 2004), Aboriginal Formosean (Tajima et al., 2003).

MtDNA haplogroups show geographic specificity within Asia (Kivisild et al., 2002; Li et al., 2007; Soares et al., 2008), therefore to determine the genetic characteristics of the Man Bac sample their mtDNA data was compared with that of populations in geographically related areas (see Table 8.3). All the haplogroups known to occur in Southeast and Northeast Asian populations, i.e, D, G, B, and F were detected in the Man Bac sample. However, most of the haplogroups that were found at Man Bac are dominant in Southeast Asian populations, except D and G, which are dominant in East Asia. The results of our phylogenetic analysis based on the Fst values show that the Man Bac population shares a relationship with south China, although they are divergent (Figure 8.2). An exact test of differentiation revealed that differences between Man Bac and other populations are statistically significant, except between north China, south China, and north Vietnam.

Table 8.3 Estimated frequencies of the mtDNA haplogroups among regional populations.

Haplogroup	Northern China (n = 125)	Southern China (n = 78)	Mainland Japan (n = 1312)	North Vietnam (n = 187)	South Vietnam (n=35)	Taiwan Aborigine (n = 640)	Philippines (n=59)	Man Bac (n = 19)
D4	35.2	14.1	32.6	34.0		1.5		33.2(D/G)
D5	6.4	5.1	4.8	2.7	14.3(D/G)	4.8	5.1(D/G)	
G	5.6	1.3	6.9	1.6		0.0		0.0
M7a	0.0	0.0	7.5	24.2	2.9	0.0	1.7	0.0
M7b	2.4	7.7	4.8	4.7	11.4	9.0	6.8	0.0
M7c	2.4	2.6	0.8	0.5	0.0	9.0	6.8	0.0
M8	6.4	2.6	1.4	0.0	2.9	0.0	0.0	0.0
M9	3.2	0.0	0.0	2.1	0.0	11.4	20.3	0.0
M10	3.2	2.6	1.3	0.0	2.9	0.4	1.7	0.0
CZ	1.6	0.0	1.8	0.3	0.0	0.0	0.0	0.0
A	4.0	0.0	6.9	8.0	0.0	0.0	0.0	8.0
B4	9.6	25.6	7.7	11.7	22.9	17.1	33.9	5.6
B5	1.6	1.3	4.3	2.4	5.7	5.9	5.1	5.6
F	7.2	23.1	5.3	2.1	11.4	26.7	6.8	50.0
N9a	3.2	1.3	4.6	0.3	5.7	1.2	0.0	0.0
N9b	0.0	0.0	2.1	4.3	0.0	0.0	0.0	0.0
Y	1.6	0.0	0.4	0.5	0.0	1.4	0.0	0.0
R	1.6	2.6	0.1	0.0	17.1	2.9	3.4	0.0
Other	4.8	10.1	6.7	0.5	2.8	8.7	8.4	5.6

The references for the population: Northern Chinese (Yao et al., 2002); Southern Chinese (Yao et al., 2002); Mainland Japanese (Tanaka et al., 2004); South Vietnam (Oota et al., 2002); Philippine (Tajima et al., 2004); Taiwan aborigines (Trejaut et al., 2005).

The distribution of mtDNA haplogroups among these areas will provide some suggestions about the population history of Man Bac. Haplogroups F and B are dominant in contemporary Southeast Asian populations; in contrast, the frequency of haplogroups D and G is relatively high in East Asian populations. It is noteworthy that both haplogroups appear in high proportions in the Man Bac series. It is suggested that southward population expansion during prehistoric times resulted in an admixture between these migrants and the local or indigenous Southeast Asian population in the region, leading to the formation of the basic genetic pattern seen in the modern northern Vietnamese population.

CONCLUSION

Inferences based on the results of the DNA analysis carried out here are limited due to the limited number of DNA sequences that could be successfully determined. Nonetheless, the establishment of kin relationships among numerous individuals buried in a single site provides extremely valuable information regarding the past social structure of the community. Furthermore, if it were possible to collect DNA data from the people inhabiting a single region over a prolonged period, it may be

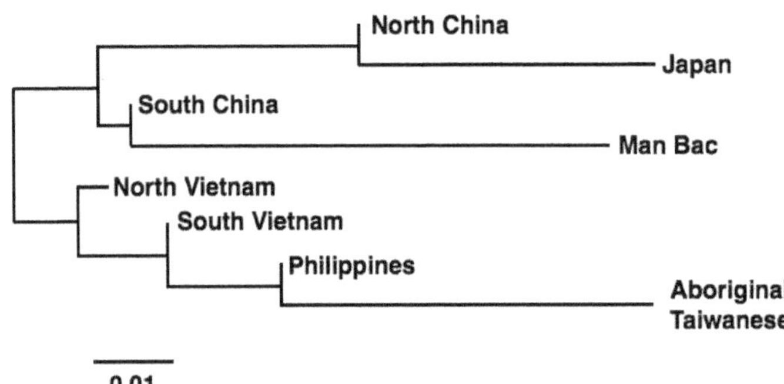

Figure 8.2 Neighbor joining tree based on the Fst values determined for 8 populations.

possible to deduce the movement of groups and their population dynamics. Since hot and humid conditions are unfavourable to the preservation of DNA, the possibilities are low for finding well-preserved DNA in a region like tropical Vietnam. However, the present study demonstrates the possibility that sufficient amounts of DNA are retained in human skeletal samples from tropical regions. It is very important that ancient DNA work continues in this region of the world.

SUMMARY

Man Bac is one of the largest neolithic sites in Vietnam. Due to its geographical and chronological position, the site is thought to play an important role in the evolution of modern-day Vietnamese. To investigate the genetic composition of the Man Bac community and to address questions regarding their potential genetic relationships with other Asian populations at a molecular level, we analysed HVR1 and HVR2 of mitochondrial DNA (mtDNA) from 35 samples excavated from this site. Some 34 out of 70 PCR amplifications were successfully analysed. The distribution of mtDNA haplotypes at the site indicated the existence of a number of different maternal lineages. The mtDNA sequence can be tentatively assigned to respective haplogroups according to specific mutations observed in the HVR 1 and 2 regions. The Man Bac sample showed affinities to Southeast and East Asian populations. The frequencies of these haplogroups indicates that a southward population expansion during the ancient past resulted in the admixture of these people with an indigenous Southeast Asian population and led to the formation of the basic pattern seen in modern northern Vietnamese.

LITERATURE CITED

Adachi N, Shinoda K, Umetsu K, Matsumura H. 2008. Mitochondrial DNA analysis of Jomon skeletons from the Funadomari site, Hokkaido, and its implication for the origins of native American. Am J Phys Anthropol 138;255-265.

Alzualde A, Izagirre N, Alonso S, Albarran C, Azkarate A, de la Rua C. 2006. Insight into the "isolation" of the Basques: mtDNA lineages from the historical site of Aldaieta (6th-7th centuries AD). Am J Phys Anthropol 130:394-404.

Andrews RM, Kubacka I, Chinnery PF, Lightowlers RN, Turnbull DM, Howell N. 1999. Reanalysis and revision of the Cambridge reference sequence for human mitochondrial DNA. Nature Genet 23: 147.

Bandelt HJ. 2005. Mosaic of ancient mitochondrial DNA: positive indicators of nonauthenticity. Europe. J Hum Genet 13:1106-1112.

Casas MJ, Hagelberg E, Fregel R, Larruga JM, Gonzalez AM. 2006. Human mitochondrial DNA diversity in an archaeological site in al-Andalus: Genetic impact of migrations from north Africa in medieval Spain. Am J Phys Anthropol 131:539-551.

Cooper A, Poinar HN. 2000. Ancient DNA: Do it right or not at all. Science 289:1139.

Excoffier L, Laval G, Schneider S. 2005. Arlequin ver.3.0: An integrated software packages for population genetics data analysis. Evol Bioinform Online 1.

Forster P. 2004. Ice Ages and the mitochondrial DNA chronology of human dispersals: a review. Phil Trans R Soc Lond B 359:255-264.

Ingman M, Kaessmann H, Paabo S, Gyllensten U. 2000. Mitochondrial genome variation and the origin of modern humans. Nature 408:708-713.

Irwin J, Saunier JL, Strouss KM, Diegoli TM, Strurk KA, O'Callaghan JE, Paintner CD, Hohoff C, Brinkmann B, Parsons TJ. 2008. Mitochondrial control region sequences from a Vietnamese population sample. Int J Legal Med 122:257-259.

Kivisild T, Tolk H-V,Parik J, Wang Y, Papiha SS, Bandelt HJ, Villems R. 2002. The emerging limbs and twigs of the East Asian mtDNA tree. Mol Biol Evol 19:1737-1751.

Kumar S, Tamura K, Nei M. 2004. MEGA3: Integrated software for molecular evolutionary genetics analysis and sequence alignment. Brief Bioinformatics 5:150-163.

Li H, Cal X, Winograd-Cort ER, Wen B, Cheng X, Qin Z, Liu W, Liu Y, Pan S, Qian J, Tan C-C, Jin L. 2007. Mitochondrial DNA diversity and population differentiation on Southern East Asia. Am J Phys Anthropol 134:481-488.

Maca-Mayer N, Cabrera VM, Arnay M, Flores C, Fregel R, Gonzalez AM, Larruga JM. 2005. Mitochondrial DNA diversity in 17th-18th century remains from Tenerife (Canary Islands). Am J Phys Anthropol 127:418-426.

Mayer, S, Weiss G, von Haeseler A. 1999. Pattern of nucleotide substitution and rate of heterogeneity in hyper variable region I and II of human mtDNA. Genetics 152:1103-1110.

Oota H, Kitano T, Jin F, Yuasa I, Wang L, Ueda S, Saitou N, Stoneking M. (2002) Extreme mtDNA homogeneity in continental Asian populations. Am J Phys Anthropol 118:146–153.

Raymond M, Rousset F. 1995. An exact test of population differentiation. Evolution 49:1280-1283.

Shinoda K, Kanai S. 1999.Intracemetery genetic analysis at the Nakazuma Jomon site in Japan by Mitochondrial DNA sequencing. Anthropol Sci 107:129-140.

Shinoda K. 2003. DNA analysis of the Jomon skeletal remains excavated from Shimo-Ohta shell midden, Chiba prefecture. Report for Sohnan Research Institute for Cultural Properties 50:201-205. (In Japanese.)

Shinoda K, 2004. Ancient DNA analysis of skeletal samples recovered from the Kuma-Nishioda Yayoi site. Bull Natl Sci Mus Tokyo D 30:1-8.

Shinoda K. Adachi N, Guillen S, Shimada I. 2006 Mitochondrial DNA analysis of ancient Peruvian highlanders. Am J Phys Anthropol 131: 98-107.

Soares P, Trejaut JA, Loo J-H, Hill C, Momina M, Lee C-L, Chen Y-M, Hudjashov G, Forster P, Macaulay V, Bulbeck D, Oppenheimer S, Lin M, Richards M. 2008. Climate change and postglacial human dispersals in Southeast Asia. Mol Biol Evol 25:1209-1218.

Tajima A, Sun C S, Pan IH, Ishida T, Saitou N, Horai S. 2003. Mitochondrial DNA polymorphisms in nine aboriginal groups of Taiwan: Implications for the population history of aboriginal Taiwanese. Hum Genet 113: 24–33.

Tajima A, Hayami M, Tokunaga K, Juji T, Matsuo M, Marzuki S, Omoto K, Horai S. 2004. Genetic origins of the Ainu inferred from combined DNA analyses. J Hum Genet 49:187-193.

Trejaut JA, Kivisild T, Loo JH, Lee CL, He CL, Hsu CJ, Lee ZY, Lin M. 2005. Traces of archaic mitochondrial lineages persist in Austronesian-speaking Formosan populations. PLoS Biol 3:e247.

Tanaka M, Cabrera VM, González AM, Larruga JM, Takeyasu T, Fuku N, Guo LJ, Hirose R, Fujita Y, Kurata M, Shinoda K, Umetsu K, Yamada Y, Oshida Y, Sato Y, Hattori N, Mizuno Y, Arai Y, Hirose N, Ohta S, Ogawa O, Tanaka Y, Kawamori R, Shamoto-Nagai M, Maruyama W, Shimokata H, Suzuki R, Shimodaira H. 2004. Mitochondrial genome variation in eastern Asia and the peopling of Japan. Genome Res 14: 1832–1850.

Thomas M, Gilbert P, Willerslev E, Hansen AJ, Barnes I, Rudbeck L, Lynnerup N, Cooper A. 2003. Distribution patterns of postmortem damage in human mitochondrial DNA. Am J Hum Genet 72:32-47.

Woodward SR, King MJ, Chiu NM, Kuchar ML, Griggs CW. 1994. Amplification of ancient nuclear DNA from teeth and soft tissues. PCR Method Appl 3:244–247.

Yao, YG, Kong QP, Bandelt HJ, Kivisild T, Zhang YP. 2002. Phylogenetic differentiation of mitochondrial DNA in Han Chinese. Am J Hum Genet 70:635–651.

9
Faunal Remains at Man Bac

Junmei Sawada[1], Nguyen Kim Thuy[2] and Nguyen Anh Tuan[2]

[1] *St. Marianna University School of Medicine, Kawasaki, Japan*
[2] *The Vietnamese Institute of Archaeology*

This chapter describes the zooarchaeological findings from an analysis of the mammalian remains from Man Bac. Several hundred vertebrate remains were recovered during the excavations of Man Bac between 2005 and 2007. Mammalian and fish bones formed the main component of the recovered vertebrate assemblage. These animal bones provide primary information for an understanding of the subsistence behaviours of the Man Bac community during the neolithic and of the palaeoenvironment of the coastal plain where Man Bac is situated.

Previous studies have examined the past mammalian fauna of northern Vietnam (Vu, 1981, 1984; Vu and Nguyen, 1988; Nguyen and Vu, 2004), however, there is limited available data on the quantity and size of the mammalian archaeological assemblages. This report provides quantitative information for the mammalian assemblage as well as supplying raw data on taxonomic identification and the measurements of bones and teeth (see Appendix 9.1 and 9.2 this chapter).

MATERIALS AND METHODS

The Man Bac faunal assemblage was collected by a combination of *in situ* recovery during excavation and the intensive sieving of two excavation squares (E3 and G1). While it is believed that all vertebrate remains were recovered, realistically it is likely that some very small vertebrate remains (e.g. rats) may have been missed during excavation and recovery. All of the faunal remains were cleaned and labelled with provenance data in the form of site, date, square, layer, and spit. Taxonomic identification of the mammalian remains was based on cranial and dental morphology. Each specimen was provided with a sample number, then identified, to at least order or family, genus and species level if possible (see Appendix 9.1 this chapter). Cetacea and Muridae were identified from post-cranial bones as no cranial remains for these taxa were recovered. The modern mammalian bone collections in the Vietnam Institute of Archaeology in Hanoi, the Raffles Museum of Biodiversity Research in Singapore, and the National Museum of Nature and Science in Tokyo, were used for comparison and identification. Measurements of cranial and dental remains were taken according to Driesch (1976), the raw data of which are presented in Appendix 9.2 (this chapter).

For *Sus scrofa* (pig or boar), the dominant species at Man Bac, age-at-death was estimated using the method of Hayashi et al. (1977) based on tooth eruption and attrition of the upper and lower teeth.

RESULTS

Ten taxa were recognised, including: Muridae (rat), *Canis* sp. (dog), *Aonyx cinerea* (oriental small-clawed otter), *Viverra* sp. (civet), *Rhinoceros* sp. (rhinoceros), *Sus scrofa* (boar), *Muntiacus muntjak* (barking deer), *Cervus* sp. (deer), *Bos* sp. (cattle) and/or *Bubalus* sp. (water buffalo), and Cetacea (whale/dolphin). With the exception of the *Rhinoceros* these taxa still inhabit northern Vietnam (Lekagul and McNeely, 1988; Parr and Hoang, 2008).

Table 9.1 shows the number of identified specimens (NISP) and the minimum number of individuals (MNI) with respect to each layer. NISP and MNI were calculated based on sample-numbered remains. The total NISP is 182, and the total MNI is 37. The mammalian assemblage by percent of NISP is shown in Figure 9.1.

Sus scrofa is the dominant taxon in the Man Bac faunal assemblage (79.1% of total NISP; 54.1% of total MNI). The age composition of the *Sus* remains is shown in Table 9.2 (see also Figure 9.2), and the molar measurements are given in Table 9.3. *Sus* remains may include a few wild boar, but most *Sus* remains are considered to be domesticated. Further information on *Sus* is discussed below.

Family Cervidae (deer) has a significant presence in the assemblge and consisted of *Cervus* sp. (6.6% of total NISP; 8.1% of total MNI) and *Muntiacus muntjak* (1.1% of total NISP; 5.4% of total MNI). *Cervus* remains are similar in size to a medium-size deer, such as *C. unicolor* (sambar), *C. nippon* (sika deer), or *C. eldii* (Eld's deer), and were difficult to identify to the species level.

The Bovinae remains consisted of two molars of a large bovine. They appeared to be *Bos* sp. and/or *Bubalus* sp. There is the possibility that Bovinae were already domesticated in Vietnam during the mid Holocene (Vu, 1981). However, we could not find evidence for domestication of Bovinae in the Man Bac site, since the Bovinae remains are too few and fragmentary.

The Carnivora remains consisted of several skull fragments of *Canis* sp., and the teeth of *Viverra* sp. (*V. zibetha* (large Indian civet) or *V. magaspila* (large-spotted civet)) and *Aonyx cinerea.* Canidae remains include *Canis*, but there is no *Cuon* (Asian wild dog), a species widely distributed in Vietnam. *Canis* was domesticated in Southeast Asia during the neolithic, and *Canis* may have been bred at Man Bac.

Rhinoceros sp. remains consisted of two molars, and are similar to *Rhinoceros sondaicus* (Javan rhinoceros).

The Cetacean remains consisted of only one vertebra and fragments of one limb bone. Family, genus and species were indeterminate.

The Muridae remains consisted of a single femur of a small rat.

DISCUSSION

Domestication of *Sus scrofa*

The very high proportion of the mammalian assemblage attributable to *Sus* is very different from the faunal signatures of hunting and gathering communities, such as during the Hoabinhian period (Nguyen and Vu, 2004; Sawada and Vu, 2006). The demographic profile of the *Sus* assemblage (Table 9.2, Figure 9.2) demonstrates a very high proportion of juvenile and young-adult individuals. In general, the observed patterns in domestic *Sus* populations are characterised by an early kill-off

(Hongo and Meadow, 2000; Hongo et al., 2007), although a high proportion of young *Sus* remains alone does not necessarily equate with domestication (Albarella et al., 2006). However, the high number and young-biased age distribution of the Man Bac *Sus* series is indicative of a domesticated population. On the other hand, the morphological features of the Man Bac *Sus* assemblage are consistent with wild pigs, making it difficult to rule out the possibility that some portion of the sample is wild, rather than domesticated.

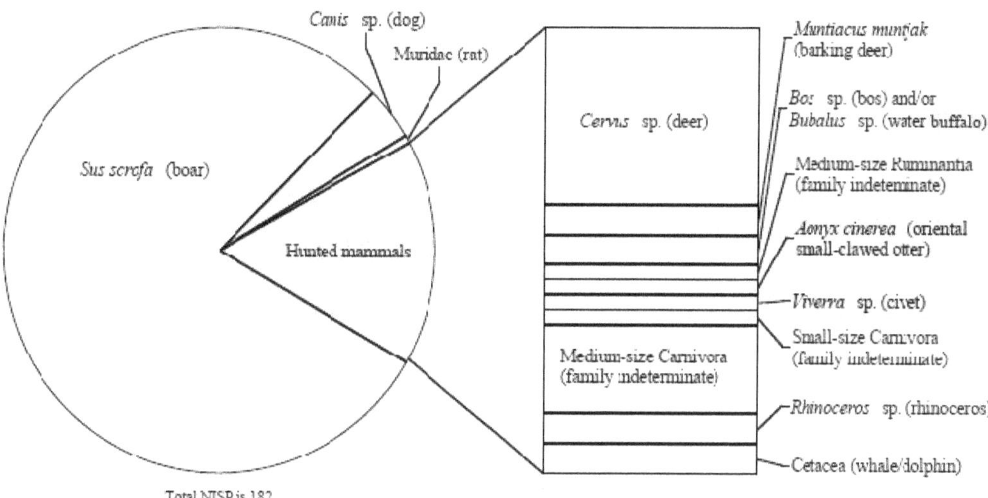

Figure 9.1 Man Bac mammalian assemblage by percent of NISP.

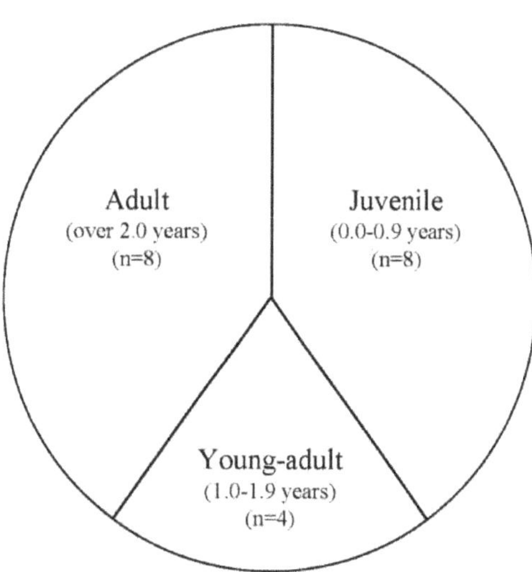

Figure 9.2 Demographic structure of the Man Bac *Sus scrofa*.

Table 9.1 Mammalian fauna of Man Bac.

Taxon	Layer I NISP (%)	Layer I MNI (%)	Layer II NISP (%)	Layer II MNI (%)	Layer III NISP (%)	Layer III MNI (%)	Total NISP (%)	Total MNI (%)
Order Rodentia								
Muridae (rat)			1 (1.7)	1 (6.3)			1 (0.5)	1 (2.7)
Order Carnivora								
Canis sp. (dog)	3 (2.8)	1 (5.9)	4 (6.9)	2 (12.5)			7 (3.8)	3 (8.1)
Aonyx cinerea (oriental small-clawed otter)	1 (0.9)	1 (5.9)					1 (0.5)	1 (2.7)
Viverra sp. (civet)	1 (0.9)	1 (5.9)					1 (0.5)	1 (2.7)
Small-size Carnivora (family indeterminate)	1 (0.9)	-					1 (0.5)	-
Medium-size Carnivora (family indeterminate)	5 (4.7)	-	1 (1.7)	-			6 (3.3)	-
Order Perissodactyla								
Rhinoceros sp. (rhinoceros)	1 (0.9)	1 (5.9)			1 (5.9)	1 (25.0)	2 (1.1)	2 (5.4)
Order Artiodactyla								
Sus scrofa (domestic/wild boar)	86 (80.4)	10 (58.8)	43 (74.1)	8 (50.0)	15 (88.2)	2 (50.0)	144 (79.1)	20 (54.1)
Muntiacus muntjak (barking deer)	1 (0.9)	1 (5.9)	1 (1.7)	1 (6.3)			2 (1.1)	2 (5.4)
Cervus sp. (deer)	6 (5.6)	1 (5.9)	6 (10.3)	2 (12.5)			12 (6.6)	3 (8.1)
Bos sp. (bos) and/or *Bubalus* sp. (water buffalo)	1 (0.9)	1 (5.9)	1 (1.7)	1 (6.3)			2 (1.1)	2 (5.4)
Medium-size Ruminantia (family indeterminate)	1 (0.9)	-					1 (0.5)	-
Order Cetacea								
Cetacea (whale/dolphin)			1 (1.7)	1 (6.3)	1 (5.9)	1 (25.0)	2 (1.1)	2 (5.4)
Total	107 (100.0)	17 (100.0)	58 (100.0)	16 (100.0)	17 (100.0)	4 (100.0)	182 (100.0)	37 (100.0)

NISP: number of identified specimens, MNI: minimum number of individuals.
NISP and MNI were calculated based on cranial and dental remains (except Muridae and Cetacea).

Table 9.2 Age composition of the *Sus* dental remains.

	< 7-8 months		7 -8 months		19-20 months		31-32 months		43-44 months		55+ months		Total	
	NISP	MNI	NISP	MNI	NISP	MNI	NISP	MNI	NISP	MNI	NISP	MNI	NISP	MNI
Layer I	3	2	10	4	5	1	2	1	2	2	0	0	22	10
Layer II	0	0	5	2	2	2	1	1	5	3	0	0	13	8
Layer III	0	0	0	0	1	1	0	0	2	1	0	0	3	2
Total	3	2	15	6	8	4	3	2	9	6	0	0	38	20

Age-at-death estimations are according to Hayashi et al. (1977).

Table 9.3 Length of molars of *Sus scrofa* (mm).

	Late Neolithic Man Bac				Modern wild [(a)]				Modern domestic [(a)]				Iron Age Noen U-Loke [(b)]			
Tooth	N	Mean	SD	Range	N	Mean	SD	Range	N	Mean	SD	Range	N	Mean	SD	Range
UM1	10	17.8	1.3	15.4 - 19.7	-	-	-	-	-	-	-	-	66	14.4	1.1	11.6 - 16.8
UM2	11	22.8	1.6	20.3 - 26.3	-	-	-	-	-	-	-	-	50	17.2	1.6	14.0 - 20.0
UM3	5	35.5	2.3	33.5 - 38.7	-	-	-	-	-	-	-	-	14	32.8	2.6	29.5 - 38.0
LM1	4	18.8	0.3	18.3 - 19.0	-	-	-	-	-	-	-	-	129	15.2	1.2	13.0 - 21.4
LM2	3	23.1	0.7	22.5 - 23.8	-	-	-	-	-	-	-	-	75	18.6	1.3	15.9 - 22.8
LM3	4	42.9	2.7	39.0 - 45.0	13	42.7	3.8	31.1 - 51.5	7	26.8	3.3	20.2 - 36.9	14	35.6	4.1	28.4 - 44.8

Abbreviations for tooth types are as follows: UM is upper molar, LM is lower molar.
(a) data from Ishiguro et al. (2008) , (b) data from McCaw (2007).

Molar dimensions of the Man Bac *Sus* series, Iron Age domestic *Sus* remains from Noen U-Loke, Thailand (data from McCaw, 2007), and the lower third molar measurements of Vietnamese modern domestic and wild pigs (data from Ishiguro et al., 2008) are shown in Table 9.3. The lower third molars of the Man Bac *Sus* series are significantly larger than both modern domestic pigs ($p<0.001$) and Noen U-Loke domestic *Sus* ($p<0.01$) using Turkey's multiple range test, while they are comparable in size to modern wild boar (Figure 9.3). The other teeth of the Man Bac *Sus* assemblage also tend to be larger than those of the Noen U-Loke remains, although there were no data for equivalent teeth of wild and Vietnamese domestic pigs.

Body, cranium and tooth size tends to decrease through domestication from wild to domestic forms (Flannery, 1983; Zeder, 2006). Ishigro et al. (2008) noted that the tooth size of Vietnamese modern wild pigs is larger than modern domestic pig teeth, with the tooth size distribution of these groups clearly separate. Figure 9.3 demonstrates that domestic *Sus* third molars in mainland Southeast Asia have reduced in size from the neolithic through to the present. Similarities in dental metrics between Man Bac *Sus* and Vietnamese modern wild pigs suggests a similarity between the two. It is not improbable that Man Bac *Sus* are at the initial stages of pig domestication in Vietnam.

Vu (1981) argued for the presence of domestic *Sus* remains at the mid Holocene Da But site of Con Co Ngua. However, Higham (1996) notes that Da But sites show no evidence for the cultivation of plants, and were likely hunter-gatherer and fishing settlements. Bellwood (2005) stated that *Sus* might have been domesticated during the neolithic in Vietnam, but clear evidence has not been found. This analysis of the Man Bac *Sus* series adds new evidence for the likelihood of *Sus* domestication in northern Vietnam by at least 3,500 BP. To clarify the timing and nature of *Sus* domestication in mainland Southeast Asia, there is a need for more work in this region.

Palaeoenvironment and Mammal Hunting

The Man Bac mammalian remains, with the exception of the Muridae and domestic *Sus/Canis*, were hunted animals: *Aonyx cinerea*, *Viverra*, *Rhinoceros*, *Muntiacus muntjak*, *Cervus*, Bovinae, and Cetacea. The habitats of these wild

Table 9.4 Primary habitats of the hunted mammals from the Man Bac site.

Taxon	Primary habitat	NISP (%)	MNI (%)
Aonyx cinerea (oriental small-clawed otter)	River and estuary	1 (4.5)	1 (7.7)
Viverra sp. (civet)	Forest	1 (4.5)	1 (7.7)
Rhinoceros sp. (rhinoceros)	Forest with a good supply of water	2 (9.1)	2 (15.4)
Muntiacus muntjak (barking deer)	Forest	2 (9.1)	2 (15.4)
Cervus sp. (deer)	Lowlands, grassland, forest	12 (54.5)	3 (23.1)
Bos sp. (bos) and/or *Bubalus* sp. (water buffalo)	Forest and grassland (*Bos*. Sp), open forest and swamp in lowlands	2 (9.1)	2 (15.4)
Cetacea (whale/dolphin)	Sea	2 (9.1)	2 (15.4)
Total hunted mammals		22 (100.0)	13 (100.0)

Habitat data is based on Lekagul and McNeely (1988) and Parr and Hoang (2008).

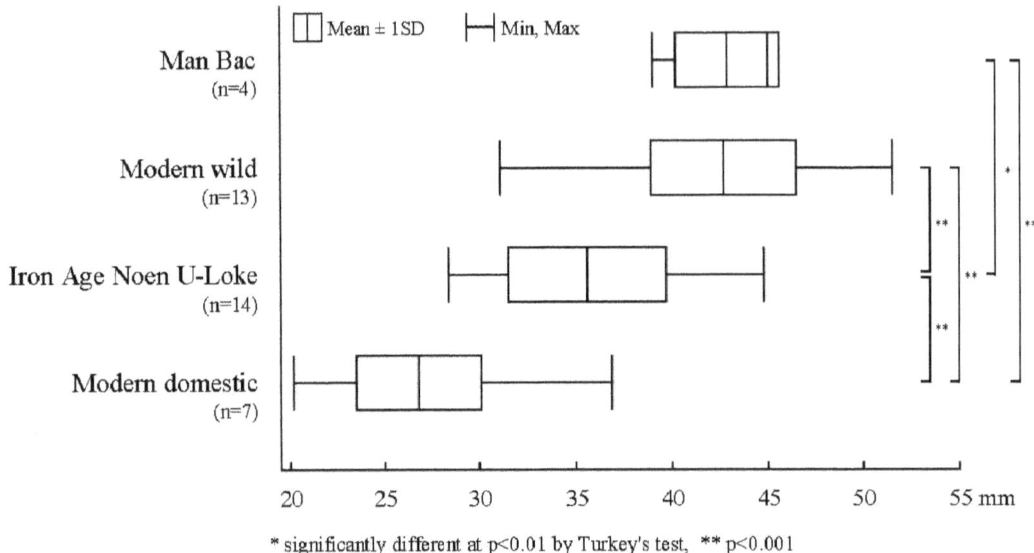

Figure 9.3 Length of lower third molars of the Man Bac *Sus* series, Iron Age domestic *Sus* remains from Noen U-Loke in Thailand (data from McCaw, 2007), and Vietnamese modern domestic and wild pigs (data from Ishiguro et al., 2008).

mammals were quite varied and included forest (*Viverra* sp., *Rhinoceros* sp., *Muntiacus muntjak*, *Cervus* sp., *Bos* sp., *Bubalus* sp.), grassland (*Cervus* sp., *Bos* sp.), watered places in lowlands (*Aonyx cinerea*, *Rhinoceros* sp., *Bubalus* sp), and the sea (Cetacea) (Lekagul and McNeely, 1988; Parr and Hoang, 2008; Table 9.4). Such varied habitats represent considerable environmental diversity in the vicinity of Man Bac during occupation of the site. Forests, grassland and lowlands can still be seen in the modern landscape near Man Bac, although there are some differences in terms of distance from the sea and probable vegetation types between the present and some 3,500 years ago.

It would appear that the occupants of Man Bac utilised a diverse range of environments for hunting and foraging. Habitat diversity aside, the behaviours and body sizes of the Man Bac mammalian series varied for different species. For instance, the head-body length of *Aonyx* is 40cm whereas that of *Rhinoceros* is over 3m (Lekagul and McNeely, 1988). Given the diversity in both the local environment and physical characteristics of the mammals, Man Bac people likely also lay claim to a diverse range of hunting skills, depending on the type of mammal targeted. Notwithstanding this however, the amount of hunted wild mammal remains (12.1% of total NISP; 35.1% of total MNI) is far less than that of the *Sus* remains. The number of species of hunted wild mammals from the Man Bac site is 7 taxa, which is rather meagre when compared to the species richness of northern Vietnam in the Holocene (Nguyen and Vu, 2004; Parr and Hoang, 2008). In contrast, the Hoabinhian pre-food production sites of northern Vietnam revealed 20 or more species of wild mammals (Nguyen and Vu, 2004; Sawada and Vu, 2006). The taxa-poor mammalian assemblage of Man Bac suggests hunting may have been more of a supplementary or secondary subsistence activity, despite the likelihood that they possessed efficient hunting skills. The initiation of domestication during the neolithic reduced the prominence of mammal hunting, and at Man Bac the key mammalian food source was domesticated (but still morphologically wild) pigs.

CONCLUSIONS

The Man Bac mammalian assemblage consisted of numerous domestic pig remains with a small number of hunted wild mammals, including several species of deer, bovids, carnivores, rhinoceros and cetaceans. The Man Bac community relied on domesticated pigs as the main mammalian food source, although they likely had sophisticated hunting skills allowing them to target a range of wild mammals in a variety of habitats in relative proximity to the site. The morphology of the pig remains suggests that they were at an initial stage of domestication. The zooarchaeological information of the Man Bac mammalian assemblage analysed in this chapter plays an important role in understanding the food-acquisition strategies of early agricultural societies in northern Vietnam.

SUMMARY

The Man Bac faunal assemblage provides primary information regarding both the ancient environment and subsistence strategies during the neolithic in northern Vietnam. Mammalian remains formed the main component of the excavated vertebrate assemblage at Man Bac which consisted of a large proportion of domestic pigs and a small number of wild mammals, including several species of deer, bovids, carnivores, rhinoceros and cetaceans. The Man Bac community utilised a range of environments and animal habitats as part of their hunting strategies. However, the relatively small proportion of hunted animals compared to domesticated pig remains suggests a reliance on pigs for their main source of meat. It is believed that Man Bac pigs represent an early stage of domestication.

ACKNOWLEDGMENTS

We thank the Raffles Museum of Biodiversity Research in Singapore and the National Museum of Nature and Science in Tokyo for access to comparative mammal collections, and Drs. Vu The Long, Nguyen Lan Cuong, Nguyen Kim Dung, Nguyen Mai Huong, Marc F. Oxenham, Hirofumi Matsumura, Mariko Yamagata, Takeji Toizumi, Hitomi Hongo, Masanari Nishimura, and Yukio Dodo for their advice and support.

LITERATURE CITED

Albarella U, Dobney K, Rowley-Conwy P. 2006. The domestication of the pig (*Sus scrofa*): new challenges and approaches. In: Zeder MA, Bradley DG, Emshwiller E, Smith BD, editors. Documenting Domestication: New Genetic and Archaeological Paradigms. Berkeley: University of California Press. p 209-227.

Bellwood P. 2005. The First Farmers: Origins of Agricultural Societies. Oxford: Blackwell Publishing.

Driesch A. 1976. A Guide to the Measurement of Animal Bones from Archaeological Sites. Cambridge: Peabody Museum Press.

Flannery KV. 1983. Early pig domestication in the fertile crescent: a retrospective look. In: Young CT, Smith PEL, Mortensen P, editors. The Hilly Flanks and Beyond: Essays on the Prehistory of Southwestern Asia. Chicago: Oriental Institute of the University of

Chicago. p 163-188.

Hayashi Y, Nishida T, Mochizuki K, Seta S. 1977. Sex and age determination of the Japanese wild boar (*Sus scrofa leucomystax*) by the lower teeth. Jpn J Vet Sci 39:165-174.

Higham CFW. 1996. The Bronze Age of Southeast Asia. Cambridge: Cambridge University Press.

Hongo H, Anezaki T, Yamazaki K, Takahashi O, Sugawara H. 2007. Hunting or management? The status of *Sus* in the Jomon period in Japan. In: Albarella U, Dobney K, Ervynck A, Rowley-Conwy P, editors. Pigs and Humans: 10,000 Years of Interaction. Oxford: Oxford University Press. p 109-130.

Hongo H, Meadow RH. 2000. Faunal remains from Prepottery Neolithic levels at Çayönü, southeastern Turkey: a preliminary report focusing on pigs (*Sus* sp.). In: Mashkour M, Choyke AM, Buitenhuis H, Poplin F, editors. Archaeology of the Near East IV A. Groningen: ARC-Publications. p 121-140.

Ishiguro N, Sasaki M, Iwasa M, Shigehara N, Hongo H, Anezaki T, Vu TL, Phan XH, Hguyen XT, Nguyen HN, Vu NT. 2008. Morphological and genetic analysis of Vietnamese *Sus scrofa* bones for evidence of pig domestication. Anim Sci J 79:655-664.

Lekagul B, McNeely JA. 1988. Mammals of Thailand (2nd edition). Bangkok: Saha Karn Bhaet Co.

McCaw M. 2007. Faunal remains. In: Higham CFW, Kijingam A, Talbot S, editors. The Origins of the Civilization of Angkor, Volume Two, the Excavation of Noen U-Loke and Non Muang Kao. Bangkok: The Thai Fine Arts Department. p 495-536.

Nguyen KS, Vu TL. 2004. Moi Truong & Van Hoa Cuoi Pleistocene Dau Holocene O Bac Viet Nam. Hanoi: Nha Xuat Ban Khoa Hoc Xa Hoi (in Vietnamese)

Parr JWK, Hoang XT. 2008. A Field Guide to the Large Mammals of Vietnam. Hanoi: People and Nature Reconciliation (Pan Nature).

Sawada J, Vu TL. 2006. Hoabinhian mammal remains from the Hang Cho site, northern Vietnam. In: Matsumura H, editor. Anthropological and Archaeological Study on the Origin of Neolithic People in Mainland Southeast Asia: Report of Grant-in-aid for International Scientific Research (2003~2005 No.15405018), p 83-86.

Vu TL. 1981. Di tich dong vat o Con Co Ngua (Thanh Hoa). Nhung Phat Hien Moi Ve Khao Co Hoc Nam 1980-1:60-6151 (in Vietnamese) .

Vu TL. 1984. So bo nghien cuu nhung xuong rang dong vat va di cot nguoi trong dot khai quat Dong Dau 1984. Nhung Phat Hien Moi Ve Khao Co Hoc Nam 1984:85-89.

Vu TL, Nguyen G. 1988. Xuong dong vat o di chi Cai Beo. Nhung Phat Hien Moi Ve Khao Co Hoc Nam 1987:49-51 (in Vietnamese).

Zeder MA. 2006. Archaeological approaches to documenting animal domestication. In: Zeder MA, Bradley DG, Emshwiller E, Smith BD, editors. Documenting Domestication: New Genetic and Archaeological Paradigms. Berkeley: University of California Press. p 209-227.

Appendix 9.1 Taxonomic identification.

Sample No.	Taxon	Skeletal part	l/r	Layer	Spit	Square	Remarks
MB05-184	Muridae (rat)	Femur	r	II	14	b1	
MB07-002	*Canis* sp. (dog)	Mandible (with I2, C, P2-P4, M1-M2)	l	II	10	f2	
MB05-041	*Canis* sp. (dog)	Maxilla (with M1 and M2)	r	I	7	d4	
MB05-104	*Canis* sp. (dog)	Maxilla (with M1)	r	II	14	e1	
MB07-013	*Canis* sp. (dog)	Maxilla (with P3, P4, M1, and M2)	l	II	11	e2	
MB05-024	*Canis* sp. (dog)	Maxilla (with P4)	r	I	6	d5	
MB05-037	*Canis* sp. (dog)	Tooth (UM1)	l	I	7	b5	
MB05-088	*Canis* sp. (dog)	Tooth (UP4)	l	II	10	c4	
MB07-047	*Aonyx cinerea* (oriental small-clawed otter)	Mandible (with P3, P4, and M1)	l	I	7	b4	
MB07-020	*Viverra* sp. (civet)	Mandible (with C and M1)	r	I	8	e1	
MB05-049	Small-size Carnivora (family indeterminate)	Tooth (LC)	l	I	7	e4	
MB05-039	Medium-size Carnivora (family indeterminate)	Mandible (ramus of mandible)	l	I	7	c1	
MB05-053	Medium-size Carnivora (family indeterminate)	Mandible (ramus of mandible)	r	I	7	a4	
MB05-115	Medium-size Carnivora (family indeterminate)	Mandible (ramus of mandible)	r	I	4	e3	
MB05-047	Medium-size Carnivora (family indeterminate)	Tooth (fragment of canine)	?	I	7	c2	
MB05-029	Medium-size Carnivora (family indeterminate)	Tooth (LI3)	l	I	6	d5	
MB07-037	Medium-size Carnivora (family indeterminate)	Tooth (UC)	r	II	11	d2	
MB05-119	*Rhinoceros* sp. (rhinoceros)	Tooth (fragment of molar)	?	I	6	a4	
MB05-120	*Rhinoceros* sp. (rhinoceros)	Tooth (LM1)	l	III	14	b2	
MB05-135	*Sus scrofa* (domestic/wild boar)	Fragment of skull	?	I	8	f6	
MB05-136	*Sus scrofa* (domestic/wild boar)	Fragment of skull	?	I	8	f6	
MB05-137	*Sus scrofa* (domestic/wild boar)	Fragment of skull	?	I	8	f6	
MB05-145	*Sus scrofa* (domestic/wild boar)	Fragment of skull	?	I	6	e2	
MB05-146	*Sus scrofa* (domestic/wild boar)	Fragment of skull	?	I	6	e2	
MB05-160	*Sus scrofa* (domestic/wild boar)	Fragment of skull	?	I	7	b5	
MB05-206	*Sus scrofa* (domestic/wild boar)	Fragment of skull	?	I	4	f3	
MB05-207	*Sus scrofa* (domestic/wild boar)	Fragment of skull	?	I	4	f3	
MB05-208	*Sus scrofa* (domestic/wild boar)	Fragment of skull	?	I	4	f3	
MB05-218	*Sus scrofa* (domestic/wild boar)	Frontal bone	?	I	6	e1	
MB05-151	*Sus scrofa* (domestic/wild boar)	Frontal bone	l	I	7	c3	
MB07-051	*Sus scrofa* (domestic/wild boar)	Incisive bone (with I2 and I3)	l	I	7	f2	Teeth unerupted
MB05-133	*Sus scrofa* (domestic/wild boar)	Mandible (angle of mandible)	l	III	18	b1	
MB05-150	*Sus scrofa* (domestic/wild boar)	Mandible (condylar process)	l	I	5	e6	
MB05-110	*Sus scrofa* (domestic/wild boar)	Mandible (with dm2, dm3, and M1)	l	II	10	c1	M1 erupting
MB07-007	*Sus scrofa* (domestic/wild boar)	Mandible (with dm3)	l	II	9	e3	
MB05-116	*Sus scrofa* (domestic/wild boar)	Mandible (with I2 and C)	r+l	I	7	c1	Female
MB05-011	*Sus scrofa* (domestic/wild boar)	Mandible (with M2 and M3)	l	I	6	f1	M3 erupting
MB07-060	*Sus scrofa* (domestic/wild boar)	Mandible (with M2 and M3)	l	I	8	d3	
MB05-118	*Sus scrofa* (domestic/wild boar)	Mandible (with P2-P4)	r	I	7	c1	
MB07-001	*Sus scrofa* (domestic/wild boar)	Mandible (with P3 and P4)	r	II	12	d4	
MB05-038	*Sus scrofa* (domestic/wild boar)	Mandible (with P4 and M1)	r	I	7	d1	P4 erupting
MB05-142	*Sus scrofa* (domestic/wild boar)	Maxilla (alveolar process)	?	I	6	b6	
MB07-046	*Sus scrofa* (domestic/wild boar)	Maxilla (alveolar process)	r	II	12	b1	
MB05-002	*Sus scrofa* (domestic/wild boar)	Maxilla (body of maxilla)	l	I	5	f5	
MB05-045	*Sus scrofa* (domestic/wild boar)	Maxilla (body of maxilla)	r	I	7	b5	
MB07-023	*Sus scrofa* (domestic/wild boar)	Maxilla (body of maxilla)	r	I	6	e3	
MB05-062	*Sus scrofa* (domestic/wild boar)	Maxilla (with C)	l	I	8	f6	Male
MB07-008	*Sus scrofa* (domestic/wild boar)	Maxilla (with C, P2, and P3)	r	II	9	b3	Male
MB05-035	*Sus scrofa* (domestic/wild boar)	Maxilla (with dm1)	l	I	7	b5	
MB05-067	*Sus scrofa* (domestic/wild boar)	Maxilla (with dm1-dm3)	r	I	9	a3	
MB05-006	*Sus scrofa* (domestic/wild boar)	Maxilla (with dm1-dm3, and M1)	r	I	6	e2	
MB07-056	*Sus scrofa* (domestic/wild boar)	Maxilla (with dm1-dm3, and M1)	r	II	10	d2	
MB05-026	*Sus scrofa* (domestic/wild boar)	Maxilla (with dm2 and dm3)	l	I	6	d6	
MB07-057	*Sus scrofa* (domestic/wild boar)	Maxilla (with dm2 and dm3)	l	II	12	b3	
MB05-034	*Sus scrofa* (domestic/wild boar)	Maxilla (with dm3 and M1)	l	I	7	b5	M1 erupting
MB07-050	*Sus scrofa* (domestic/wild boar)	Maxilla (with dm3 and M1)	r	I	7	f2	
MB07-011	*Sus scrofa* (domestic/wild boar)	Maxilla (with dm3, M1, and M2)	r	III	15	c1	
MB05-077	*Sus scrofa* (domestic/wild boar)	Maxilla (with M1 and M2)	r	II	10+11	a4	M2 erupting
MB05-114	*Sus scrofa* (domestic/wild boar)	Maxilla (with M1)	l	I	4	e3	
MB05-139	*Sus scrofa* (domestic/wild boar)	Maxilla (with M1)	r	I	7	c6	
MB05-083	*Sus scrofa* (domestic/wild boar)	Maxilla (with M2 and M3)	l	II	10	a5	
MB05-073	*Sus scrofa* (domestic/wild boar)	Maxilla (with M2)	r	II	9	a6	M2 erupting
MB05-108	*Sus scrofa* (domestic/wild boar)	Maxilla (with M3)	r	II	12	d1	
MB05-003	*Sus scrofa* (domestic/wild boar)	Maxilla (with P2)	l	I	5	f2	
MB05-102	*Sus scrofa* (domestic/wild boar)	Maxilla (with P2-P4)	l	II	13	b1	
MB05-025	*Sus scrofa* (domestic/wild boar)	Maxilla (with P3 and P4)	r	I	6	f4	
MB05-001	*Sus scrofa* (domestic/wild boar)	Maxilla (with P4 and M1)	l	I	4	f3	
MB05-036	*Sus scrofa* (domestic/wild boar)	Maxilla (with P4 and M1-M3)	l	I	7	b5	M3 erupting

Abbreviations for tooth types are as follows: I is incisor, C is canine, P is premolar, M is molar, d and unicase letters are deciduous teeth, U is upper, L is lower.

Appendix 9.1 (Continued 1).

Sample No.	Taxon	Skeletal part	l/r	Layer	Spit	Square	Remarks
MB05-111	*Sus scrofa* (domestic/wild boar)	Maxilla (with P4 and M1-M3)	l	II	13	b1	
MB07-052	*Sus scrofa* (domestic/wild boar)	Maxilla (with P4 and M1-M3)	r	II	11	d4	M3 erupting
MB05-057	*Sus scrofa* (domestic/wild boar)	Maxilla (with P4, M1, and M2)	r	I	7	b3	M2 erupting
MB05-084	*Sus scrofa* (domestic/wild boar)	Maxilla (with P4, M1, and M2)	r	II	10	c3	
MB05-225	*Sus scrofa* (domestic/wild boar)	Nasal bone	?	I	6	a4	
MB05-147	*Sus scrofa* (domestic/wild boar)	Nasal bone	l	I	6	a4	
MB05-148	*Sus scrofa* (domestic/wild boar)	Nasal bone	r	I	6	b4	
MB05-174	*Sus scrofa* (domestic/wild boar)	Temporal bone	l	I	?	cd7	
MB05-315	*Sus scrofa* (domestic/wild boar)	Temporal bone	r	II	12	d1	
MB07-010	*Sus scrofa* (domestic/wild boar)	Temporal bone	r	II	11	a'3	
MB05-031	*Sus scrofa* (domestic/wild boar)	Tooth (fragment of incisor)	?	I	6	c3	
MB05-131	*Sus scrofa* (domestic/wild boar)	Tooth (fragment of LC)	?	III	18	b1	
MB05-125	*Sus scrofa* (domestic/wild boar)	Tooth (fragment of Ldi1 or Ldi2)	?	III	15	a5	
MB05-007	*Sus scrofa* (domestic/wild boar)	Tooth (fragment of molar)	?	I	6	e2	
MB05-018	*Sus scrofa* (domestic/wild boar)	Tooth (fragment of molar)	?	I	6	b6	
MB05-048	*Sus scrofa* (domestic/wild boar)	Tooth (fragment of molar)	?	I	7	d5	
MB05-052	*Sus scrofa* (domestic/wild boar)	Tooth (fragment of molar)	?	I	7	a4	
MB05-059	*Sus scrofa* (domestic/wild boar)	Tooth (fragment of molar)	?	I	8	a3	
MB05-072	*Sus scrofa* (domestic/wild boar)	Tooth (fragment of molar)	?	II	8	d3	
MB05-082	*Sus scrofa* (domestic/wild boar)	Tooth (fragment of molar)	?	II	10+11	a2	
MB05-092	*Sus scrofa* (domestic/wild boar)	Tooth (fragment of molar)	?	II	10+11	c2	
MB05-096	*Sus scrofa* (domestic/wild boar)	Tooth (fragment of molar)	?	II	12	b3	
MB05-097	*Sus scrofa* (domestic/wild boar)	Tooth (fragment of molar)	?	II	12	b3	
MB05-106	*Sus scrofa* (domestic/wild boar)	Tooth (fragment of molar)	?	II	14	c1	
MB05-128	*Sus scrofa* (domestic/wild boar)	Tooth (fragment of molar)	?	III	12	c3	
MB05-130	*Sus scrofa* (domestic/wild boar)	Tooth (fragment of molar)	?	III	12	d2	
MB05-182	*Sus scrofa* (domestic/wild boar)	Tooth (fragment of molar)	?	II	10+11	b2	
MB05-251	*Sus scrofa* (domestic/wild boar)	Tooth (fragment of molar)	?	I	7	a4	
MB05-252	*Sus scrofa* (domestic/wild boar)	Tooth (fragment of molar)	?	I	7	a4	
MB05-253	*Sus scrofa* (domestic/wild boar)	Tooth (fragment of molar)	?	I	7	a4	
MB07-028	*Sus scrofa* (domestic/wild boar)	Tooth (fragment of molar)	?	I	6	d1	
MB05-126	*Sus scrofa* (domestic/wild boar)	Tooth (fragment of premolar)	?	III	16	a2	
MB05-132	*Sus scrofa* (domestic/wild boar)	Tooth (fragment of premolar)	?	III	18	b1	
MB05-064	*Sus scrofa* (domestic/wild boar)	Tooth (LC)	l	I	8	f3	Male
MB05-117	*Sus scrofa* (domestic/wild boar)	Tooth (LC)	l	I	7	c1	Female
MB05-023	*Sus scrofa* (domestic/wild boar)	Tooth (Ldi2)	l	I	6	e4	
MB05-093	*Sus scrofa* (domestic/wild boar)	Tooth (Ldi2)	l	II	10	f6	
MB05-074	*Sus scrofa* (domestic/wild boar)	Tooth (Ldi2)	r	II	9	b3	
MB05-123	*Sus scrofa* (domestic/wild boar)	Tooth (Ldi2)	r	III	15	c3	
MB07-042	*Sus scrofa* (domestic/wild boar)	Tooth (Ldi2)	r	II	9	e3	
MB07-044	*Sus scrofa* (domestic/wild boar)	Tooth (Ldi2)	r	II	13	b3	
MB05-046	*Sus scrofa* (domestic/wild boar)	Tooth (Ldm3)	r	I	7	e6	
MB05-090	*Sus scrofa* (domestic/wild boar)	Tooth (LI1)	l	II	10+11	c1	
MB05-103	*Sus scrofa* (domestic/wild boar)	Tooth (LI1)	l	II	13	a6	
MB07-015	*Sus scrofa* (domestic/wild boar)	Tooth (LI1)	l	I	8	d1	
MB07-043	*Sus scrofa* (domestic/wild boar)	Tooth (LI1)	l	II	9	b3	
MB05-076	*Sus scrofa* (domestic/wild boar)	Tooth (LI1)	r	II	9	f6	
MB05-015	*Sus scrofa* (domestic/wild boar)	Tooth (LI2)	l	I	6	b4	
MB05-075	*Sus scrofa* (domestic/wild boar)	Tooth (LI2)	l	II	9	d3	
MB05-065	*Sus scrofa* (domestic/wild boar)	Tooth (LI2)	r	I	8	b5	
MB05-124	*Sus scrofa* (domestic/wild boar)	Tooth (LI2)	r	III	15	a5	
MB05-030	*Sus scrofa* (domestic/wild boar)	Tooth (LI3)	r	I	6	c3	
MB05-129	*Sus scrofa* (domestic/wild boar)	Tooth (LM1)	l	III	12	a5	
MB05-070	*Sus scrofa* (domestic/wild boar)	Tooth (LM1)	r	I	9	a2	Unerupted
MB05-091	*Sus scrofa* (domestic/wild boar)	Tooth (LM1)	r	II	10+11	c2	
MB07-048	*Sus scrofa* (domestic/wild boar)	Tooth (LM2)	l	I	10	d4	
MB07-058	*Sus scrofa* (domestic/wild boar)	Tooth (LM2)	l	III	14	a'3	
MB05-010	*Sus scrofa* (domestic/wild boar)	Tooth (LM2)	r	I	6	f3	
MB07-049	*Sus scrofa* (domestic/wild boar)	Tooth (LM3)	l	I	10	d4	
MB07-059	*Sus scrofa* (domestic/wild boar)	Tooth (LM3)	l	III	14	a'3	
MB05-080	*Sus scrofa* (domestic/wild boar)	Tooth (LM3)	r	II	10+11	a2	
MB05-021	*Sus scrofa* (domestic/wild boar)	Tooth (LP2)	r	I	6	b6	
MB05-058	*Sus scrofa* (domestic/wild boar)	Tooth (M3 fr)	?	I	7	b3	
MB05-040	*Sus scrofa* (domestic/wild boar)	Tooth (UC)	l	I	7	e1	Female
MB07-004	*Sus scrofa* (domestic/wild boar)	Tooth (UC)	r	I	7	c1	Male
MB07-025	*Sus scrofa* (domestic/wild boar)	Tooth (Udi1)	l	II	13	f4	
MB05-019	*Sus scrofa* (domestic/wild boar)	Tooth (Udm2)	l	I	6	b6	
MB05-004	*Sus scrofa* (domestic/wild boar)	Tooth (Udm3)	l	I	5	e6	

Appendix 9.1 (Continued 2).

Sample No.	Taxon	Skeletal part	l/r	Layer	Spit	Square	Remarks
MB05-033	*Sus scrofa* (domestic/wild boar)	Tooth (Udm3)	r	I	7	b5	
MB05-014	*Sus scrofa* (domestic/wild boar)	Tooth (UI1)	l	I	6	b4	
MB05-050	*Sus scrofa* (domestic/wild boar)	Tooth (UI1)	l	I	7	a3	
MB07-039	*Sus scrofa* (domestic/wild boar)	Tooth (UI1)	l	I	7	c3	
MB05-022	*Sus scrofa* (domestic/wild boar)	Tooth (UI2)	r	I	6	e1	
MB05-008	*Sus scrofa* (domestic/wild boar)	Tooth (UI3)	l	I	6	e2	
MB05-155	*Sus scrofa* (domestic/wild boar)	Tooth (UI3)	r	I	6	a3	
MB05-017	*Sus scrofa* (domestic/wild boar)	Tooth (UM1)	l	I	6	b6	Unerupted
MB05-032	*Sus scrofa* (domestic/wild boar)	Tooth (UM1)	l	I	7	b5	
MB05-066	*Sus scrofa* (domestic/wild boar)	Tooth (UM1)	l	I	8	f2	Unerupted
MB05-028	*Sus scrofa* (domestic/wild boar)	Tooth (UM2)	l	I	6	c2	
MB05-060	*Sus scrofa* (domestic/wild boar)	Tooth (UM2)	l	I	8	a3	
MB05-127	*Sus scrofa* (domestic/wild boar)	Tooth (UM2)	l	III	17	c2	
MB07-053	*Sus scrofa* (domestic/wild boar)	Tooth (UM2)	l	II	7	a'6	
MB07-054	*Sus scrofa* (domestic/wild boar)	Tooth (UM2)	l	II	11	a1	
MB05-051	*Sus scrofa* (domestic/wild boar)	Tooth (UM2)	r	I	7	a4	Unerupted
MB05-107	*Sus scrofa* (domestic/wild boar)	Tooth (UM3)	l	II	14	a1	
MB05-122	*Sus scrofa* (domestic/wild boar)	Tooth (UM3)	l	III	14+15	a2b2	
MB05-054	*Sus scrofa* (domestic/wild boar)	Tooth (UM3)	r	I	7	a3	Unerupted
MB05-061	*Sus scrofa* (domestic/wild boar)	Tooth (UP1)	r	I	8	a3	
MB05-020	*Sus scrofa* (domestic/wild boar)	Tooth (UP2)	l	I	6	b6	
MB05-013	*Sus scrofa* (domestic/wild boar)	Tooth (UP2)	r	I	6	b4	Unerupted
MB05-079	*Sus scrofa* (domestic/wild boar)	Tooth (UP2)	r	II	10+11	a4	
MB05-078	*Sus scrofa* (domestic/wild boar)	Tooth (UP3)	r	II	10+11	a4	
MB05-068	*Sus scrofa* (domestic/wild boar)	Tooth (UP4)	l	I	9	d1	
MB05-105	*Sus scrofa* (domestic/wild boar)	Tooth (UP4)	l	II	14	b1	
MB05-081	*Sus scrofa* (domestic/wild boar)	Tooth (UP4)	r	II	10+11	a2	
MB05-109	*Muntiacus muntjak* (barking deer)	Antler	?	II	11	a1	
MB05-112	*Muntiacus muntjak* (barking deer)	Frontal bone and antler	l	I	5	d5	
MB05-172	*Cervus* sp. (deer)	Antler	?	I	7	d5	
MB05-087	*Cervus* sp. (deer)	Mandible (with dm3)	r	II	10+11	b1	
MB05-156	*Cervus* sp. (deer)	Occipital bone	m	I	7	b5	
MB05-043	*Cervus* sp. (deer)	Tooth (fragment of premolar)	?	I	7	f4	
MB05-044	*Cervus* sp. (deer)	Tooth (fragment of premolar)	?	I	7	f4	
MB05-101	*Cervus* sp. (deer)	Tooth (LM3)	r	II	13	b2	
MB05-094	*Cervus* sp. (deer)	Tooth (UM1)	r	II	11	e3	
MB05-009	*Cervus* sp. (deer)	Tooth (UM2)	l	I	6	e2	Unerupted
MB05-095	*Cervus* sp. (deer)	Tooth (UM2)	r	II	11	e3	
MB05-089	*Cervus* sp. (deer)	Tooth (UM3)	r	II	10	f4	
MB05-056	*Cervus* sp. (deer)	Tooth (UP2)	l	I	7	b3	
MB05-086	*Cervus* sp. (deer)	Tooth (UP2)	r	II	10+11	a1	
MB07-005	*Bos* sp. (bos) and/or *Bubalus* sp. (water buffalo)	Tooth (fragment of molar)	l	II	11	d2	
MB05-055	*Bos* sp. (bos) and/or *Bubalus* sp. (water buffalo)	Tooth (LP3)	l	I	7	b3	
MB05-027	Medium-size Ruminantia (family indeterminate)	Tooth (fragment of molar)	?	I	6	c2	
MB05-330	Cetacea (whale/dolphin)	Limb bone (shaft)	?	III	15	a1	
MB05-171	Cetacea (whale/dolphin)	Vertebra	m	II	12	a6	

Appendix 9.2 Raw data measurements of the Man Bac mammal remains (mm).

Sus scrofa (domestic/wild boar)

Sample No.	Skeletal part	l/r	LP2-LP4 length	UM1 length	UM2 length	UM3 length	LM1 length	LM2 length	LM3 length
MB05-006	Maxilla (with dm1-dm3, and M1)	r		19.70					
MB05-010	Tooth (LM2)	r						23.84	
MB05-011	Mandible (with M2 and M3)	l						22.54	45.00
MB05-017	Tooth (UM1)	l		19.33					
MB05-028	Tooth (UM2)	l			20.26				
MB05-032	Tooth (UM1)	l		17.69					
MB05-036	Maxilla (with P4 and M1-M3)	l			26.34				
MB05-038	Mandible (with P4 and M1)	r					18.93		
MB05-051	Tooth (UM2)	r			23.39				
MB05-054	Tooth (UM3)	r				33.89			
MB05-057	Maxilla (with P4, M1, and M2)	r		15.35					
MB05-060	Tooth (UM2)	l			23.26				
MB05-066	Tooth (UM1)	l		17.19					
MB05-070	Tooth (LM1)	r					18.97		
MB05-077	Maxilla (with M1 and M2)	r		17.50	21.11				
MB05-080	Tooth (LM3)	r							43.90
MB05-083	Maxilla (with M2 and M3)	l				33.47			
MB05-107	Tooth (UM3)	l				34.10			
MB05-110	Mandible (with dm2, dm3, and M1)	l					18.28		
MB05-111	Maxilla (with P4 and M1-M3)	l		16.75	23.27	37.10			
MB05-118	Mandible (with P2-P4)	r	39.73						
MB05-122	Tooth (UM3)	l				38.72			
MB05-127	Tooth (UM2)	l			22.36				
MB05-129	Tooth (LM1)	l					18.88		
MB07-011	Maxilla (with dm3, M1, and M2)	r		18.49	22.95				
MB07-048	Tooth (LM2)	l						22.90	
MB07-049	Tooth (LM3)	l							43.84
MB07-050	Maxilla (with dm3 and M1)	r		18.18					
MB07-052	Maxilla (with P4 and M1-M3)	r		17.66	22.98				
MB07-053	Tooth (UM2)	l			22.85				
MB07-054	Tooth (UM2)	l			21.58				
MB07-060	Mandible (with M2 and M3)	l							38.97
Mean			-	17.78	22.76	35.46	18.77	23.09	42.93
SD			-	1.25	1.56	2.32	0.33	0.67	2.69

Cervus sp. (deer)

Sample No.	Skeletal part	l/r	UM1 length	UM2 length	LM3 length
MB05-094	Tooth (UM1)	r	22.32		
MB05-095	Tooth (UM2)	r		27.16	
MB05-101	Tooth (LM3)	r			31.79

10
Fish Remains at Man Bac

Takeji Toizumi[1], Nguyen Kim Thuy[2], Junmei Sawada[3]

[1] *Institute of Comparative Archaeology, Waseda University, Japan*
[2] *The Vietnamese Institute of Archaeology*
[3] *School of Medicine, St. Marianna University, Japan*

Many fish remains were recovered from excavations at Man Bac during the 2005 and 2007 seasons. This chapter focuses on the identification of fish remains recovered in the 2004-5 season, with some general observations made on the 2007 assemblage. In addition, a discussion of the aquatic palaeoenvironment surrounding the site and the fishing activities of its inhabitants is outlined here. The analysis was carried out at the Institute of Archaeology, Hanoi in 2008. The elements considered for identification were maxillaries, premaxillaries, dentaries, angulars, quadrates, vertebrae and other identifiable elements. These specimens were identified through comparison with skeletal specimens of modern fishes.

MATERIALS AND RESULTS

Identifications

The identification results are shown in Table 10.1. A total of 722 specimens were available for analysis. Separating them by strata; specifically Layer I, II, and III, yielded 561, 121, and 40 specimens respectively, with most of the specimens coming from Layer I, and the number decreasing in the lower layers.

As with the mammalian assemblage (see Chapter 9) the Man Bac fish assemblage was collected by a combination of *in situ* recovery during excavation and the wet sieving of two excavation squares (squares E3 and G1).

Among the material analysed, 692 specimens were identified to the level of order or lower, and 4 taxa of Chondrichthyes (Elasmobranchii) plus 10 taxa of Osteichthyes (Teleostei) were identified. In addition, there were 25 unidentified Osteichthyes specimens (Figure 10.1, No. 13-17). *Acanthopagrus* sp. (black seabreams) were the most numerous (54% of total MNI), followed by *Lates calcarifer* (barramundi), Siluriformes (catfishes), Rajiformes (rays), Lamnidae / Lamniformes (sharks), and Serranidae (groupers). This pattern is basically the same from Layer I to Layer III (Table 10.2, Figure 10.2).

Brief Description of the Dominant Taxa

Because a comparison with modern fish specimens was insufficiently detailed, the identification of the Elasmobranchii (Sharks and Rays) remains uncertain. Most of the shark vertebrae are from Carcharhinidae or similar types (Figure 10.1, No. 2),

Table 10.1 Fish remains from the 2004-2005 excavation season at Man Bac.

taxon	common name	element	Layer I	Layer II	Layer III	Total
			L / R	L / R	L / R	L / R
Lamnidae	Mackerel sharks	vertebra	4		1	5
Lamniformes ?	Sharks	vertebra	36	10	2	48
Myliobatididae ?	Eaglerays ?	teeth	146	27	2	175
Rajiformes A	Rays (type A)	vertebra	34	18	3	55
Rajiformes B	Rays (type B)	vertebra	13	2		15
Rajiformes	Rays	caudal spine	5	1	1	7
Elasmobranchii	Sharks or Rays	vertebra	20	3		23
Clupeidae	Sardines or Shads	caudal vertebra	1			1
Siluriformes	Catfishes	pectoral spine	7 / 9	2 / 2	/ 1	9 / 12
		fin spine(fragment)	12	5	4	21
		cleithrum	/ 1			0 / 1
Siluriformes ?	Catfishes ?	caudal vertebra	2			2
Mugilidae	Mullets	opercle	/ 1			0 / 1
Mugilidae ?	Mullets ?	opercle		/ 1		0 / 1
		abdominal vertebra	2			2
Lates calcarifer	Barramundi	maxillary	1 / 1			1 / 1
		premaxillary	2 / 5	/ 1		2 / 6
		dentary	6 / 6	1 /	1 /	8 / 6
		angular	5 / 3	2 /		7 / 3
		quadrate	1 / 6	1 /	1 / 1	3 / 7
		preopercle		/ 1	1 /	1 / 1
		cleithrum	1 / 1	/ 1		1 / 2
		1st vertebra	3			3
		abdominal vertebra	15	3	2	20
		caudal vertebra	3			3
Serranidae (middle)	Groupers (middle)	maxillary	1 /			1 / 0
		dentary		1 /		1 / 0
		preopercle	1 /			1 / 0
Serranidae (middle) ?	Groupers (middle) ?	quadrate	/ 1			0 / 1
Serranidae (large)	Groupers (large)	maxillary	1 /			1 / 0
		premaxillary	3 / 1	1 /		4 / 1
		cleithrum	/ 1			0 / 1
		angular	/ 1			0 / 1
Serranidae (large) ?	Groupers (large) ?	quadrate	1 / 1	/ 1		1 / 2
		abdominal vertebra	1			1
		caudal vertebra	3	2	1	6
Lates calcarifer or Serranidae ?	Barramundi or Groupers	basioccipital		1		1
		angular	/ 1			0 / 1
		quadrate	/ 1			0 / 1
		opercle	/ 1			0 / 1
		cleithrum	1 /			1 / 0
		caudal vertebra	7			7
Carangidae	Jacks	caudal vertebra	1			1
Sciaenidae ?	Croakers ?	maxillary	1 /			1 / 0
		premaxillary	/ 1			0 / 1
		angular	1 /			1 / 0
Acanthopagrus	Black Seabream	maxillary	2 / 1	1 / 1		3 / 2
		premaxillary	31 / 30	7 / 4	2 / 3	40 / 37
		dentary	13 / 16	2 / 4		15 / 20
		angular	4 / 6	2 /		6 / 6
		palatine	/ 1	/ 1		0 / 2
		opercle	1 / 2	1 /		2 / 2
		anal spine	31	9	13	53
Sparidae ?	Seabreams ?	1st vertebra	1			1
		abdominal vertebra	4			4
		caudal vertebra	6			6
Platycephalidae	Flatheads	dentary			/ 1	0 / 1
		caudal vertebra	1			1
Teleostei (unidentified) A	-	unknown	4	1		5
Teleostei (unidentified) B	-	unknown	1			1
Teleostei (unidentified) C	-	caudal vertebra	7	1		8
Teleostei (unidentified) D	-	dentary	/ 1			0 / 1
Teleostei (unidentified) E	-	dentary	1 /			1 / 0
Teleostei (unidentified) others	-	vertebra	9			9
Teleostei (unidentifiable)	-	vertebra	4			4
		urostyle	1			1
	Total		561	121	40	722

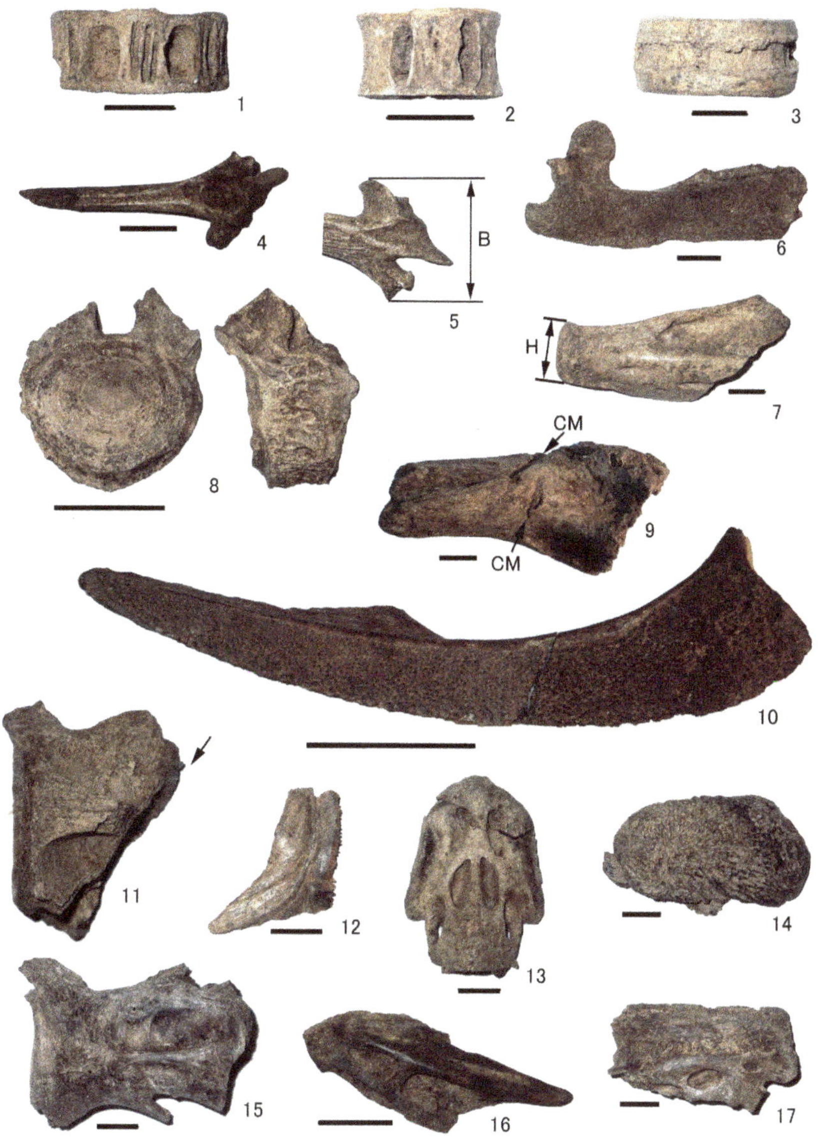

Figure 10.1 Fish remains from Man Bac.

1 Lamnidae vertebra, 2 Carcharhinidae? vertebra, 3 Rajiformes A vertebra, 4-5 Siluriformes pectoral spine (B: greatest breadth of the proximal end), 6-7 *Lates calcarifer* [6 premaxillary, 7 dentary (H: height of the anterior end)], 8 and 11 Serranidae (large) ? [8 abdominal vertebra, 11 caudal vertebra (artificially cut?)] , 9-10. Serranidae (large) [9 maxillary (CM: cut mark), 10 premaxillary], 12. Serranidae (middle) preopercle, 13 Teleostei (unidentified) A, 14 Teleostei (unidentified) B, 15 Teleostei (unidentified) C caudal vertebra, 16. Teleostei (unidentified) D dentary, 17 Teleostei (unidentified) E dentary. scale bar: 8 and 10: 5cm, others: 1cm

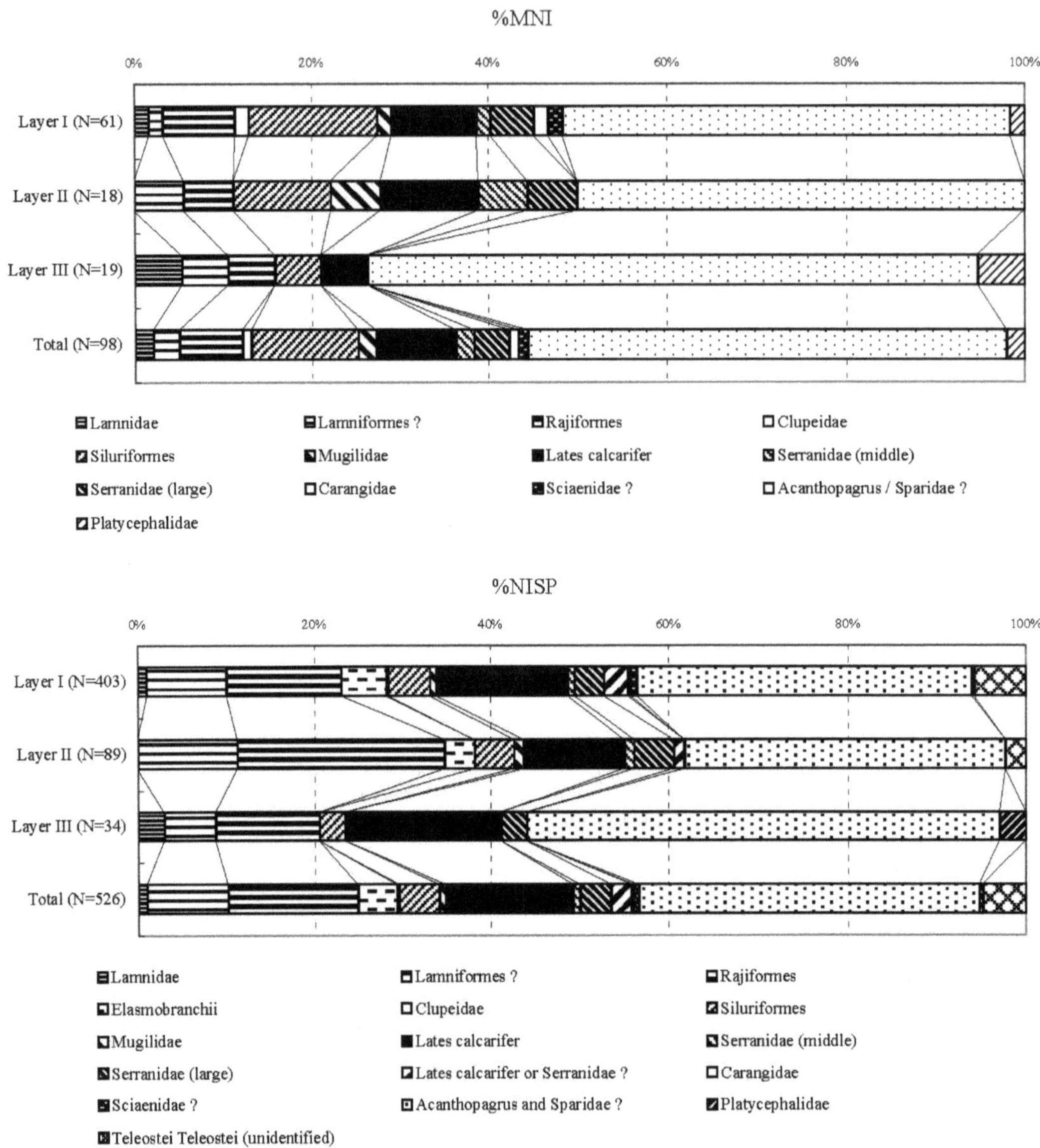

Figure 10.2 Assemblage of fish remains from Man Bac 2004-2005. upper: %MNI, lower: %NISP.

but there are also some from Lamnidae (Mackerel sharks, Figure 10.1, No.1). Most of these vertebrae are small to medium in size with diameters around 2 cm.

The Rajiformes teeth are probably from Myliobatididae (Eaglerays) and there is a variety of sizes with the largest specimen having a width of 4 cm. Among the Rajiformes vertebrae, there are those that exhibit a pulley-like shape (Rajiformes A, Figure 10.1, No.3), and those that don't (Rajiformes B). Most of the Rajiformes A vertebrae are small with diameters of 9 to 13 mm, but there are also some medium to large ones with diameters of 18 to 36 mm. The size of the Rajiformes B vertebrae is essentially the same as the A.

As for the Siluriformes, the Man Bac collection at hand and comparative research on modern specimens are insufficient, so difficult to identify to taxonomic family. The greatest breadth of the proximal end of the pectoral spine (Figure 10.1, No. 5) is 7 to 18 mm, which at least identifies them as relatively large fishes. *Lates calcarifer* is also limited to relatively large fishes. The height of the anterior end of the dentary (Figure 10.1, No. 7) is 14.5 to 19.1 mm.

Table 10.2 Assemblage of fish remains from Man Bac 2004-2005.

taxon	common name	NISP				MNI			
		Layer I	Layer II	Layer III	Total	Layer I	Layer II	Layer III	Total
Lamnidae	Mackerel sharks	4	0	1	5	1	0	1	2
Lamniformes ?	Sharks	36	10	2	48	1	1	1	3
Rajiformes A	Rays (type A)	34	18	3	55				
Rajiformes B	Rays (type B)	13	2	0	15	5	1	1	7
Rajiformes	Rays	5	1	1	7				
Elasmobranchii	Sharks or Rays	20	3	0	23	-	-	-	-
Clupeidae	Sardines or Shads	1	0	0	1	1	0	0	1
Siluriformes	Catfishes	17	4	1	22	9	2	1	12
Siluriformes ?	Catfishes ?	2	0	0	2				
Mugilidae	Mullets	1	0	0	1	1	1	0	2
Mugilidae ?	Mullets ?	2	1	0	3				
Lates calcarifer	Barramundi	59	10	6	75	6	2	1	9
Serranidae (middle)	Groupers (middle)	2	1	0	3	1	1	0	2
Serranidae (middle) ?	Groupers (middle) ?	1	0	0	1				
Serranidae (large)	Groupers (large)	6	1	0	7	3	1	0	4
Serranidae (large) ?	Groupers (large) ?	7	3	1	11				
Lates calcarifer or Serranidae ?	Barramundi or Groupers	11	1	0	12	-	-	-	-
Carangidae	Jacks	1	0	0	1	1	0	0	1
Sciaenidae ?	Croakers ?	3	0	0	3	1	0	0	1
Acanthopagrus	Black Seabream	138	32	18	188	31	9	13	53
Sparidae ?	Seabreams ?	11	0	0	11				
Platycephalidae	Flatheads	1	0	1	2	1	0	1	2
Teleostei (unidentified) A	-	4	1	0	5	-	-	-	-
Teleostei (unidentified) B	-	1	0	0	1	-	-	-	-
Teleostei (unidentified) C	-	7	1	0	8	-	-	-	-
Teleostei (unidentified) D	-	1	0	0	1	-	-	-	-
Teleostei (unidentified) E	-	1	0	0	1	-	-	-	-
Teleostei (unidentified) others	-	9	0	0	9	-	-	-	-
Teleostei (unidentifiable)	-	5	0	0	5	-	-	-	-
Total		403	89	34	526	62	18	19	99

* NISP: number of identified specimens; MNI: minimum number of individuals.

* Teeth of Rajiformes and fin-spine fragments of Siluriformes are not included in NISP.

There are both medium and large specimens of Serranidae, with the large type being more numerous. Among the measurable specimens of the large type, the longest premaxillary is 197 mm (Figure 10.1, No.10). The greatest breadth of an abdominal vertebra is 71 mm (Figure 10.1, No. 8), which is the largest size for Osteichthyes. The medium type has an estimated body length of 30 to 50cm, which is significantly smaller than the large type (Figure 10.1, No.12). Most of the *Acanthopagrus* sp. are adult fish.

Materials Recovered by Sieving

The identification results for the materials from Square E3 (1mm sieve) in Layer I are presented in Table 10.3. Seven specimens each of Rajiformes and *Acanthopagrus* sp., and 1 specimen each of Clupeidae, Mugilidae and Sparidae were identified. In addition, there were 5 unidentified specimens of Teleostei, Clupeidae, Mugilidae and unidentified Teleostei, which were small-sized fishes, unlike those seen in the materials collected through *in situ* excavation.

Materials Recovered from the 2007 Excavation

Many fish remains were also recovered in the 2007 excavation. As with the 2004-5 season, fish remains were collected by a combination of *in situ* recovery during excavation and the intensive sieving (1mm) of selected squares in Trench 2, layer 1.

In general, the same specimens of fish are seen in both seasons of excavation. The identification results of the sieved assemblage are shown in Table 10.4. Nine specimens of Rajiformes, 4 specimens each of *Acanthopagrus* sp. and Sparidae, plus 1 specimen each of Cyprinidae, Mugilidae, and Serranidae were identified. In addition, there were 4 unidentified specimens of Teleostei (possibly Cyprinidae or Siluformes). Cyprinidae, Mugilidae and the unidentified Teleostei are all small-sized fishes.

Table 10.3 Fish remains collected by sieving from square E3 in layer I (2004-5).

taxon	common name	element	L/R	N
Rajiformes A	Rays type A	vertebra		1
Elasmobranchii	Sharks or Rays	vertebra		6
Clupeidae	Sardines or Shads	caudal vertebra		1
Mugilidae ?	Mullets ?	abdominal vertebra		1
Acanthopagrus	Black Seabream	premaxillary	L	4
		premaxillary	R	1
		dentary	R	1
		angular	R	1
Sparidae ?	Seabreams ?	caudal vertebra		1
Teleostei (unidentified)	-	vertebra		5
Total				22

Table 10.4 Fish remains collected by sieving from area H2 in layer I (2007).

taxon	common name	element	L/R	N	remarks
Myliobatididae ?	Eaglerays ?	teeth		2	
Rajiformes A	Rays (type A)	vertebra		7	
Elasmobranchii	Sharks or Rays	vertebra		3	
Cyprinidae	minnows or carps	abdominal vertebra		1	
Mugilidae ?	Mullets ?	abdominal vertebra		1	
Serranidae ?	Groupers ?	caudal vertebra		1	
Lates calcarifer or Serranidae ?	Barramundi or Groupers	angular	L	1	
Acanthopagrus	Black Seabream	premaxillary	L	1	
		dentary	R	1	
		angular	L	1	
		opercle	R	1	
Sparidae ?	Seabreams ?	abdominal vertebra		2	
		caudal vertebra		2	
Teleostei (unidentified)	-	caudal vertebra		1	Cyprinidae ?
		quadrate	R	1	Siluriformes ?
		abdominal vertebra		1	Siluriformes ?
		caudal vertebra		1	Siluriformes ?
Teleostei (unidentifiable)	-	vertebra		2	
Total				30	

DISCUSSION

Palaeoenvironmental Reconstruction Based on Fish Remains

The habitat types of the identified fishes are listed in Table 10.5. Most of the fishes identified are types that inhabit marine (littoral) or brackish waters (Figure 10.3) (Masuda et al., 1980; Masuda and Kobayashi, 1994). In particular, *Acanthopagrus* sp., which comprised more than half of the fish remains, mainly inhabits embayments and lagoons with relatively low salinity, or brackish waters such as estuaries and mangrove wetlands (Masuda et al., 1980; Masuda and Kobayashi, 1994). *Lates calcarifer* inhabit a variety of areas, from marine (littoral) waters to downstream areas of relatively large rivers. The specific habitats of the sharks and rays are uncertain due to insufficient identification, but it is certain that they are marine species. Because these fish remains were recovered in large numbers, there is no doubt that marine embayments or lagoons existed near the site.

There are both freshwater and marine species of Siluriformes (Masuda et al., 1980; Masuda and Kobayashi, 1994), but since the excavated remains have not been identified to the family level so far, their habitats are uncertain. As for aquatic animals besides fish, the remains of large Trionychidae (soft-shelled turtles) are common. Since Trionychidae live in freshwater, it can be assumed that there was a freshwater environment of a certain size near the site.

In summary, we can ascertain that there was a series of aquatic environments, from marine embayments or lagoons to freshwater ponds in the vicinity of the site (Figure 10.4). Compared to the current landscape surrounding Man Bac, and the aquatic environment along the northern Vietnam coast, the area was likely to have been similar to present day Ha Long Bay and the downstream basins of the rivers flowing into the bay. On the other hand, Serranidae inhabit rocky or coral reefs facing the open sea (Masuda et al., 1980; Masuda and Kobayashi, 1994), so there is a possibility that such an environment also existed in portions of the coastal area near the site.

Characteristics of Fishing Activities Estimated From Fish Remains

The dominant fishes caught at this site consist of *Acanthopagrus* sp., sharks and rays, *Lates calcarifer*, Siluriforme and Serranidae. Based on their habitats, as described above, it is likely that the main fishing grounds ranged from embayments or lagoons to brackish waters such as estuaries near the site (see Table 10.6). *Lates calcarifer* and Siluriformes were possibly caught in freshwater environments, but this is not certain. Combined with the low frequency of freshwater fish such as Cyprinidae, it is likely that freshwater fishing activities were of relatively limited importance.

Since Serranidae have a different habitat from other fishes, and most of their remains recovered were of an extremely large size, they may have been imported. Among the artefacts recovered from the site, those possibly used as fishing tools are bone pointed tools, bone harpoons and stone net sinkers. In particular, bone pointed tools were numerous (Dung 2006). Further research is required to clarify the relationship between these tools and the fishes identified, but it seems quite likely that the many bone pointed tools were used to catch *Acanthopagrus* sp. (Toizumi, 1988, 2000).

The tendency towards large individuals of Siluriforme, *Lates calcarifer* and Serranidae is a unique characteristic of the assemblage. Since most of the analysed materials were collected through *in situ* excavation, there is a possibility that bones from smaller fish are missing due to sampling issues, but even considering that, the tendency towards large fish is clear. It can be hypothesised that these fishes were caught using tools such as spears, or hooks and lines, with a strong selectivity for larger individuals. Serranidae, in particular, were probably caught by angling (hook

Table 10.5 Habitats of fishes identified at Man Bac.

taxon	common name	total %MNI in Man Bac 2005	habitat: marine / freshwater	habitat: remarks
Lamnidae	Mackerel sharks	2.0	marine (open sea - littoral)	
Lamniformes ?	Sharks	3.0	marine (open sea - littoral)	
Rajiformes	Rays	7.1	marine (open sea - littoral)	
Clupeidae	Sardines or Shads	1.0	marine (open sea - littoral) - brackish	
Cyprinidae	minnows or carps	-	freshwater - brackish	
Siluriformes	Catfishes	12.1	marine (littoral) - freshwater	
Mugilidae	Mullets	2.0	marine (littoral) - brackish	
Lates calcarifer	Barramundi	9.1	marine (littoral) - freshwater	
Serranidae	Groupers	6.1	marine (littoral)	rocky bottom
Carangidae	Jacks	1.0	marine (open sea - littoral)	
Sciaenidae ?	Croakers ?	1.0	marine (littoral) ?	
Acanthopagrus	Black Seabream	53.5	marine (littoral) - brackish	
Platycephalidae	Flatheads	2.0	marine (littoral)	sandy bottom

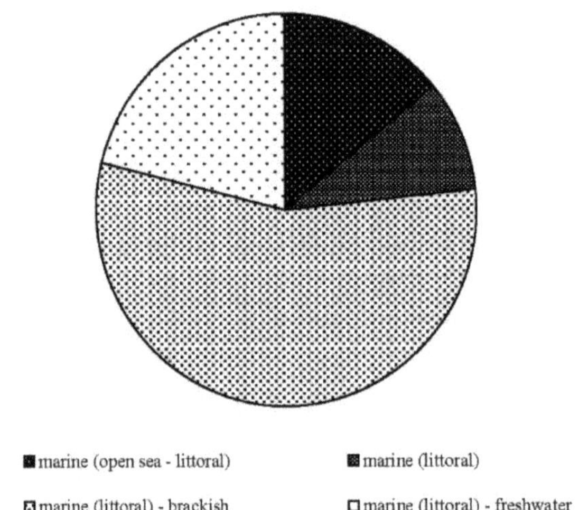

Figure 10.3 Habitats of fishes identified at Man Bac 2004-2005 (%MNI).

marine			brackish	freshwater
open sea	littoral	littoral (bay / lagoon)		
	Serranidae			
	sharks and rays			
		***Acanthopagrus* sp.**		
			Lates calcarifer	
			Siluriformes	
				Trionychidae

Figure 10.4 Habitats of the dominant fishes at Man Bac (Trionychidae are also shown).

and line) since they inhabit relatively deep waters (see Clark and Szabó 2009; Ono and Clark 2010). Although no fishing hooks have been found at the site, considering Siluriformes, *Lates calcarifer* and Serranidae are all carnivores with a large mouth and that large individuals were caught, it is possible that some of the bone pointed tools, which are relatively short and pointed at both ends, were used as "gorges".

It needs to be noted that in the sieved samples, bones of smaller fishes, including types of fishes not present in the *in situ* recovered assemblage such as Clupeidae and Cyprinidae, were identified. Therefore, it is likely that these smaller fishes were also caught to a certain extent. If further research is done on these smaller fish remains, the observations described above may have to be revised.

Among the specimens of large Serranidae, a maxillary and several vertebrae with cutting traces by a large axe-like blade were found (Table 10.7, Figure 10.1, Nos. 9 and 11). Many stone axes were excavated at the site (Dung 2006), and were possibly used for butchering these large fishes.

Table 10.6 Estimated fishing grounds and technologies used for the dominant fish taxa caught at Man Bac.

taxon	fishing ground	technology
sharks	marine (details unknown)	unknown (hook and line / spears ?)
rays	marine (details unknown)	unknown
Siluriformes	marine (bay / lagoon) - freshwater	hook and line / spears
Lates calcarifer	marine (bay / lagoon) - freshwater	hook and line / spears
Serranidae	marine (littoral)	hook and line / (spears?)
Acanthopagrus sp.	marine (bay / lagoon) - estuaries	spears / (hook and line?)

Table 10.7 Modified fish remains from Man Bac.

taxon	common name	layer	sample bag No.	element	modification
Rajiformes A	Rays (type A)	I	98	vertebra	perforated ?
		I	146	vertebra	perforated
		I	101	vertebra	perforated
		II	237	vertebra	perforated
		II	219	vertebra	perforated
		II	217	vertebra	perforated
		II	215	vertebra	perforated
		I	94	vertebra	perforated ?
Lates calcarifer	barramundi	I	127	abdominal vertebra	artificially cut ?
Serranidae (large)	Groupers (large)	I	158	maxillary	cut marks
Serranidae (large) ?	Groupers (large) ?	I	151	caudal vertebra	artificially cut ?
		I	75	caudal vertebra	artificially cut ?
		II	182	caudal vertebra	artificially cut ?
		II	180	caudal vertebra	artificially cut ?
Acanthopagrus sp.	Black Seabream	III	329	anal spine	polished at the tip

SUMMARY

The dominant fishes caught by the Man Bac community include *Acanthopagrus* sp., sharks and rays, *Lates calcarifer*, Siluriformes and Serranidae. The large size of Siluriformes, *Lates calcarifer* and Serranidae is a unique characteristic and it is likely that they were caught with spears, or hooks and lines. It is likely that the main fishing grounds ranged from marine embayments or lagoons to brackish waters such as estuaries, with freshwater fishing activities being relatively limited.

ACKNOWLEDGEMENTS

We would first like to express our deep thanks to the following people for their help with analysing the material: Dr. Nguyen Kim Dung, Dr. Nguyen Lan Cuong, and Mr. Manabu Uetsuki.

LITERATURE CITED

Dung NK. 2006. Preliminary report on the Vietnamese - Japanese - Australian archaeological excavation at Man Bac site. In: Matsumura editor. Anthropological and Archaeological Study on the Origin of Neolithic People in Mainland Southeast Asia. Report of Grant-in-Aid for International Scientific Research (unpublished). p 88-128.

Clark G, Szabó K. 2009. The fish bone remains. In The Early Prehistory of Fiji, Clark GR, Anderson AJ (eds). ANU E Press: Canberra; 213–230. Terra Australis 31.

Masuda H, Araga C, Yoshino T. 1980. Coastal Fishes of Southern Japan. Tokyo: Tokai Daigaku Shuppankai.

Masuda H, Kobayashi Y. 1994. Grand Atlas of Fish Life Modes. Tokyo: Tokai Daigaku Shuppankai.

Ono, R. and Clark, G. 2010. A 2500-year record of marine resource use on Ulong Island, Republic of Palau. International Journal of Osteoarchaeology DOI: 10.1002/oa.1226.

Toizumi T. 1988. A method of seasonality estimation based on seasonal cycle of refuse deposition observed in the shell middens. Bull Natn Mus Japan Hist 29: 197-233. (in Japanese with English summary)

Toizumi T. 2000. Prehistoric fishery in the Final Jomon period around Atumi peninsula, central Japan. Zoo-archaeol 14: 23-38. (in Japanese)

11

Man Bac: Regional, Cultural and Temporal Context

Marc F. Oxenham[1] and Hirofumi Matsumura[2]

[1] *School of Archaeology and Anthropology, Australian National University*
[2]*Department of Anatomy, Sapporo Medical University, Japan.*

One of the most important issues facing us is the interpretation of Man Bac and its placement in a cultural and temporal context. The archaeological context of Man Bac places it firmly within a cultural complex identified as the Phung Nguyen, dated to around 1,800-1,400 BCE (Nguyen et al. 2004). In recent years, this cultural complex has been referred to as either Late neolithic or earliest Bronze Age (Nguyen et al. 2004, Oxenham et al. 2008). The term 'neolithic' has been traditionally used in Southeast Asia to characterise communities with presumed or proven agriculture and pottery, but without metal (Bellwood, 1992: 94). However, the presence or absence of pottery is not necessarily useful by itself in identifying 'the neolithic', with good examples of pottery manufacture by hunter-gatherers in late Pleistocene and early Holocene northeast Asia, including Initial and Early Jomon Japan (Habu, 2004). There was also a lack of pottery in many of the earliest clearly agricultural contexts in many parts of the world, for instance in the Levant (Lev-Yadun et al., 2000), and in Mesoamerica and the Andes (Bellwood 2005). Conversely, the earliest evidence for pottery in South America (c. 6,000 BCE in the lower Amazon and c. 5,000 BCE in northern Columbia) occurred in the absence of agriculture (Bellwood 2005: 158).

In northern Vietnam at least, pottery appeared first among Mid-Holocene hunter-gatherer communities (e.g. Da But, see Patte, 1965), well before any evidence for agriculture. Similarly, the Early Neolithic in China, dating to some 16,000-8,000 years BCE and named as such due to the presence of pottery, appears devoid of any clear evidence for agriculture. Furthermore, while clear support for domestication appears in the Middle Neolithic (8,000-5,000 BCE) of the Middle and Lower Yangtze basin, the same time period (still termed the Middle Neolithic) in southern and southwest China lacks evidence for agriculture (Zhang and Hung 2008). It was not until the early phase of the Late Neolithic (5,000-3,500 BCE) in southern China that clear evidence for pig domestication first occurred (Zhang and Hung 2008:313). During the late phase of the Late Neolithic (3,500-2,500 BCE), Zhang and Hung (2008:313-314) regard the Guangxi-Guangdong region (termed Lingnan by Chinese archaeologists) plus Fujian as a recipient of major farming dispersals from the more northerly Middle and Lower Yangtze basin. In the Terminal Neolithic, 2,500-2,000 BCE,

> the number of settlements dramatically increased in the Lingnan-Fujian region and southwest China. Locally, this was the full blossoming of the Neolithic in this area, at a time when regional populations are estimated to have exceeded in size those of the middle and lower Yangtze
> Zhang and Hung (2008:314)

The earliest clear material cultural links involving pottery between southern China and northern Vietnam can be seen in the presumed hunter-gatherer archaeological sites attributed to the Da But and Dingsishan (Phases 1 to 3) cultural complexes in Vietnam and Guangxi respectively (Zhang and Hung 2010). The dating of Da But and allied sites is problematic, but they are believed to have flourished between 4,500 and 2,500 BCE (see Nguyen 2005). Da But sites are generally characterised by large shell midden deposits, edge ground pebble ('Bacsonian') axes and very coarse cord-marked pottery (Oxenham et al. 2005). The Da But burial practices are unique in northern Vietnam, in that a significant number of the inhumations are tightly flexed squatting burials with limited grave goods, and no pottery (Oxenham 2000). The largest excavated Da But cemetery site is at Con Co Ngua, which included 96 burials that exhibited very low levels of oral disease but a high frequency of serious accidental trauma, possibly related to their hunter-gatherer subsistence activities (Oxenham et al. 2001, Oxenham et al. 2006, Oxenham 2006). The presence of pottery in the Da But complex, and at least one large cemetery, suggests some level of sedentism (Oxenham 2000). As with the convention adopted in China, Vietnamese archaeologists see the Da But as a "neolithic" culture, ostensibly due to the presence of pottery and edge ground axes. Nguyen et al. (2004) have gone as far as to characterise the Da But as Middle neolithic, and the pre-Bacsonian Hoabinhian period in general, which extends back to at least 16,000 BCE, as early neolithic, despite the absence of pottery prior to Da But and Con Co Ngua.

The value and archaeological meaning of the term "Neolithic" in China and northern Vietnam is thus very unclear, particularly in the context of its use as a form of short-hand for the presence of pottery. The meaning of the term is clearly tested in the use of "Middle Neolithic" to describe clear agricultural communities in the Yangtze Basin on the one hand, and contemporaneous hunter-gatherer communities in southern China on the other (see Zhang and Hung 2008, for instance). In Chinese archaeology this terminological problem stems from the imposition of a rigid chronological framework, based on the better known developments over the last 10,000 years in the Yangtze basin, on to other parts of China that experienced very different cultural and technological developments at very different periods of time. Using this system to characterise the Da But in Vietnam would lead to it being labelled the Early Phase of the Late neolithic. Man Bac (as part of the Phung Nguyen) would become the Terminal neolithic, even though the Phung Nguyen complex to which it belongs contains the earliest evidence for agriculture and animal domestication in northern Vietnam.

In would seem that the real value of the term "neolithic" lies in its use as a short hand for a major change in human subsistence and associated behaviour, i.e. the development of food production (animal and plant domestication). Pottery by itself is not a good signature for animal and plant domestication, as pointed out above.

"Neolithic" as a chronological marker, let alone a behavioural indicator, has limited, if any value, in northern Vietnam, or southern China for that matter. If pottery using cultures associated with otherwise hunter-gatherer economies require some form of special signification, perhaps it would be better to use a term without profound agricultural connotations - "Pre-Neolithic Pottery using Cultures" (PNPC) may be a value-free substitute.

Much of the discussion so far has aimed to secure an appropriate regional, cultural and temporal context for Man Bac. It can be seen that Man Bac was part of the agriculturally driven demographic expansion that occurred in what is now southern China between 2,500 and 2,000 BCE. The Da But complex represents the southern periphery of the expansion of Late Pre-Neolithic Pottery using Cultures (PNPC), as represented by Dingsishan, that seem to have emerged in northern Vietnam between 4,500-4,000 BCE. The Phung Nguyen, including Man Bac, represents the southern periphery of a subsequent expansion of Neolithic (here meaning food producing) communities originating in the geographical region of southern China.

The agricultural developments, and their dating, in southern China prior to the emergence of the Phung Nguyen appear to provide a good framework or model for understanding sites like Man Bac, which is dated to between 1,500-1,800 BCE. But do the findings of this monograph support such a model?

Several bioarchaeological lines of evidence from Man Bac support the view that the site is representative of a demographic expansion from southern China into northern Vietnam some time shortly after 2,000 BCE. In Chapter 2 it was argued that Man Bac appeared to be a reasonably representative sample in terms of its demographic profile. Moreover, a relatively high level of fertility was evident, consistent with an earlier investigation by Bellwood and Oxenham (2008), with high JA and MCM values indicative of high population growth. High levels of fertility, and evidence for population growth, are consistent with the model for demographically driven expansion from southern China, involving the arrival of new immigrant food producers in the Man Bac region.

Cranio-morphometric analyses (Chapter 3) demonstrate: (1) a significant morphological, and by extension genetic, gulf between late Pleistocene to mid Holocene Pre-Neolithic Pottery using Cultures, such as those from Da But and Con Co Ngua, and later Metal Period populations in northern Vietnam; (2) that Man Bac is morphometrically mosaic, and by implication genetically heterogeneous, with some individuals showing the earlier Hoabinhian/Bacsonian morphology, others being indistinguishable from the Metal Period morphology, and some a hybrid between the two. Qualitative and quantitative analyses of the teeth (Chapter 5) revealed a slightly different, albeit not entirely inconsistent, picture to the cranio-morphometrics. The greater dental size variance, relative to Metal Period samples, suggests a greater genetic heterogeneity in the Man Bac sample. Moreover, the close affinity of the Man Bac to Metal Period samples in terms of odontometric proportions, but the intermediate position of Man Bac between earlier Da But and Hoabinhian samples on the one hand, and later Metal Period series on the other, speaks to the highly variable nature of the Man Bac population. The most parsimonious explanation for such morphological and genetic heterogeneity is that the Man Bac population was undergoing a major and rather rapid genetic transition. It would seem clear the source of this transition is to be found within a population of food producing immigrants into the region.

Interestingly, the analysis of Man Bac cranial non-metric traits (Chapter 4) did not find the sample particularly heterogeneous, but rather found that Man Bac was mostly closely affiliated with Metal Period and modern Southeast Asians, as well as with the Chinese Weidun Neolithic sample from the lower Yangtze, dated to between 4,000-3,000 BCE. This finding is consistent with Zhang and Hung's (2008) suggestion that there was a major dispersal of farming communities from the lower Yangtze south into the Lingnan-Fujian region between 3,500-2,500 BCE. It is from such immigrant populations that the subsequent demographic expansion into northern Vietnam occurred. Why Dodo's (Chapter 4) non-metric analysis does not indicate the level of heterogeneity found by Matsumura (Chapters 3 and 5) is unclear. It may be simply that a greater contribution of immigrant genes existed in the Man Bac sample than was indicated by Matsumura's analyses. Shinoda's ancient mDNA analysis (Chapter 8) clearly points to a major genetic input to the Man Bac community from a more northerly source, although genetic heterogeneity rather than a predominantly modern Southeast Asian or East Asian signal is suggested. Whatever the reason, Dodo's analysis is still consistent with the general model for the expansion of farming communities from southern China into northern Vietnam after 2,000 BCE.

Two final strands of evidence were dealt with in this monograph: human palaeohealth and zooarchaeology. Earlier work comparing the palaeohealth of the Da But complex (Con Co Ngua specifically) and a Metal Period series from northern Vietnam suggested an increase in infectious disease and/or a decline in immuno-competence in the latter period, arguably associated with agricultural activities (Oxenham et al. 2005). This is consistent with bioarchaeological studies globally, that show a generalised trend of declining health with the adoption and/or intensification of agriculture (see an earlier summary in Larsen 1995). However, Southeast Asian assemblages do not, for the most part, follow the trend of declining oral health seen in many other parts of the world with the adoption/intensification of agriculture (see Oxenham et al. 2006). The level and patterning of infectious disease in Man Bac has not been assessed, although oral and physiological health has. Man Bac exhibits the highest rate of caries seen in any Southeast Asian assemblage, albeit only marginally greater than that observed in the contemporary site of Khok Phanom Di in Thailand. Moreover, very high levels of physiological stress are indicated by the elevated frequencies of LEH and cribra orbitalia. Given the apparent caries-neutral effects of a rice diet (Tayles et al. 2009), a cereal apparently absent at Man Bac, but present at Khok Phanom Di, it is intriguing that two of the earliest neolithic assemblages in the region also exhibit the highest rates of caries, by far. In addition to the likely effects of elevated levels of fertility on oral health in both series (see Chapter 2), it is possible that root crops were an important staple at both sites. Tayles (1999) has suggested the possible availability of tubers and bananas at Khok Phanom Di, while Zhang and Hung (2010) suggest that tubers played a significant role, in addition to rice agriculture, in Neolithic subsistence in Lingnan-Fujian. If tubers, rather than rice, played a significant role in subsistence during the beginnings of food production in Southeast Asia, only to be replaced by rice as agriculture became more established, this may explain the elevated rates of caries in the earliest agricultural populations in the region.

But are tubers particularly cariogenic foodstuffs? Despite being inferred to be cariogenic (e.g. see Turner 1979), little is known of the cariogenic or prophylactic properties of root crops such as taro (see Toverud et al. 1952:485). Modified tubers such as *poi* (fermented taro) have been implicated in elevated rates of caries in Hawaii (Miller 1974), but we have been unable to locate experimental evidence to confirm this. Nonetheless, some processed tubers, cassava flour at least, are believed to be cariogenic (Rosalen et al. 1997). A study of variation in caries incidence in Papua New Guinea suggests that the particular mineral contents of foods such as sago and taro are most important in the prevalence of caries (Barmes et al. 1970). Unfortunately, Man Bac is silent on the presence or otherwise of root crops such as taro, and clearly more research is needed to determine the effect, positive or negative, that tubers have on dental health.

The final set of biological evidence comes from the animal remains (Chapter 9), which showed not only that food production in the form of pig domestication occurred at Man Bac, but that pigs contributed significantly in terms of a terrestrial food source. The view of Sawada et al. that we are seeing the initial stages of pig domestication at Man Bac is intriguing. It could be that limited stocks of domesticated immigrant pigs were being supplemented by locally domesticated indigenous pigs. Apart from pig domestication, Sawada and colleagues also point out that the Man Bac community targeted a range of taxa in their hunting activities, albeit taxonomically impoverished in comparison to earlier hunter-gatherer communities in the region. Whatever the role of animal and plant domestication in terms of subsistence contributions at Man Bac, the diverse range of animals and habitats exploited, highlights the importance of such activities in their day to day lives. Moreover, the fish remains (see Chapter 10) also indicate a considerable variety of exploited habitats, with an apparent emphasis on targeting very large fish with a sophisticated fishing technology.

The habitat diversity and breadth of hunting and fishing technologies and behaviours suggested by the mammalian and fish remains is intriguing. Not only is a very good knowledge of the local environment indicated, but a long history of hunting and fishing behaviours is suggested. How do we reconcile the model of rapid demographic expansion of agricultural communities from Lingnan-Fujian into northern Vietnam, with a community such as Man Bac, retaining sophisticated hunting and fishing skills that also required an intimate local knowledge of the environment? The biologically mosaic nature of the Man Bac human community provides an answer. Man Bac is clearly representative of interaction, possibly at many levels, of in-coming food producing migrants and an indigenous population. The new migrants brought domesticated plants and pigs to the table, while the indigenous populations with whom they were integrating likely brought a sophisticated and intimate knowledge of the local environment, including an ancient tradition of hunting and fishing. Man Bac is one of those rare archaeological instances of a community in transition, both in terms of its human genetic makeup, and with respect to major behavioural shifts in subsistence life-ways.

LITERATURE CITED

Barmes DE, Adkins BL, and Schamschula RG. 1970. Etiology of caries in Papua-New Guinea: associations in soil, food and water. Bull Wld Hlth Org 43:769-784.

Bellwood P. (1992) Southeast Asia before history. In: Tarling N., editor. The Cambridge History of Southeast Asia: Volume One Frome early Times to c. 1800. Cambridge University Press: Cambridge. pp. 55-136.

Bellwood P. 2005. First farmers; The Origins of Agricultural Societies. Blackwell; Main St, Maiden, MA.

Bellwood P, Oxenham M. 2008. The expansions of farming societies and the role of the Neolithic Demographic Transition. In: Bocquet-Appel J-P, Bar-Yosef O, editors. The Neolithic Demographic Transition and its Consequences. Dordrecht: Springer.

Habu J. (2004) Ancient Jomon of Japan. Cambridge University Press; Cambridge.

Larsen CS. 1995. Biological changes in human populations with agriculture. Annual Review of Anthropology 24:185-213.

Lev-Yadun, S., A. Gopher & S. Abbo. 2000.The cradle of agriculture. Science 288:1602-1603.

Miller CD. 1974. The influence of foods and food habits upon the stature and teeth of the ancient Hawaiians. In CE Snow (edited). Early Hawaiians;an Initial Study of Skeletal Remains from Mokapu, Oahu. Lexington, KY: University of Kentucky Press, pp. 167-175.

Nguyen KS, Pham MH, and Tong TT. 2004. Northern Vietnam from the Neolithic to the Han Period. In I Glover and P Bellwood (editors) Southeast Asia from Prehistory to History, London and New York: Routledge Curzon, p:177-201.

Nguyen V. 2005. The Da But Culture evidence for cultural development in Vietnam during the middle Holocene. Indo-Pacific Prehistory Association Bulletin 25(3):89-93.

Oxenham MF. 2000. Health and Behaviour During the Mid-Holocene and Metal Period of Northern Viet Nam. Unpublished PhD Thesis, Northern territory University, Darwin, NT, Australia.

Oxenham MF. 2006. Biological responses to change in prehistoric Vietnam. Asian Perspectives 45:212-239.

Oxenham MF, Walters I, Nguyen LC and Nguyen KT. 2001. Case Studies in Ancient Trauma: Mid-Holocene through Metal Periods in Northern Viet Nam. In Henneberg M, and Kilgariff J, editors. *The Causes and Effects of Biological Variation.* Australasian Society for Human Biology: University of Adelaide, p. 83-102.

Oxenham MF, Nguyen KT, Nguyen LC. 2005. Skeletal evidence for the emergence of infectious disease in bronze and iron age northern Vietnam. *Am J Phys Anthropol 126:359-376.*

Oxenham MF, Nguyen LC, Nguyen KT. 2006. The oral health consequences of the adoption and intensification of agriculture in Southeast Asia. In: Oxenham M, Tayles N, editors. Bioarchaeology of Southeast Asia. Cambridge: Cambridge University Press. p 263-289.

Oxenham MF, Matsumura H, Domett K, Nguyen KT, Nguyen KD, Nguyen LC, Huffer D, Muller S. 2008. Health and the experience of childhood in late Neolithic Vietnam. Asian Perspectives 47:190-209.

Patte E. 1965. Les ossements du kjokkenmodding de Da But. Bulletin du Service Ethnologie de Indochine 10: 1-87.

Rosalen PL, Volpato MC, and Ruenis AP. 1997. Cariogenic [correction of Carcinogenic] potential of a typical cassava flour from the Amazonian region of Brazil. Indian J Dent Res. 8(3):72-6.

Tayles N. 1999. The excavation of Khok Phanom Di: a prehistoric site in Central Thailand. Volume IV: the people. The Society of Antiquaries of London: Oxbow Books.

Tayles N, Domett K, Halcrow S. 2009. Can dental caries be interpreted as evidence of farming? The Asian experience. In: Kope T, Meyer G, Alt KW, editors. Comparative Dental Morphology. Basel: Karger. p 162-166.

Toverud G, Finn SB, Cox GJ, Bodecker CF, and JH Shaw. 1952. A Survey of the Literature of Dental Caries. National Academy of Sciences: Washington D.C.

Turner CG (II). 1979. Dental anthropological indications of agriculture among the Jomon people of central Japan. Am J Phys Anthropol 51(4):619-635.

Zhang C, and Hung HC. 2008. The Neolithic of Southern China–Origin, Development, and Dispersal. Asian Perspectives 47(2): 299-329.

Zhang C, and Hung HC. 2010. The emergence of agriculture in Southern China. Antiquity 84: 11-25.

Appendix 1

Man Bac Burial Descriptions

Damien G. Huffer[1] and Trinh Hoang Hiep[2]

[1] *School of Archaeology and Anthropology, Australian National University*
[2] *The Vietnamese Institute of Archaeology*

All of the burial descriptions below are individually identified using the MB Year M Burial # format, where MB = Man Bac, Year = Field season the burial was excavated in, M = Mộ (grave), and Burial # = Sequential number assigned to the burial by order of discovery during that season. NB: For the 2007 season, H (Hộ) 1 corresponds to the north-south oriented grid to the west of the 2005 excavation, while H2 corresponds to the east-west oriented grid to the south of the 2005 excavation. Regarding the descriptive terminology, 'orientation' refers to the position of the long axis of the body (head through torso) with respect to cardinal directions. For instance, an orientation of north-south means the long axis of the body was oriented north-south, with the head north (the first direction mentioned). 'Facing' refers to the direction in which the head is facing (generally either up or to either side, which is designated with the appropriate cardinal direction). See Chapter 2 for the methods employed to determine age-at-death and sex. Photos accompany descriptions of the better preserved burials only.

For the 2005 and 2007 excavation seasons arbitrary 10cm spits were the chief excavation unit, with each spit corresponding to a level. In 2005 and 2007 cultural units I and II tended to be free of burials, while cultural unit III contained the majority of burials. In 2005 the following levels were used: 1-8 (cultural unit I), 9-12 (II), 13-20 (III). For 2007 Square 1 (H1): 1-8/9 (I), 9-13/14 (II), and 15-19 (III). For 2007 (H2): 1-8/9(I), 9-12 (II), and 13-21 (III). Actual burial depths below original surface for 2005 are approximately 1.4-2.0m; 2007 (H1) 1.0-1.9m; and 2007 (H2) 1.6-2.1m.

1999

Sex and age-at-death estimates for the 1999 sample was carried out by H. Matsumura. No pottery illustrations are available for the 1999 burials (Appendix 2).

MB99M1:

Square A5, A6. A 16-18 month old infant, poorly preserved, missing most the os coxae, most ribs, the right ulna, the distal right radius, and most carpals, tarsals, metacarpals and metatarsals. It is oriented east-west, supine, and facing north. A row of stones are positioned parallel to the body on the right side, with one positioned at the feet. These are possibly all that remains of a stone circle constructed around this individual at interment. One small globular pot is to the right of the face.

MB99M1

MB99M2:

Square C6, D6. A young adult female, 18-20 years old, poorly preserved, missing all elements below the pelvis, the right os coxa, both hands, several ribs, lower thoracic and lumbar vertebrae, left clavicle, and part of the cranium. *In situ* orientation is east-west, burial is supine and facing up. Grave goods consist of a globular pot placed to the right of the face. The 2nd vessel is a footed cup with curvilinear/incised motifs, placed to the left of the left distal humerus.

MB99M3:

Square A4, B4. A young adult female, 18-20 years old, only missing the left metacarpals and rib fragments. Orientation is northwest-southeast, burial is positioned supine and facing north. Grave goods consist of one vessel placed beyond the feet: it is footed with walls thinning towards the rim and straight with a decorative motif consisting of bands of cordmarking surrounding a middle band of burnishing (removed prior to photography).

MB99M3

MB99M4:

Square C5, C6. A moderately to poorly preserved infant skeleton >1 year old. Most of the cranium, several ribs, the left wrist and hand, most of the pelvic bones, and the patella were not recovered. From those elements recovered *in situ*, orientation was observed to be east-west, burial position is supine, facing upwards. Three stones are present near the body; one at the head, one at the feet, one right of the pelvis, but no grave goods were recovered.

MB99M4

MB99M5a:

Square D3. A nearly complete skeleton of an approximately 4-5 year old subadult. Burial orientation is east-west, position is supine, facing upwards. Part of the cranium, many ribs, the left clavicle, left forearm, most of both os coxae, the right metacarpals, and neither of the patellae were recovered. This burial overlaps MB99M5b and both were removed en-bloc. One small globular pot was placed to the left of the cranium.

MB99M5a and MB99M5b

MB99M5b:

Square D3, E3. A partially complete skeleton of an adult male approximately 30-50 years old. The entire torso, from clavicle to pelvis is overlapped by MB99M5a. This individual is flexed, oriented southwest-northeast, interred on its right side, facing southeast. The arms are flexed at the elbow at approximately 45^0, as are the legs, and the hands are next to or underneath the head. The only grave good recorded as definitely belonging to this burial is a small globular pot, placed to the right of the cranium.

2001

Sex and age-at-death estimates for the 1999 sample was carried out by H. Matsumura. No pottery illustrations are available for the 2001 burials (Appendix 2).

MB01M1:

Square C2. A nearly complete skeleton of a subadult approximately 9 years old. Only the right clavicle, a few right ribs, the distal left radius and ulna and the left hand are missing. *In situ* orientation is northeast-southwest, burial position is supine, and facing direction is south. A large, redware, paddle-impressed globular pot was interred inverted in front of the face; soot blackened on the base. The second vessel is a large, redware, footed bowl, placed just at the back of the cranium. It has diagonal cross-hatching between the shoulder carination and the base of the bowl, and also exhibits soot blackening.

MB01M1

MB01M2:

Square B2. A very poorly preserved infant skeleton (age indeterminate), represented by only a few fragments of the left and right tibia and fibula, the left proximal femoral epiphysis, and a single metatarsal. *In situ* orientation is northeast-southwest, burial position and facing direction are indeterminate. No grave goods were recovered.

MB01M3:

Square C2. A partially preserved skeleton of an approximately 8 month old infant. It was recovered from the west bulk of the southwest corner, and consists of only a few skull fragments, the mandible (lacking dentition), both clavicles and scapulae, the sternum, ulnae and radii, the os coxae, femora, and the proximal tibiae and fibulae. Orientation, burial position and facing position were indeterminate. No grave goods recovered.

MB01M4a:

Square C2. A nearly complete skeleton of an infant approximately 8 months old, recovered from the south bulk of the southwest corner. Most of the cranium, right ulna and radius, both hands, feet, the vertebrae and sacrum are missing. Orientation is east-west, burial position is supine, but facial direction is unrecorded. One paddle-impressed and footed globular pot was recovered at the feet.

MB01M4b:

Square C2. This burial consists solely of the scapula of an infant. It was recovered intermixed with the bones of burial MB01M4a.

MB01M5:

Square C2. A partially recovered skeleton of an adult male between 50-60 years old. All elements below the pelvis (as well as the right phalanges and left hand) were not recovered due to the skeleton extending into the west section. The individual is oriented east-west, burial position is supine, and facing direction is upwards. One redware, paddle-impressed, globular pot was recovered to the right of the cranium.

MB01M5

MB01M6:

Square A1. This burial consists solely of the tibiae and fibulae, metatarsals and some tarsals of an infant of indeterminate age, protruding from the east bulk of the northeast corner. Orientation is roughly northeast-southwest. No grave goods were recorded.

MB01M7:

Square B2. This burial consists solely of the proximal half of the left humerus and left scapula of an infant of indeterminate age. No grave goods were recovered.

MB01M8:

Square A2. This burial consists solely of several cranial fragments of a subadult of indeterminate age. Numerous cowrie shells and thin-cut shell discs surrounded these fragments, suggesting the inclusion of one or more shell necklaces as grave goods. Orientation, burial position and facing direction are indeterminate.

MB01M9:

Square A2. A partially excavated skeleton of an adult female (?). Only the skull, left distal humerus, left scapula, clavicles and 1st-3rd cervical vertebrae were recovered protruding from the west section. Tooth wear suggests an older individual. A few cowrie shells were recovered around the head, suggesting deliberate placement.

MB01M10:

Square B2. A partially excavated skeleton of an adult male approximately 40 years old. The skull, both proximal humeri, clavicles, scapulae, and 1st-3rd cervical vertebrae recovered *in situ*, protruding from the west section. Burial orientation is east-west, burial position is supine, and facing direction is upwards. No grave goods recovered in association with those elements represented within the excavation grid.

2005

Sex and age-at-death estimates for the 2005 sample was carried out by M. Oxenham and K. Domett. For illustrations of burial pots, see Appendix 2.

MB05M1:

Square A2; level 11-12(II). Subadult 18 months +/-5 months. The remains of this individual are distributed inside and surrounding a large redware, undecorated, globular pot. To the south of this vessel, there a small globular pot (not illustrated in Appendix 2). The recovery of this burial within the artefact rich 'Cultural Layer II', made discerning the interment boundary, and extent of deliberate grave goods difficult. However, a single green nephrite cylindrical bead was recovered inside the fill of the redware vessel in association with the cranial fragments.

MB05M2:

Square F5; level 14(III). Neonate. *In situ* orientation is east-west, position is supine, facing upwards. One cowrie shell is to the left of the skull, and one undecorated redware bowl is at the feet.

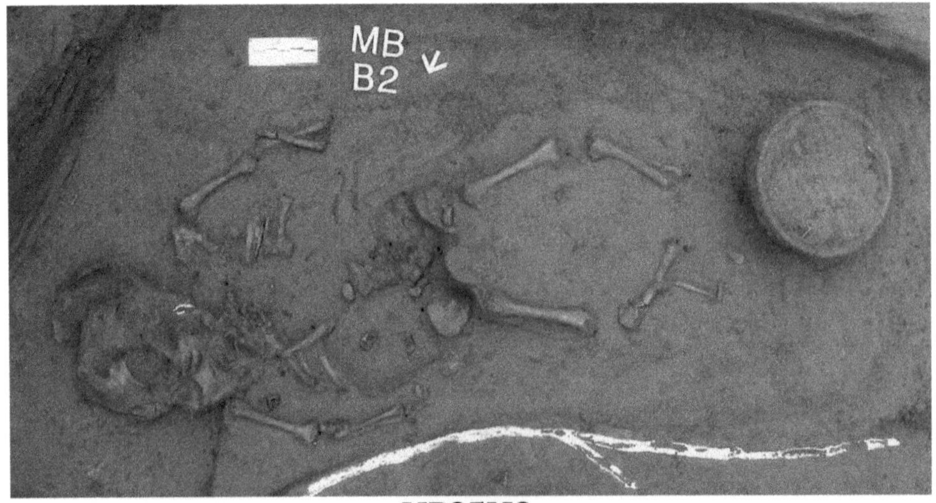

MB05M2

MB05M3:

Square C4, C5; level 14(III). Infant skeleton, 6 months +/- 2 months. *In situ* orientation is east-west, position in supine, facing up. Two small, redware, cord marked globular pots are positioned north of the skull, parallel to the body, and tilted at a 45^0 angle. Freshwater bivalve shells are contained within one vessel.

MB05M3

MB05M4:

Square D3, D4; level 14(III). Age estimate is 2 years +/- 6 months. Burial orientation and positioning are indeterminate, nor were any grave goods recovered.

MB05M5:

Square D4; level 15(III). Infant, 18 months +/-3months. *In situ* orientation is east-west, burial positioning is supine, facing upwards. A large cross-ribbed globular pot was interred on its side just to the right of the skull.

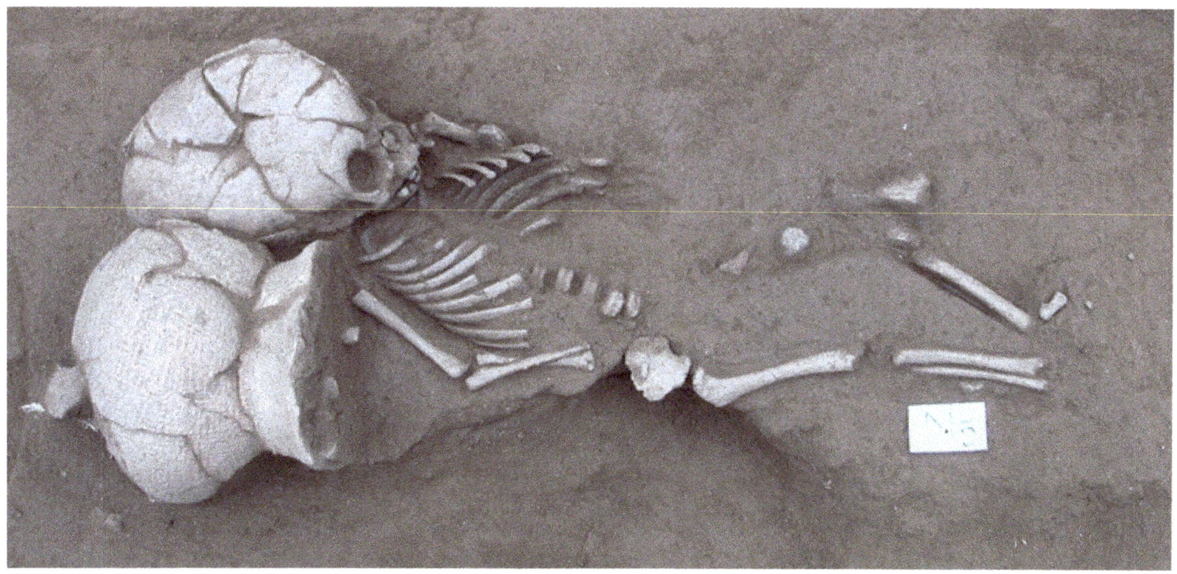

MB05M5

MB05M6:

Square D3; level 16(III). A ~1.5 year old child. Those elements superior to the pelvis were removed via disturbance by burial MB05M13. *In situ* orientation is east-west, burial position is supine, facing direction indeterminate. No grave goods recovered (see MB05M8 or MB05M13).

MB05M7:

Square D5; level 15(III). Neonate. *In situ* orientation is east-west, but burial position and facing direction are indeterminate. No grave goods recovered.

MB05M7

MB05M8:

Square B5, C5; level 15(III). An infant ~6 months old. A cross-ribbed globular pot is positioned between the legs. Orientation, burial position and facing direction are indeterminate.

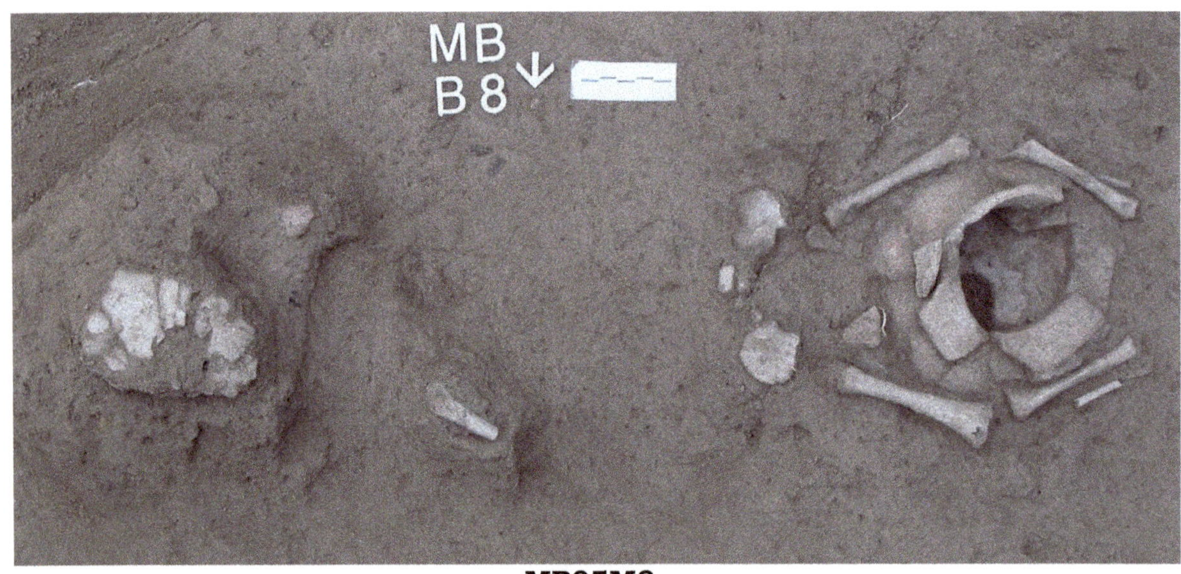

MB05M8

MB05M9:

Square C5; level 16(III). An adult female 40-49 years old. *In situ* orientation is east-west, position is supine, facing south. Some of the bones are lightly tinged green from copper mineralisation in the surrounding soil matrix. Two medium sized globular pots are placed to the left of the body. The first is cord marked and placed next to the left elbow. The second is cross-ribbed and further from the body, southwest of the left scapula.

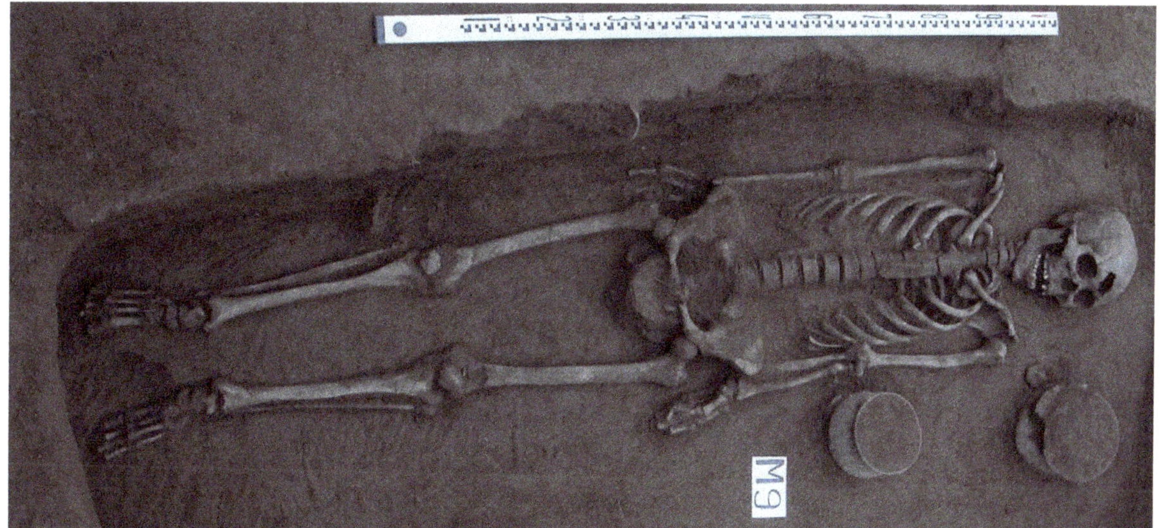

MB05M9

MB05M10:

Square F6; level 14-16(III). A subadult skeleton, 9 years +/-9months. *In situ* orientation is east-west, position is supine, facing northeast. Grave goods consist of five vessels and a separate ring-foot grouped either side of the cranium, two to the left and three to the right.

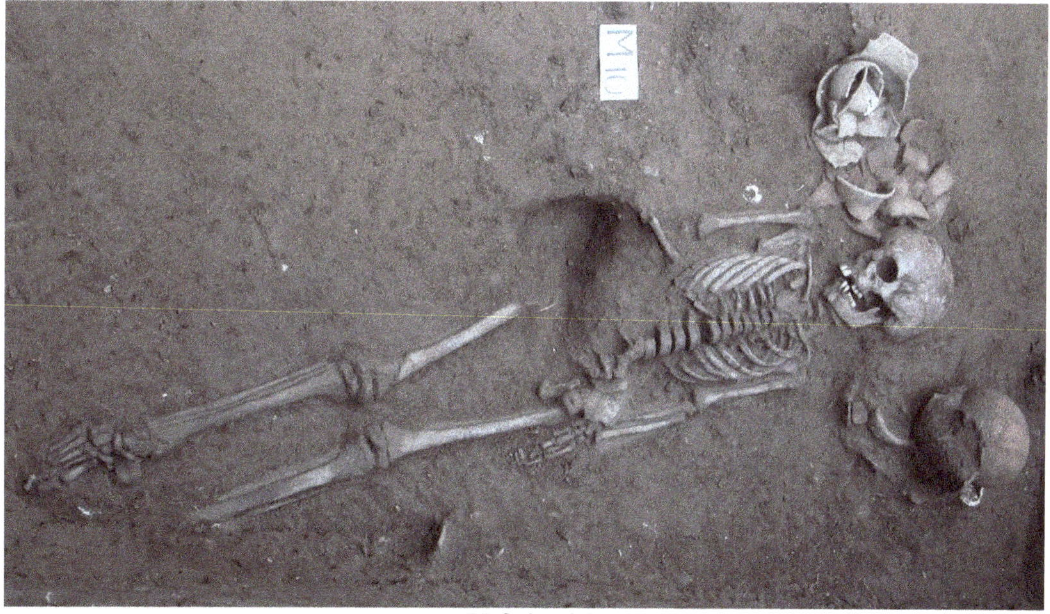

MB05M10

MB05M11:

Square A4, B4; level 16(III). A young adult male skeleton, 18-25 years old. *In situ* orientation is east-west, position is supine, facing south. A single cross-ribbed globular pot is placed right of the cranium, and an untanged polished stone adze is just superior of the right os coxa, to the right of L4.

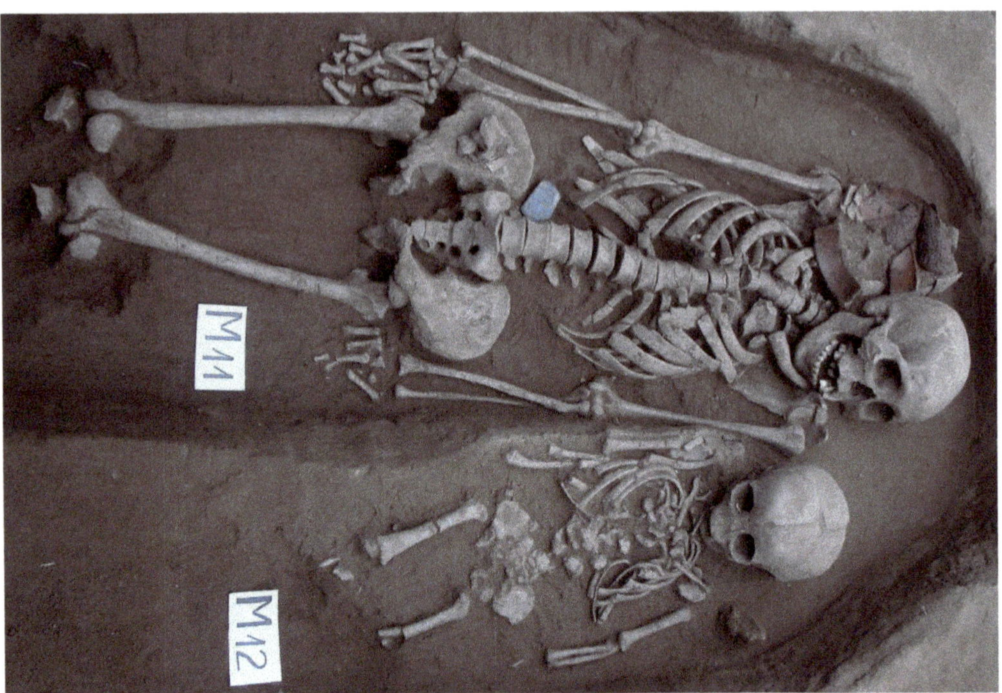

MB05M11 and MB05M12

MB05M12:

Square B4; level 16(III). An infant 2 years +/- 6 months, buried directly to the south of MB05M11. The distal left femur and hands are also missing. Most dentition is missing, but both dm^1 have erupted and are present. *In situ* orientation is east-west, position is supine, facing upwards. A small cross-ribbed globular pot was recovered just to the right of the cranium, superior to the right shoulder. A small piece of unworked sandstone was recovered just left of the cranium, but its use as a deliberate grave good is unclear.

MB05M13:

Square D3, E3; level 16(III). A ~16 year old young adult. *In situ* orientation is east-west (slightly northeast-southwest), position is supine, facing south. Sections of the cranium, femora, os coxae and humeri are green tinged, likely from natural copper mineralisation. Grave goods include a large cross-ribbed globular pot just right of the cranium, and a parallel ribbed globular pot placed lateral to the right elbow. Also included is a fragment of a grey sandstone grinding stone to the right of L4/5, and a pointed tool made from the distal end of an ungulate femur, recovered underneath the left humerus.

MB05M13

MB05M14:

Square C6, D6; level 16(III). A subadult between 2-5 years old. Some light green staining was recorded on the ribs. *In situ* orientation is west-east, position is supine, facing upwards. A single undecorated globular pot (soot blackened on the base) was recovered slightly northwest of the cranium.

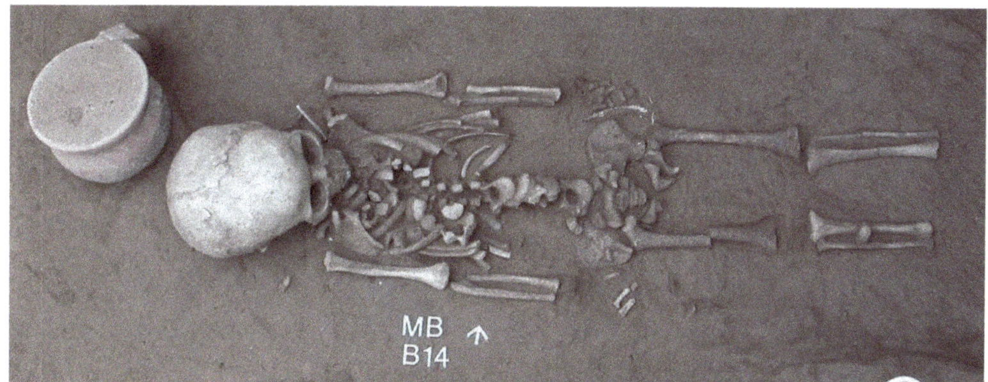

MB05M14

MB05M15:

Square A6, B6; level 15(III). A young adult female(?) approximately 17-18 years old. The grave has suffered extensive post-burial disturbance. Morphologically it is female, while all recovered long bone epiphyses (including clavicle) are unfused (sex estimation is thus problematic). Most of the dentition is present and the 3rd mandibular molars are just beginning to erupt. *In situ* burial position indicates a flexed burial, southeast-northwest oriented, recovered lying on its left side, and facing west/southwest. Only one, direct rimmed cord marked shallow bowl was recovered. It was placed just left of the skull and appeared to be interred at a 45 degree angle.

MB05M16a:

Square A5, B5, A6, B6; level 15-16(III). A mature adult female, 40-49 years old. The skull of this individual was revealed protruding from the west bulk in this square during the 2005 excavation, and was subsequently removed. The rest of the body was recovered during the first stage of the 2007 excavation. *In situ* orientation is slightly northeast-southwest, position is supine, facing west. No clear grave goods were identified, although a small ceramic potting anvil was identified in the southern end of the grave shaft, where the feet would have been. This burial was excavated and removed over two seasons.

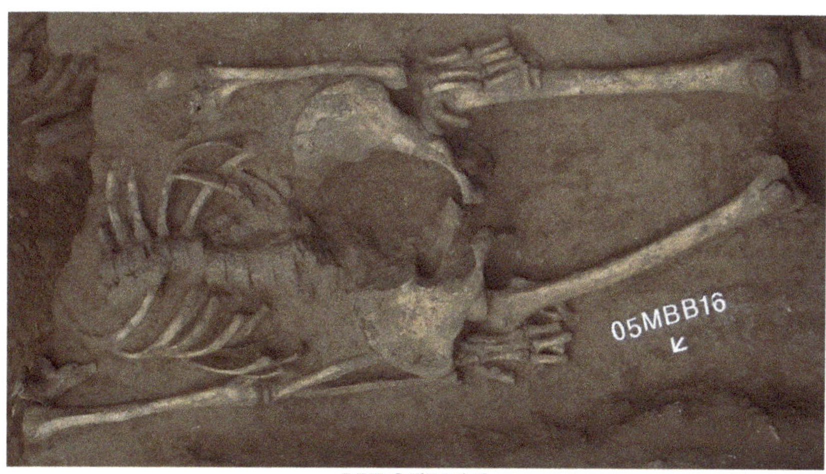

MB05M16

MB05M16b.

This individual was recognised by M. Oxenham during post-excavation analysis of the remains of MB05M16a (the female above). The following material defines this individual: (L15, III): 1 x right maxillary di^2 unerupted (crown only formed); (L16 (III), b4): 1 x right tibial diaphysis; 1 x fibula diaphysis, 1 x rib fragment, 2 x vertebral arches. All material is consistent with a neonate. It is possible that this infant was directly associated with MB05M16 as an unborn or recently born child (although M16a is a mature female).

MB05M17:

Square E1; level 17(III). An adult individual of indeterminate age and sex. East-west orientation is suggested by the position of the long bone fragments *in situ*, but this is not certain. Burial position and facing are indeterminate. This is the first burial of the 2005 season to be recovered from within the deeper stratum characteristic of a high concentration of shell and a low PH, covering the west half of the grid. Grave goods consist of three decorated cups with flaring sides, and decoration mimicking Phung Nguyen vessels, but they lack the burnishing or dark, finely grained temper seen on clear Phung Nguyen "imported" vessels within other graves. The positioning of these objects is also indeterminate.

MB05M18:

Square A3, A4; level 18(III). An infant, 18 months +/- 3 months. It was recovered *in situ* against the western bulk, at the feet of burials MB05M11 and M12, and its interment is likely responsible for the loss of all elements below the patellae from those burials. *In situ* orientation is north-south, position is supine, facing east. A single globular cord marked pot is lateral of the right forearm.

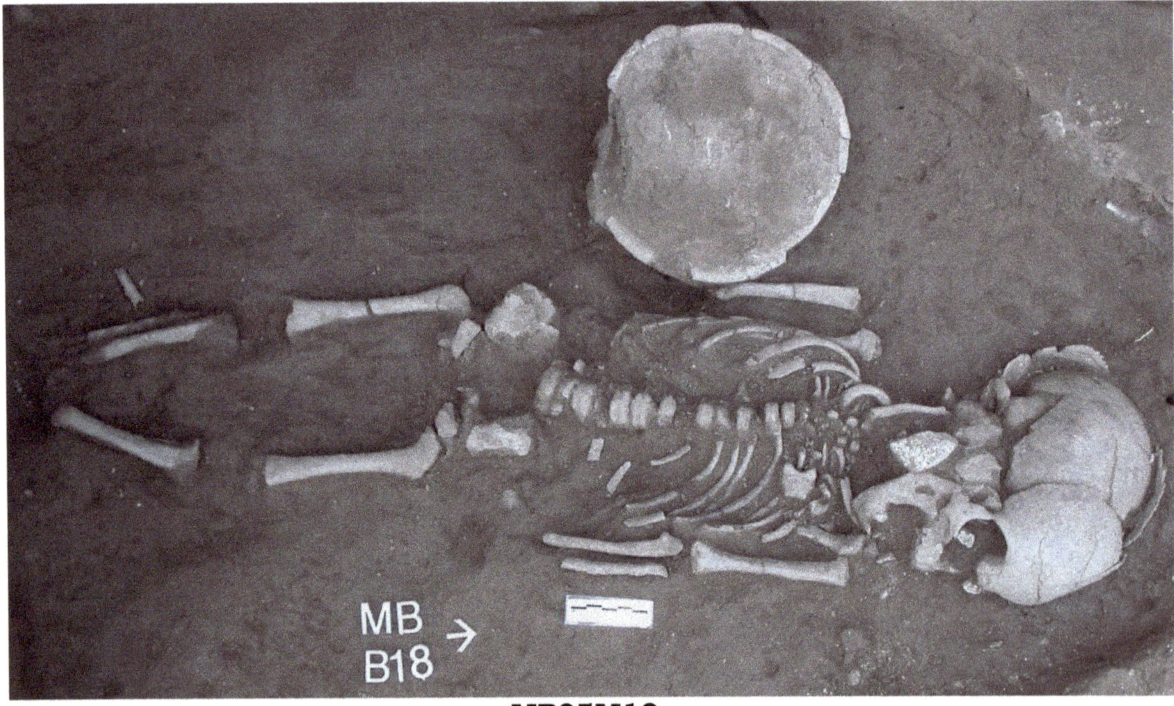

MB05M18

MB05M19:

Square F1; level 19(III). An adult of indeterminate age and sex, recovered from within the same shell/beach sand stratum as MB05M17, but approximately 15cm deeper. A small globular pot with unusual external decoration that resembles relief netting was next to the right tibia.

MB05M20:

Square C1, D1; level 18-20(III). A young adult male (?), 15-29 years old. *In situ* orientation is east-west, position is supine, facing north. The body was also interred on an angle of at least 5 degrees, so that it slopes from level 18 into level 20. Three ceramic vessels are included with this burial. Two are globular pots with cross ribbing, and the third is an undecorated open bowl. One globular bowl is placed north-east of the cranium, the other is in front of the face.

MB05M20

MB05M21:

Square B1, B2; level 18(III). Infant ~6 months old. *In situ* orientation is southeast-northwest, position is supine, facing upwards. The only grave good is a small footed undecorated bowl placed to the right of the cranium.

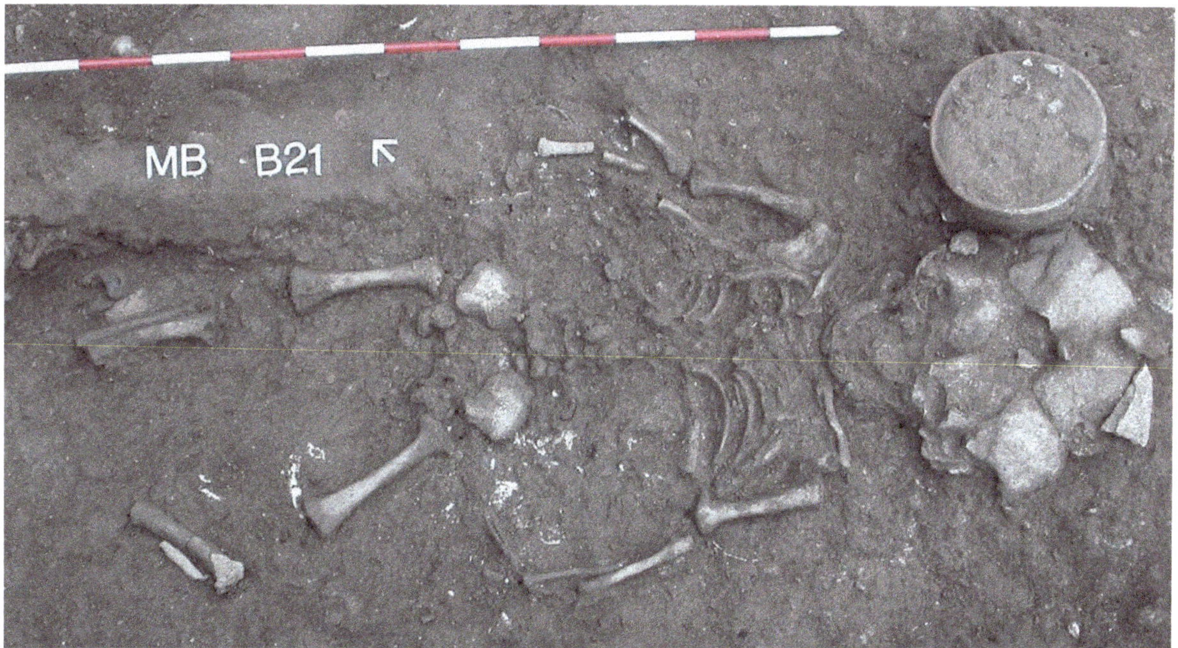

MB05M21

MB05M22:

Square D6; level 16(III). Infant ~18 months old, recovered against the south section, extending into it slightly. Orientation is indeterminate, as are body and facing positioning. No grave goods were recovered.

MB05M23:

Square C2; level 18(III). Infant ~ 15 months old. *In situ* orientation is approximately northeast-southwest, position is supine. A single globular pot is located between the legs at the knees.

MB05M23

MB05M24:

Square D2, E2; level 19(III). An 8 years +/- 9 months subadult. *In situ* orientation is east-west, position is supine, facing south. The grave good assemblage for this individual is unique amongst the 2005 burials. A cowrie shell necklace is *in situ* superior to the 2nd cervical vertebrae, and two bivalve shells are held, one in each hand. A small globular pot is just superior to the right shoulder. A pedestalled bowl with an everted incised rim was reconstructed from fragments next to the first vessel, north of the cranium.

MB05M24

MB05M25:

Square A6; level 16(III). A subadult 5 years +/- 9 months. *In situ* orientation is east-west, position is supine, facing upwards. The only grave goods recovered are a large pedestalled bowl decorated with cross combed and burnished decoration to the right of the scapula and a small nephrite adze.

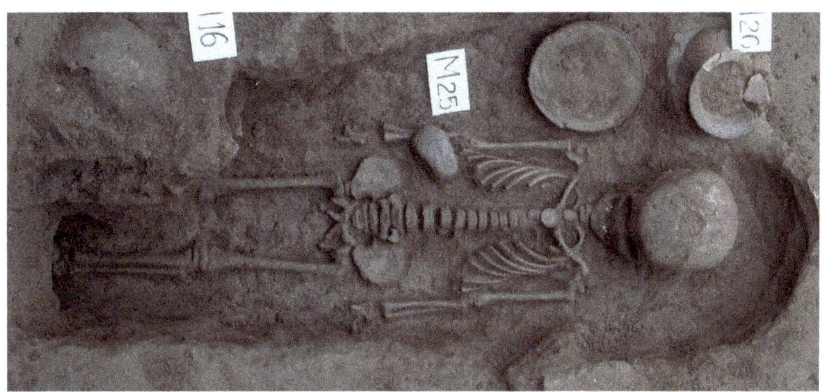

MB05M25

MB05M26:

Square A5; level 16(III). A 4-5 +/- 1 year old subadult (Oxenham assigned this age estimate based on a similar degree of tooth wear seen on independently aged MB05M25 & MB07H2M15). This individual is only represented by a few cranial fragments and several deciduous teeth found in direct association with a small globular pot. The remains are identified as a separate burial, as they appear too deep to be associated with burial M25 or M16, and too shallow to be associated with M29.

MB05M27:

Square B1; level 19(III). Adult of indeterminate age-at-death and sex. Represented by the right patella, tibia, fibula and metatarsals (protruding from the north section). This burial was not excavated.

MB05M28:

Square A2, B2; level 18(III). An adult female (?) 15-29 years old. More exact age estimates are complicated by the poor preservation and fragile nature of the skeletal material, buried within the sand/shell matrix. *In situ* orientation is east west (slightly northeast-southwest), position is supine, facing direction is indeterminate. A number of unique grave goods were recovered with this individual. Non-ceramic grave goods included a segment of shell disc necklace located in the mandibular region, and a grey polished stone adze placed superior to the right elbow. Seven ceramic vessels were recovered. Two are footed cups with incised and punctate decorative motifs, and there is a segment of a third. One small pedestalled bowl was recovered, and two globular pots (only one is illustrated in Appendix 2). Finally, a unique mortuary vessel with distinct Phung Nguyen incised (although unburnished) motifs was reconstructed. It is small, with a very small ring-foot base, a narrow vessel aperture, and a cylindrical shape to its body. It is unique within the Man Bac assemblage.

MB05M28

MB05M29:

Square A5, B5, B6; level 19-20(III). A robust adult male 30-39 years old. As with burial M16, the cranium was recovered during the 2005 excavation, while the rest of the body was revealed and removed during the 2007 excavation. *In situ* orientation is northeast-southwest, position is supine, facing upwards. This individual was interred with five globular pots clustered around the head and on top of the upper torso, positioned at various angles. In addition, a green nephrite Phung Nguyen style T-section bracelet is *in situ* around the right forearm, but broken into two segments and once repaired. A small black stone is just lateral of the left radial midshaft. This burial was excavated and removed over two seasons.

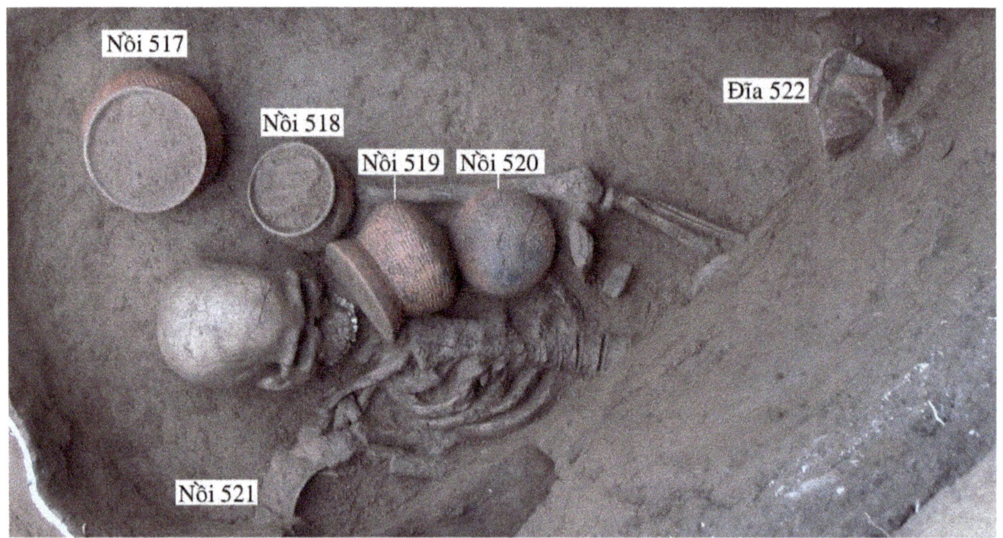

MB05M29 cranium

MB05M29 post cranium

MB05M30:

Square C/D 5/6; level 17 An infant ~6 months old. *In situ* orientation is northeast-southwest, position is supine, facing south. A small globular pot was recovered superior to the cranium, and two small grey nephrite beads were recovered to the right of the cervical vertebrae.

MB05M31:

Square C/D 5/6; level 18. A mature adult male 20-29 years old. *In situ* orientation is northeast-southwest, position is supine, facing upwards. Three very small, disk-cut grey nephrite beads, as well as a cluster of 4 small cylindrical black nephrite beads, are directly on top of the ribs and sternum (with two just lateral of the right iliac crest). Two clusters of cowrie shells are underneath each wrist, and an intact, T-sectioned black nephrite bracelet encircles the right wrist. A globular pot is located superior to the cranium, and a second such vessel was reconstructed from fragments between the legs at the knees. A cord marked footed bowl was recovered to the left of the left tibia. Finally, a small clay projectile pellet was recovered from the fill.

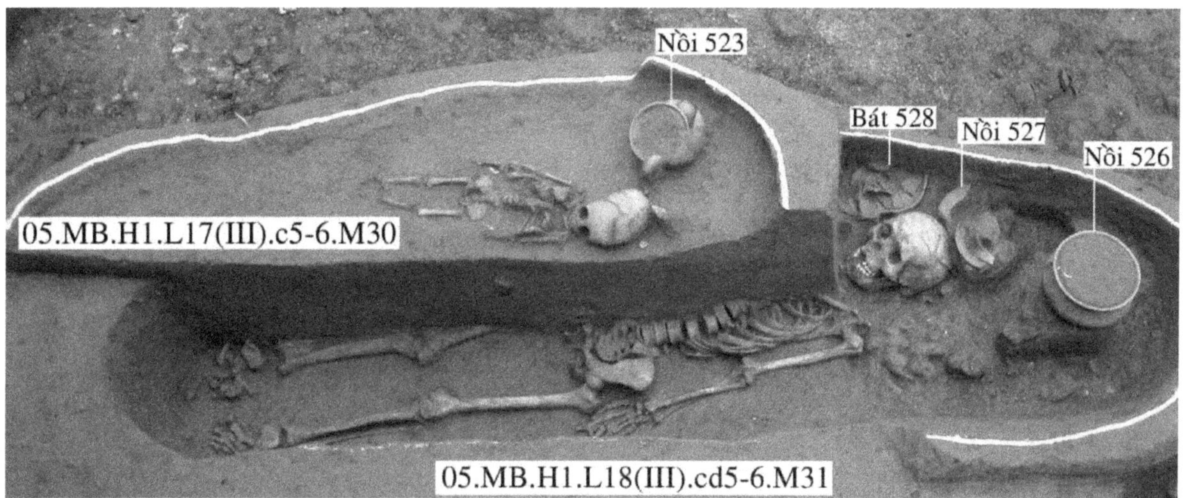

MB05M30 and MB05M31

MB05M32:
Level 18. An adult male (?) 15-29 years old. Burial orientation and positioning are indeterminate. There was a pig (*Sus scrofa*) mandibular canine located inside the mandibular arch. There were four pottery vessels associated with this burial, and also fragments of two other vessels. There was a green nephrite adze in the grave also.

MB05M33:
One pot only, no clear association with human skeletal remains at the time of discovery.

MB05M34:
Square A/B5; level 17. A mature adult female 40-49 years old. *In situ* orientation is approximately northeast-southwest (only slightly off a direct east-west coordinate), position is supine, facing north. A single, redware globular pot is just lateral of the right shoulder (not illustrated in Appendix 2).

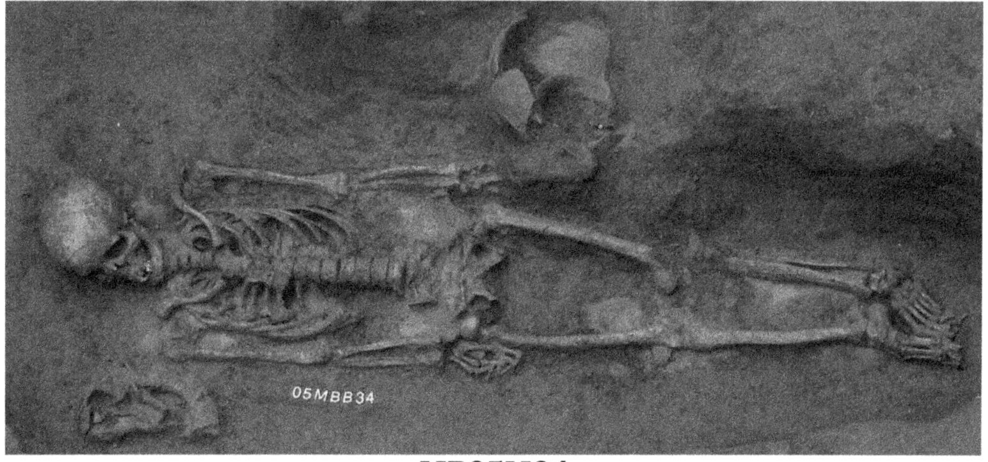

MB05M34

MB05M35:

One pot only, no clear association with human skeletal remains at the time of discovery.

MB05M36:

Square A1; level 20(III). A subadult 3 years +/- 6 months old. It was recovered directly underneath the feet of burial MB05M27, and the cranium thus belonged within the 2005 excavation grid. The entire left arm is underneath the fragments of a redware, highly fired, globular pot. A bivalve shell is underneath the right hand, and at least five grey nephrite cylindrical beads are located on the chest; between the left and right 4th, 5th, and 6th ribs and superior to the left 3rd rib.

2007H1

Sex and age-at-death estimates for the 2007 sample was carried out by M. Oxenham and K. Domett.

MB07H1M1:

Square B5, B6; level 10(II). A subadult 12 years +/- 6 months old. Importantly, this burial represents a substantially later interment, being recovered from within Cultural Layer II. *In situ* orientation is northeast-southwest, position is supine, but the maxillary region is too fragmentary to determine facing direction. No clear grave goods are associated with this burial, although a cluster of sherds are on top and to the left of the cranium. Because this burial was placed within the midden layer, clear interment boundaries are indeterminate.

MB07H1M1

MB07H1M2:

Square A2; level 13-14(II). Only the proximal half of a tibia and associated fibula fragment of a neonate represent this burial. No dentition is present, nor any clear grave goods: this individual is also interred within Cultural Layer II.

MB07H1M3:

Square A4, A5; level 16(III). A nearly complete skeleton of an older subadult >12 years < 18 years. Only a few cranial fragments and distal tarsal and carpal phalanges are missing. *In situ* orientation is slightly northeast-southwest, position is supine, facing south. A large bivalve shell is covering the left hand, and a cluster of cross-ribbed globular pot fragments are on top of and to the right of the head.

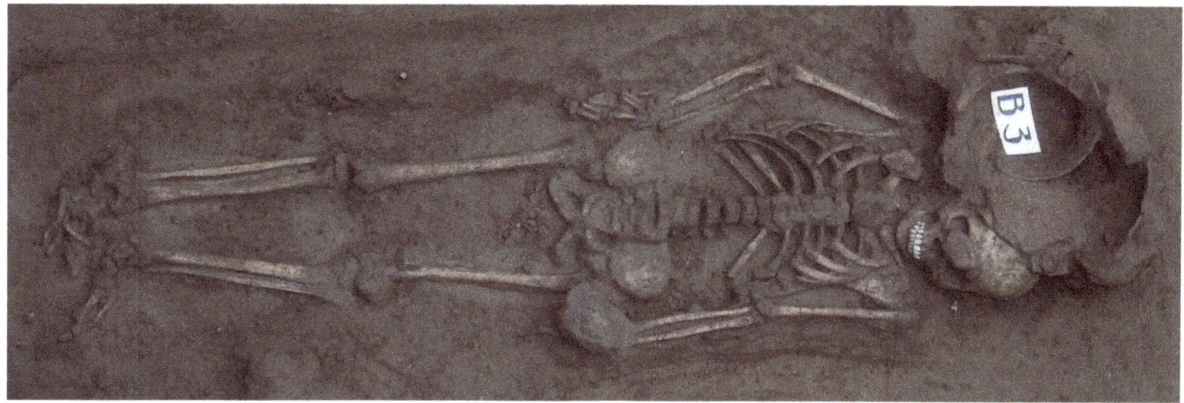

MB07H1M3

MB07H1M4:

Square A4, B5; level 17(III). A mature adult female (30+ years old) in good condition. *In situ* burial orientation is northeast-southwest, position is supine, facing upwards. Importantly for the determination of superpositioning, the skull of burial MB07H1M10 is revealed (supine, facing up) between the legs of M4 at the pelvis. A small, highly fired, redware footed bowl is to the left of the left tibia (but far enough away so that provenance is uncertain), and two further highly fired redware globular pots (both fragmented) are northwest of the right ulna/radius. However, their association with this burial as opposed to MB07H1M10 is also uncertain. They are not illustrated in Appendix 2.

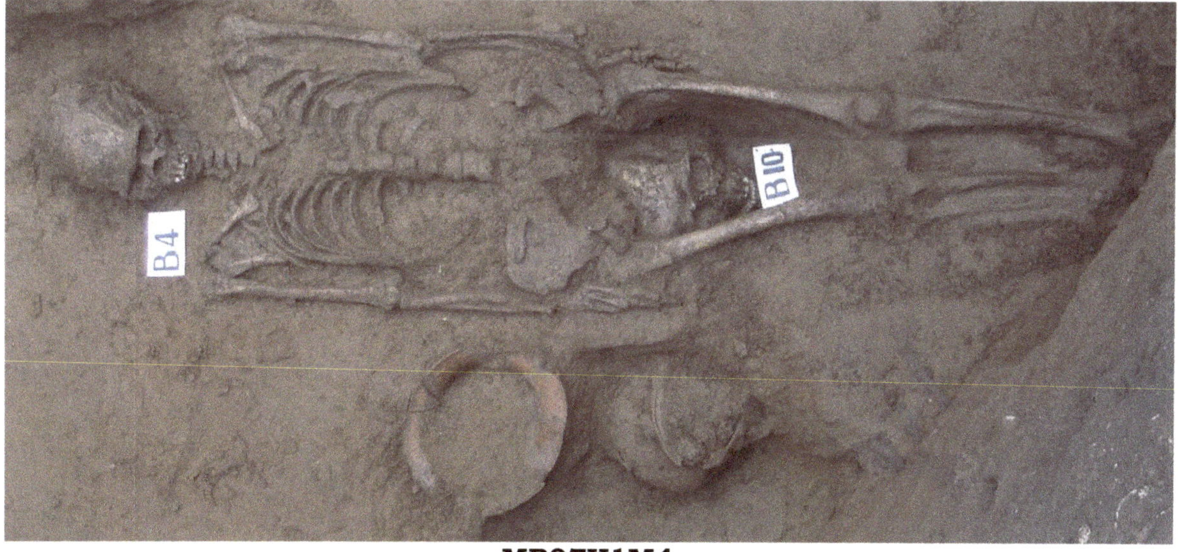

MB07H1M4

MB07H1M5:

Square A6(H1), B1(H2); level 17(III). A mature adult male (40-49 years old). The hands are prone and articulated. *In situ* orientation is northeast-southwest, position is supine, facing upwards. Grave goods consist of two small vessels. One is a shallow undecorated bowl and the other is a small globular pot. Both are placed together directly right of the head. A cluster of sherds are left of the left wrist (only one vessel is illustrated in Appendix 2). Importantly, this burial intrudes into burial MB07H1M13, removing much of the upper body.

MB07H1M5 and MB07H1M13a

MB07H1M6:

Square B4; level 15(III). A subadult ~6-9 months old. *In situ* orientation is northeast-southwest, position is supine, facing up. A single globular pot is recovered just to the right of the head.

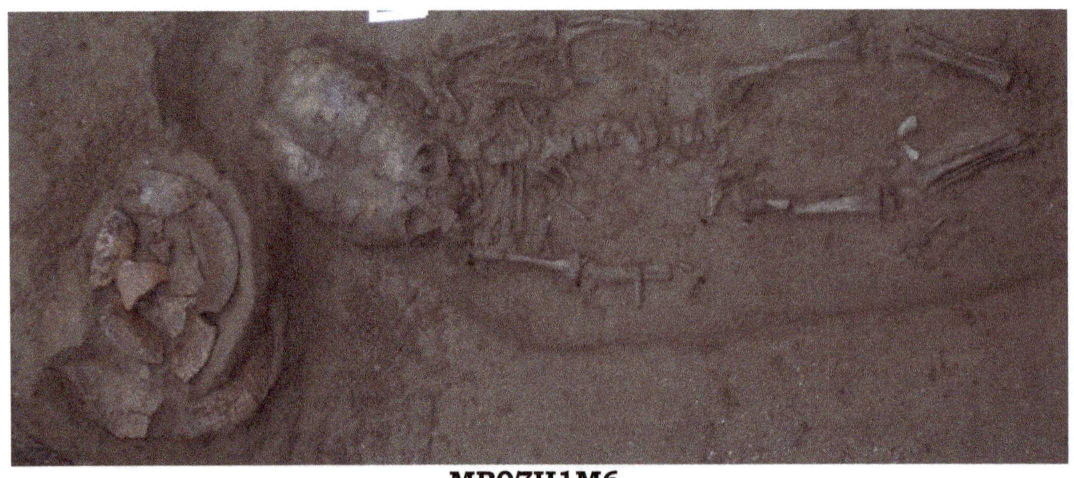

MB07H1M6

MB07H1M7:

Square A5, A6; level 17(III). A subadult 1 year +/- 3 months old. Extensive disturbance of the skeleton was caused by the intrusion of MB07H1M3. *In situ* orientation is northeast-southwest, position is supine, facing direction is indeterminate. No grave goods were recovered with this individual.

MB07H1M8:

Square B2, B3; level 18(III). A mature adult male 30-39 years old. *In situ* orientation is northeast-southwest, position is supine, facing north. A single cord marked pot was reconstructed from fragments lateral to the right shoulder.

MB07H1M8

MB07H1M9:

Square A2; level 18(III). An adult male approximately 20-29 years old. This burial is partially flexed and interred on its right side directly against the eastern bulk which abuts the western edge of the 2005 excavation. *In situ* orientation is north-south, position is flexed, facing westwards. A parallel ribbed globular pot and a pedestalled dish with finely combed decoration are directly in front of the face. A cross-ribbed globular pot was also included (not illustrated in Appendix 2).

MB07H1M9

MB07H1M10:

Square A5, B6; level 18-19(III). A mature adult male approximately 40-49 years old. *In situ* orientation is northeast-southwest, position is supine, facing up. Six vessels are associated with this individual. The first is a large cross-ribbed globular pot with a ring-foot base and vertical bands of incised lines on the shoulder. It is approximately 30cm east of the cranium. The second is a very large cross-ribbed globular pot approximately 35cm northwest of the cranium. A medium sized cross-ribbed globular pot with a notched rim is approximately 28cm west/northwest of the cranium. A very rare Phung Nguyen motif decorated pedestal base is covering the distal right femur. It exhibits curvilinear bands of surface burnishing delineated by incision, over cross-hatched combed decoration. Another globular pot is placed between the legs at the tibial midshaft. It is soot-blackened on the base. Finally, another cross-ribbed globular pot, also with a soot-blackened base, is located next to the left foot. Six black nephrite cylindrical beads are recovered between the ribs.

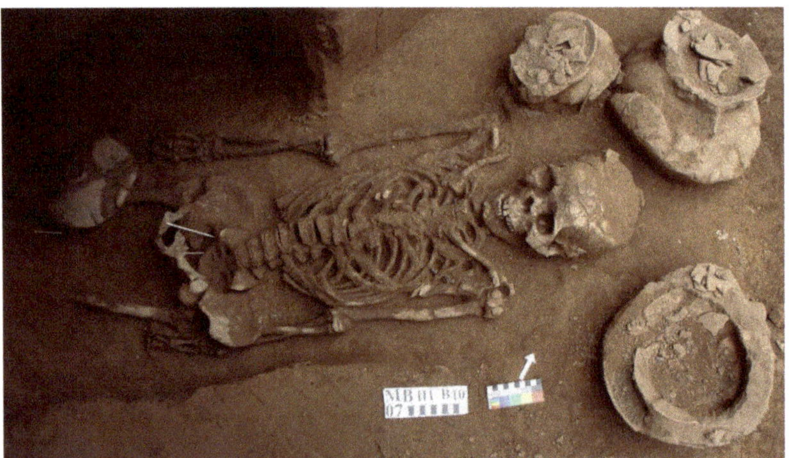

MB07H1M10

MB07H1M11:

Square B2, B3; level 19(III). A mature adult female 50+ years old. A window was excavated into the west bulk of square B3 to recover portions of this individual below the pelvis, and next to the feet, another adult calvarium was uncovered, but not excavated. *In situ* orientation is northeast-southwest, position is supine, facing north. A small globular pot is located approximately 18cm east of the cranium (not illustrated in Appendix 2).

MB07H1M11

MB07H1M12:

Square B4, B5; level 18(III). Neonate. *In situ* orientation and facing direction are indeterminate, but burial position is supine. No grave goods recorded.

MB07H1M13a:

Square B6(H1), A1(H2); level 16(III). An adult (indeterminate sex) aged 30+ years old. The cranium is absent, but a mandible of appropriate age and dimensions was recovered in level 15(III), square A1H2. Both hands and feet are intact and *in situ*. Orientation is east-west (slightly northwest-southeast), burial position is supine. No clearly associated grave goods. See top left of illustration of MB07H1M5 which cut through this burial.

MB07H1M13b

This individual was recognised by M. Oxenham during post-excavation analysis of the remains of MB07H1M13a (above). 1 x permanent I^1, di^2 and dm^1 are consistent with a subadult ~10 years old. No other burials in the vicinity of MB07H1M13b are consistent with this age.

MB07H1M14

This individual was recognised by M. Oxenham during post-excavation analysis. It consists of an isolated and undesignated mandible located in Layer 17(III) in square ab1, most NW corner of the excavation (north of burials 11 and 8). No other adult burial anywhere near this position is missing a mandible. Sex in indeterminate and age-at-death is ~30+ years.

2007H2

MB07H2M1:

Square F1, G1 (extends ~1m into NE corner of E bulk); level 16-17(III). A mature adult male 40-49 years old. *In situ* orientation is east-west (slightly southeast-northwest), position is supine, facing up. No obvious grave goods, although a few shells and redware sherds appear in the vicinity of the body; likely due to taphonomic processes.

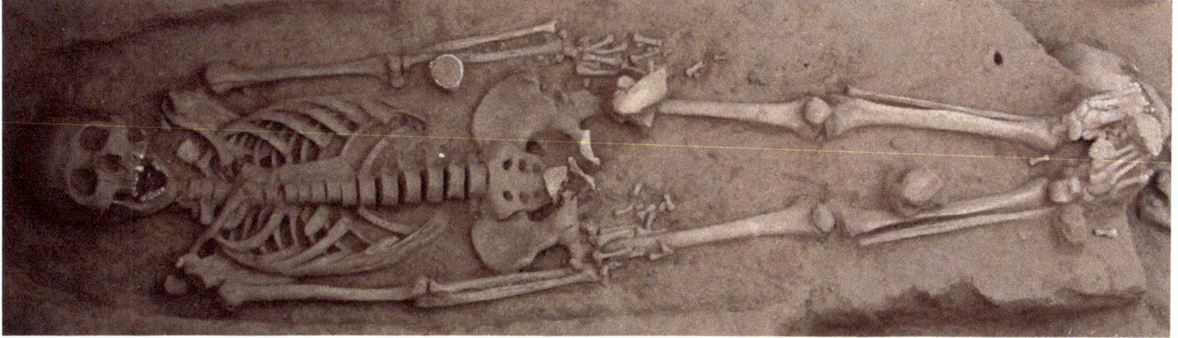

MB07H2M1

MB07H2M2:

Square E3, D3; level 18(III). A subadult > 12 years < 18 years. Intrusive damage (from a pit or post hole) fractured the left humerus (moving the proximal half out of position), disarticulated the left clavicle, and crushed the left scapula. *In situ* orientation is east-west, position is supine, facing north/northwest. Four vessels are present as grave goods, and a bivalve shell is underneath the right hand. A very tall cross-ribbed jar with a small ring-foot is placed ~15cm east of the cranium. One globular pot is on top of the right elbow, and a second it placed on its side next to the right foot. Finally a small grey polished stone adze is located to the west of upper torso.

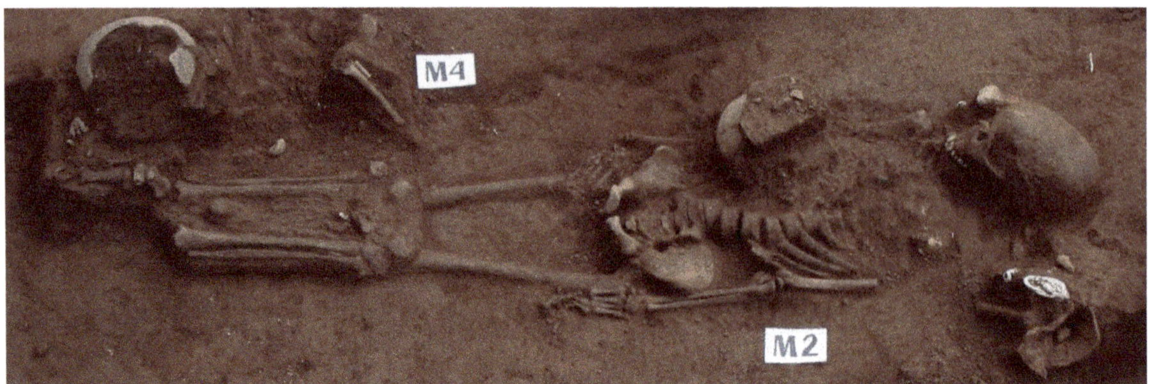

MB07H2M2

MB07H2M3:

Square D3; level 18(III). Neonate. Not enough elements *in situ* to judge burial orientation or facing position, but what remains suggests a supine burial. This individual lies parallel to the proximal tibia/distal femur of H2M2 at ~28cm distance. No clear grave goods.

MB07H2M4:

Square D3; level 18(III). Infant burial ~18 months old. Orientation is determined to be east-west, burial position is supine, facing direction is indeterminate. No clear grave goods associated.

MB07H2M5:

Square A2, B2, C2; level 18(III). An adult female 20-29 years old. *In situ* orientation is east-west, position is supine, facing up. A large, highly fired, cord marked, redware globular pot is placed over the left elbow. A single redware sherd is clenched between the jaws on the right side, propping the mouth open. A small cup was also placed in this burial.

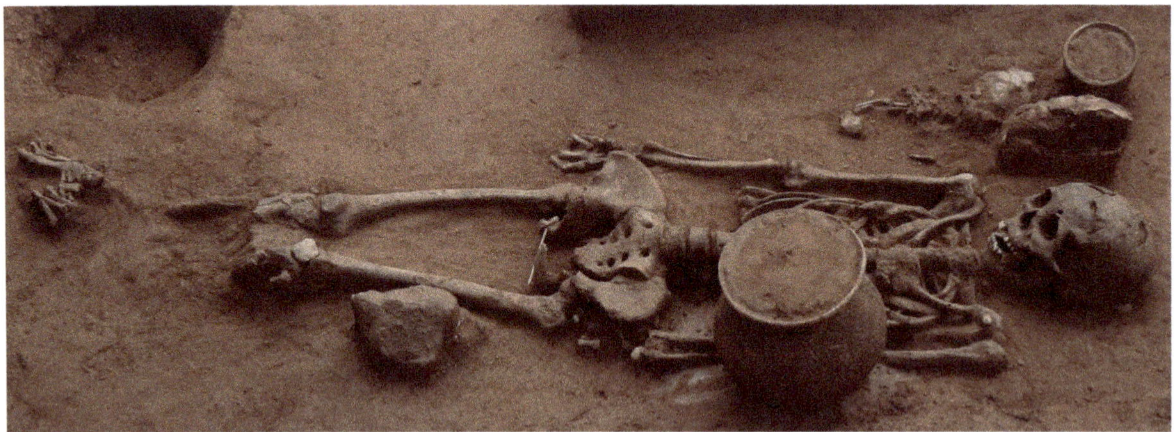

MB07H2M5 and MB07H2M9

MB07H2M6:

Square A3; level 19-20(III). A subadult 2 years +/- 6 months old. *In situ* orientation is east-west, position is supine, facing direction is indeterminate. Two vessels were recovered, the first one just south of the cranium, and the other just to the north of the cranium. Both are small, highly fired, cord marked redware globular pots.

MB07H2M6

MB07H2M7:

Square D3, E3; level 18(III). An infant 18 month +/- 5 months. *In situ* orientation is east-west (slightly northeast-southwest in relation to H2M2), position is supine, facing south. Grave goods include a marine bivalve shell lateral and parallel to the left distal femur, another marine bivalve shell fragment just left of the lumbar vertebrae, and two globular pots just north of the cranium.

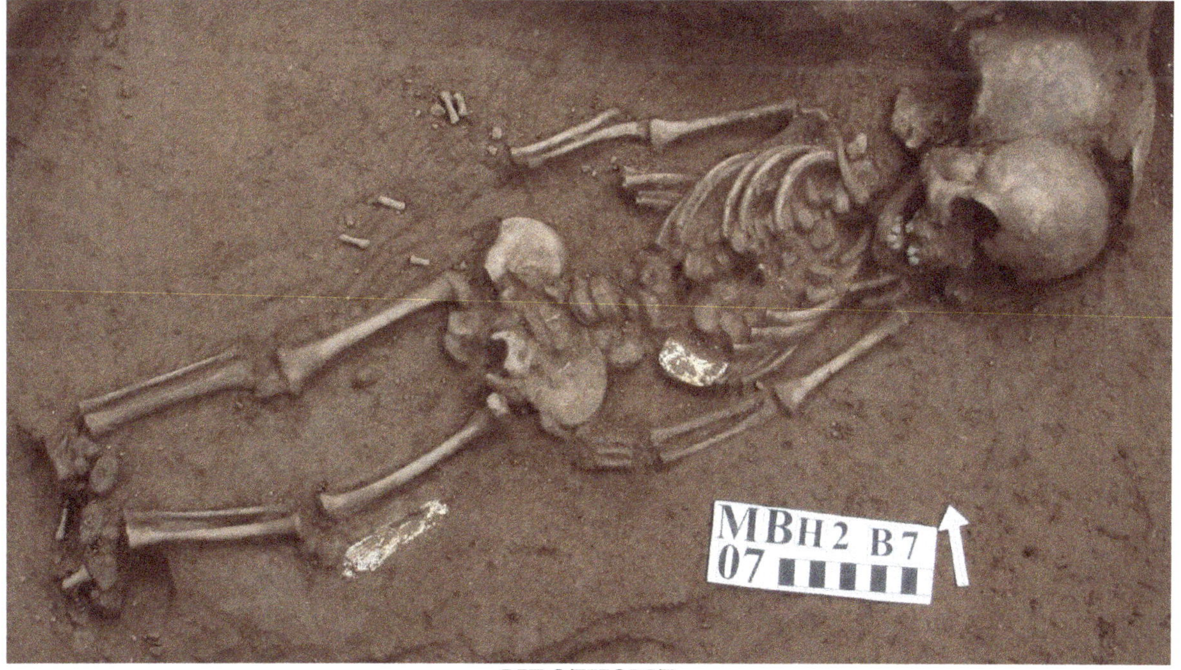

MB07H2M7

MB07H2M8:

Square E2, F2; level 18(III). A subadult 18 months +/-5 months old. *In situ* orientation is east-west, position is supine, facing south. Importantly, this individual is lying directly on top of the chest of burial H2M10, an adult male. Grave goods include marine bivalve shells held in both hands, a cluster of six shells (cowrie, clam, and gastropod shells) at the level of the left hand. part of a decorated pedestal base resting between the iliac blades, and two black stone beads found together in the upper right thorax region.

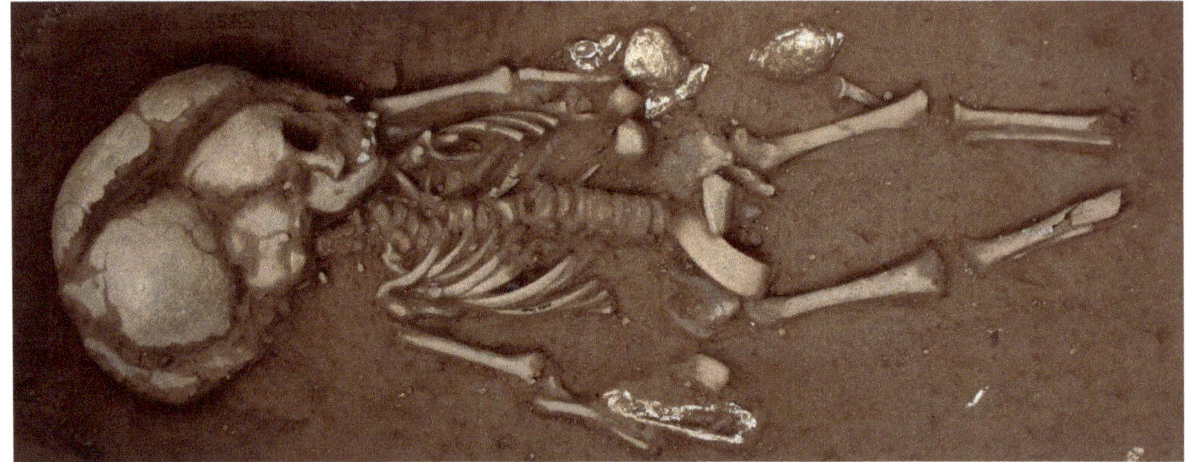

MB07H2M8

MB07H2M9:

Square B2; level 18(III). Neonate. *In situ* orientation is indeterminate, position is supine, facing indeterminate. Grave goods include two vessels directly east and south of the cranium respectively. The vessel to the east is a small cup with traces of parallel ribbing. The vessel to the south is a fragmented globular pot with cross ribbing (not reconstructed or illustrated in Appendix 2). The grave cut of burial MB07H2M5 bisected this burial, see the top right of the illustration of MB07H2M5.

MB07H2M10:

Square E2, F2; level 18(III). An adult male 30-39 years old. *In situ* orientation is southeast-northwest, position is supine, facing north. Importantly, burial H2M8 is interred directly on top of the ribs of this individual. His head is at the feet of H2M1. Two pottery vessels were included in this burial.

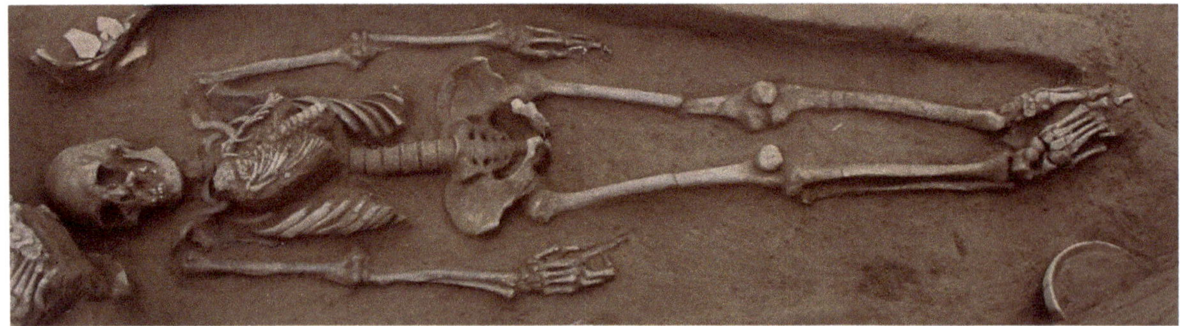

MB07H2M10

MB07H2M11:

Square D1; level 18(III). Only the calvarium of this individual, an adult of indeterminate sex and age, was exposed projecting from the western bulk of the square, as well as one half of a small globular pot. Unexcavated.

MB07H2M12:

Square C2, D2; level 19(III). A mature adult female 50+ years old. *In situ* orientation is slightly northeast-southwest, position is supine, facing north. A single globular pot is directly right of the right elbow (not illustrated in Appendix 2).

MB07H2M12

MB07H2M13:
Square D3; level 19(III). A subadult 4 years +/- 9 months old. *In situ* orientation is east-west, position is supine, facing upwards (although the skull is fragmented). Grave goods include two small, cylindrical, black nephrite beads at the sternal ends of the clavicles (one each), a third black nephrite bead just superior to the left shoulder, and a cluster of three cowrie shells just medial to the left wrist.

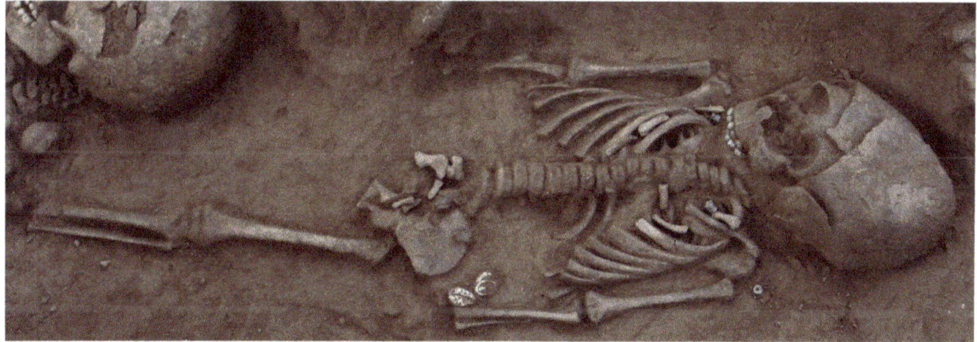

MB07H2M13

MB07H2M14:
Square D3; level 19(III). Neonate. *In situ* orientation is northeast-southwest, position is supine, facing upwards. One medium-sized globular pot is just left of the head (not illustrated in Appendix 2).

MB07H2M14

MB07H2M15:

Square A3; level 19(III). A subadult 4 years +/- 5 months old. *In situ* orientation is east-west, position is supine, facing north. Unusually, the hands are supinated and the feet are close together. No grave goods recorded.

MB07H2M15

MB07H2M16:

Square D3; level 19(III). A 18 month +/- 5 months old. *In situ* orientation is east-west, position is supine, facing south. A small cross-ribbed globular pot is to the right, beyond the cranium and approximately one-half of a burnished, Phung Nguyen motif decorated plate is positioned between the legs (not illustrated). A bivalve shell is underneath the left hand, and cluster of cowrie shells are just medial to the left wrist. Finally, eight small cylindrical black nephrite beads are scattered around the neck region.

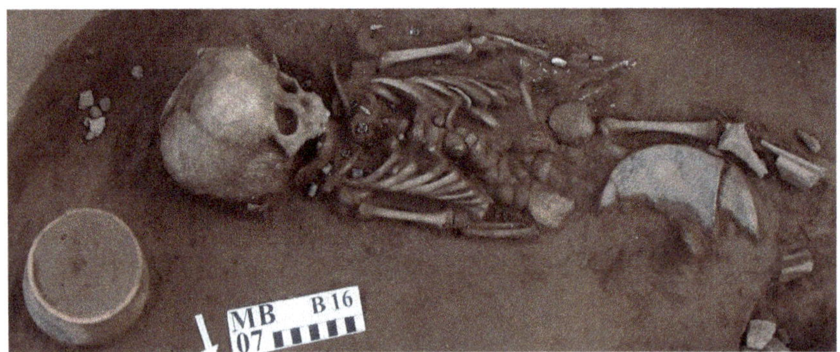

MB07H2M16

MB07H2M17:

Square C3, D3; level 19(III). A subadult 12-18 years old. *In situ* orientation is east-west, burial position is supine, and facing direction is north. Grave goods consist of a small cross-ribbed globular pot just to the right, beyond the cranium, and a small, cylindrical, green nephrite bead just lateral of L3.

MB07H2M17

MB07H2M18:

Square F2, E3; level 19(III). A young adult female > 18 years < 24 years old. *In situ* orientation is slightly northeast-southwest, position is supine, facing up. Grave goods include a bivalve shell in the left hand, a small globular pot to the right of the cranium and a cup with a Phung Nguyen style "S" decorative motif. It does not possess a clear ring foot base, and it is located to the right of, and adjacent to, the globular pot (not illustrated in Appendix 2).

MB07H2M18

MB07H2M19:

Square F4, E4, D4; level 19(III). A mature adult male > 20 years < 24 years old. Many elements have a greenish-black mineralised staining. *In situ* orientation is east-west, position is supine, facing south (judged by the *in situ* position of the mandible). At least six, possibly seven, vessels are associated with this individual. Two footed vessels, one with a Phung Nguyen style 'leaf' motif, and the other cord marked, are located at the right shoulder. A large globular pot is right of the right knee. At least two other globular pots are placed together at the feet. Another is 18cm beyond and to the left of the left foot. Bivalve shells are present underneath both hands, and a small, cylindrical nephrite bead is underneath the right 3rd rib.

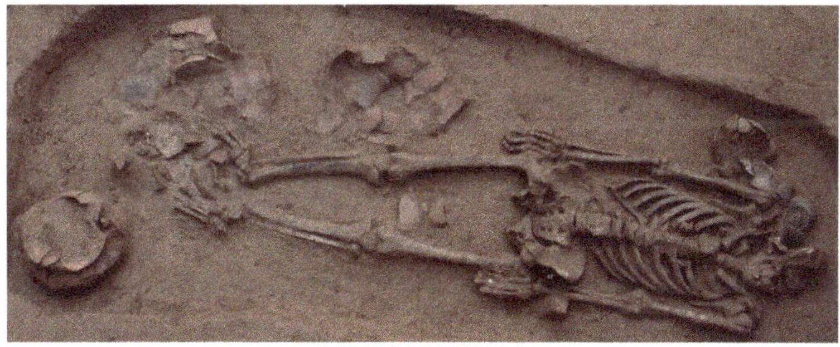

MB07H2M19

MB07H2M20:

Square C1; level 19(III). A subadult 6 months +/- 2 months old. *In situ* orientation is east-west, position is supine, facing upwards. No grave goods recovered.

MB07H2M20

MB07H2M21:

Square A2; level 19(III). A subadult 9 months +/- 2 months old. *In situ* orientation is approximately east-west, position is supine, facing direction is indeterminate. No grave goods were recovered.

MB07H2M21

MB07H2M22:

Square B3, B4, C3, D3; level 19-20(III). An adult female approximately 30-39 years old. *In situ* orientation is east-west, position is supine, facing north. Two ribbed globular pots (one cross-ribbed and one mostly parallel) are located beyond the cranium and left of the left distal fibula (one each). One green nephrite bead is recovered from between the 5th and 6th right rib, while two small black nephrite beads are recovered underneath the right mandibular ramus.

MB07H2M22

MB07H2M23:

Square A1; level 19(III). A partially uncovered cranium of an adult was exposed projecting from the southern bulk of this square. Unexcavated and no grave goods noted.

MB07H2M24:

B1, C1, D1; level 19-20(III). A mature adult female (?) approximately 40-49 years old. *In situ* orientation is east-west, position is supine (with legs close together), facing south. A large cross-ribbed globular pot is to the right of the cranium, while another such vessel, also crushed (not illustrated in Appendix 2), is just beyond and to the left of the cranium. Three small, cylindrical, black nephrite beads are present between the right 5th and 6th rib (one), and on the sternum (two).

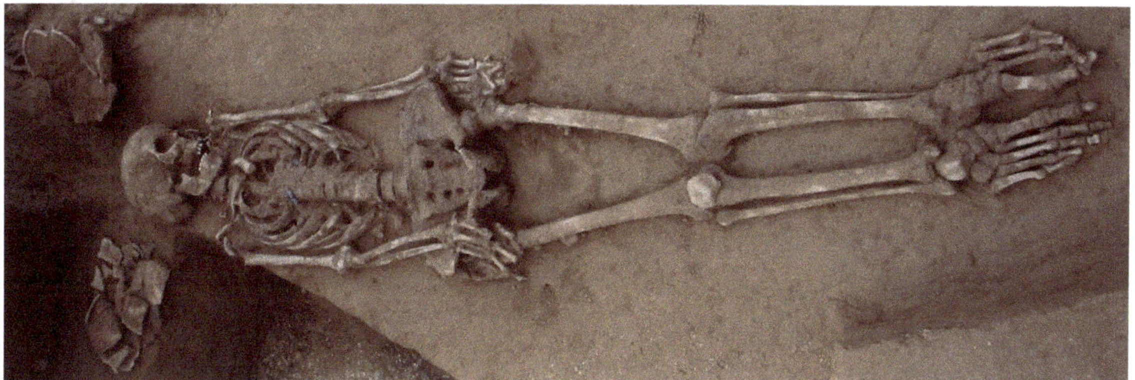

MB07H2M24

MB07H2M25:

Square A6; level 19(III). Only the calvarium of this individual, an adult, was exposed projecting from the southern bulk of this square. Unexcavated and no grave goods noted.

MB07H2M26:

Square D1, C1; level 19(III). An infant 18 months +/- 5 months old. *In situ* orientation is northeast-southwest, position is supine, facing south. Two redware vessels are nested together just beyond and to the left of the cranium. One is a small bowl (not illustrated), and the other (outer) vessel is a cross-ribbed globular pot, soot blackened, with parallel notches around the rim. Another small cross-ribbed globular pot is to the left of the right tibia/fibula.

MB07H2M26

MB07H2M27:

Square A4, B4, C4; level 21(III). An adult male 30-39 years old. *In situ* orientation is northeast-southwest, position is supine, facing south. The calcaneus, talus, 2nd-4th metatarsal, and a cluster of tarsal phalanges from another adult individual were recovered beyond the cranium, between it and a small globular pot. The fragments of another globular pot are covering the right elbow. A string of five, articulated, black nephrite beads (suggestive of a necklace) extend in a line directly below the mandible. A sixth black nephrite bead is located immediately laterally of the left 10th rib.

MB07H2M27

MB07H2M28:

Square F1, F2; level 18(III). Neonate. A small section of the calvarium was removed during the interment of H2M10, and the large redware, globular pot resting on the top right of the cranium, is associated with that burial. *In situ* orientation is east-west (head to the west), position is supine, facing up. A small, upturned, undecorated, footed bowl is leaning against the right side of the face (not illustrated in Appendix 2), and two fragile bivalve shells are recovered from where the hands would have rested if preserved.

MB07H2M28

MB07H2M29:

Square A1, B1; level 21(III). A subadult 7 years +/- 9 months old. *In situ* orientation is east-west, position is supine, facing direction is indeterminate as the cranium was not preserved. No clear grave goods recorded, however, four sherds from what appears to be a bowl are located directly on top of the sacrum.

MB07H2M29

MB07H2M30:

Square F1, F2; level 21(III). An adult male 30-39 years old. Windows into the southern and eastern bulks were required to excavate the head and foot regions. *In situ* orientation is east-west, position is supine, facing upwards (slightly north). Numerous grave goods are associated with this burial. Two cowrie shells are positioned on the skull, one in the left orbit, and the other next to the right. Several black nephrite cylindrical beads (at least five) are located between the ribs in the sternal region, as are numerous very small grey nephrite cut-disc beads. A cluster of cowrie shells lie underneath both wrists, and a bivalve shell is held in the left hand. A large black nephrite Phung Nguyen style T-section bracelet encircles the left wrist. Covering the left proximal tibia is a vertically crushed large globular pot, and the fragments of a burnished and very finely cord marked Phung Nguyen style bowl, once on a pedestal. Just left to the left knee is a complete pedestalled plate with incised decoration infilled with rows of punctate stamping. Another small globular pot is beyond the cranium (not visible in the photo), and a very tall footed and cross-ribbed jar is recovered further to the right of the cranium (not visible in the photo), within the excavated window.

MB07H2M30

MB07H2M31:

C3, D3; level 21(III). A subadult 4 years +/- 9 months old. *In situ* orientation is northeast-southwest, position is supine, facing south. Traces of a bivalve shell were recorded in association with the right hand, but no other clear grave goods were present.

MB07H2M31

MB07H2M32:

Square D4, E4, E3, F3, E2, F2; level 21(III). A young adult male < 25 years old. This is the most elaborate burial recovered from Man Bac to date. *In situ* orientation is slightly northeast-southwest, position is supine, facing up. Numerous grave goods are associated with this individual. A bivalve shell lies on top of the left os coxa in the acetabular region. Two clusters of cowrie shells are approximately where the wrists would be given complete articulation. A black nephrite T-sectioned Phung Nguyen style bracelet is *in situ* around the right proximal ulna/radius. At least six black nephrite cylindrical beads are clustered around the neck region, as are two small clusters of very small grey nephrite discs, one group of which remain articulated as a string. Three vessels surround the head. The first, adjacent to the cranium to its left, is a small cross-ribbed globular pot. Directly right of the cranium are fragments of a second, as well as an incised and stamped pedestalled plate. Substantially further east of the cranium is a cluster of five vessels. These included a pedestalled plate, a cross-ribbed globular pot, two carinated footed vessels with everted rims and incised decoration over parallel ribbing (see descriptions in Appendix 2), and a tall footed jar. Even though this cluster is somewhat distant from the body, there are no other burials nearby to which they would more logically belong.

MB07H2M32

Appendix 2
The Man Bac Burial Pottery – An Illustrated Corpus of the Whole Vessels from the Burials in Cultural Unit III.

Nguyen Kim Dung[1], Mariko Yamagata[2], Shinya Watanabe [2] and Peter Bellwood[3]

[1]*The Vietnamese Institute of Archaeology*
[2]*Department of Archaeology, Waseda University, Japan*
[3]*School of Archaeology and Anthropology, Australian National University*

Although the archaeology of Man Bac will be dealt with in separate publications, the opportunity is taken here to illustrate the complete vessels excavated in association with the burials recovered in 2005 and 2007. The vessel photographs were taken by Nguyen Kim Dung in the Institute of Archaeology in Hanoi and compiled into Adobe Illustrator figures by Peter Bellwood. The vessel measurements were recorded by Peter Bellwood and Watanabe Shinya. They are numbered as in the previous burial descriptions by Damien Huffer and Trinh Hoang Hiep. Only the vessels from the burials in cultural unit III are illustrated here.

Microscopic analysis of soil samples from grave fills indicates that all burial pits in cultural unit III were filled and sealed before the accumulation of the cultural unit II above. The grave fills contain no charcoal or other evidence of human habitation activity, and are basically sterile. Hence, the unit III burials were not dug down from unit II above, but predate it. The sterility of the grave fills also suggests that the seven radiocarbon dates from the site, all on charcoal and listed in Table 1, postdate the burials and their contained pottery offerings. The burial assemblage, therefore, should date to before 1900 BC.

Table 1. Radiocarbon dates from Man Bac, with calibrated ages.

Lab. number	Provenance	Material	$\delta^{13}C$ (‰)	C14 Age BP	Calibrated age BC (OxCal 4.1)
*IAAA-102758	2007 Level 7	charcoal	-25.88 ± 0.45	3370 ± 30	1745-1538 (95.4%)
IAAA-102759	2007 Level 11	charcoal	-23.93 ± 0.43	3380 ± 30	1751-1608 (94.3%)
IAAA-102760	2007 Level 14	charcoal	-27.35 ± 0.57	3490 ± 30	1895-1700 (95.4%)
IAAA-102761	2007 Level 17	charcoal	-24.87 ± 0.55	3560 ± 30	2016-1775 (95.4%)
**AA-69831	2005 Level 18	charcoal	-26.0	3393 ± 36	1775-1608 (92.6%)
AA-69832	2005 Level 18	charcoal	-27.7	3341 ± 38	1737-1524 (95.4%)
IAAA-102762	2007 Level 19	charcoal	-27.27 ± 0.54	3570 ± 30	1984-1876 (80.6%)

* Institute of Accelerator Analysis Ltd, Kanagawa, Japan (provided by Hirofumi Matsumura)
** Institute of Archaeology, Hanoi: see Khao Co Hoc, 2007, volume 1, page 101.

The burial vessels from Man Bac consist mostly of small globular pots with everted rims, impressed with grooves made in one or two directions - they are only rarely cord-marked. Usually, it is quite difficult to decide whether the grooves were beaten on with a wooden paddle carved with parallel lines, or made by dragging a comb of some kind across the surface. Decoration of this type is referred to below as "ribbed", either parallel or crossed (i.e. made in one or two directions). None of the impression or combing convincingly resembles basketry, but one vessel has small square impressions that resemble a net, applied in relief (2005 M19).

Most of the pedestalled dishes and two large footed beakers with flaring sides have combed and stamped decoration of the style referred to in northern Vietnam as "Phung Nguyen" (e.g. Pham 2004: Fig. 8.11), represented at perhaps its highest levels of skill in the site of Xom Ren, about 100 km upstream from Hanoi in Phu Tho Province in the Red River valley (Han 2009). Most of the Man Bac vessels, including the globular pots, have tempers of laterite sand or calcareous sand (probably derived from shells and limestone) that are presumed to be of local origin, but several of the "Phung Nguyen" vessels have much finer tempers that suggest exotic origins, possibly from further inland in the Red River valley. Such observations will need to be confirmed by further ceramic research.

In terms of regional comparisons, the Man Bac unit III burial pottery appears to be most closely related to contemporary assemblages to the north, in Guangxi and Guangdong, rather than to more southerly assemblages from the lower Mekong valley or northeastern Thailand (Carmen Sarjeant is currently examining these regional relationships for her PhD at ANU). For instance, the globular vessels with everted rims, the presences of ribbed/combed and cord marked decoration, the uses of incision and combing to delineate curvilinear motifs, and the absence of complex geometrically stamped patterns, all combine to align the Man Bac pottery closely with that from Assemblage F in the site of Sham Wan, Lamma Island, Hong Kong (Tsui and Meacham 1978:127-33). The Sham Wan C14 dates suggest a date in the third millennium BC for Assemblage F, with the more complex geometrically stamped wares that belong to the later Assemblage C dating to circa 2000 BC or later. Man Bac shows little affinity with Sham Wan Assemblage C, and lacks the complex forms of geometric stamping such as lozenges, herring-bone patterns, circles and rectangular meanders that characterise it.

However, it should also be noted that the style of incised and stamped decoration that is referred to as Phung Nguyen in northern Vietnam has parallels across very large areas of Mainland Southeast Asia and southern China (Rispoli 2008), including the lower Mekong valley and northeastern Thailand. In this regard, population origins and the movements of ceramic style features need not always have been in unison. Research is ongoing on these issues.

In the photographs that follow, each vessel is shown sufficiently large for any surface decoration or other details to be visible. This means that the vessels are not illustrated to a single scale. Heights and mouth (rim) diameters are indicated by the letters H and D, followed by the dimensions in centimetres (in 0.5cm intervals). Where red slipping is clearly visible it is mentioned, and many vessels have faint traces of red slip, suggesting that the original proportion of this type of surface decoration was much higher than is visible now.

LITERATURE CITED

Han Van Khan 2009. *Xom Ren.* Hanoi: Nha Xuat Ban Dai Hoc Quoc Gia Ha Noi. (in Vietnamese)

Pham Minh Huyen 2004. The Metal Age in the north of Vietnam. In I. Glover and P. Bellwood (eds), *Southeast Asia,* pp. 189-201. Abingdon (UK): RoutledgeCurzon.

Rispoli, F. 2008. The incised and impressed pottery of Mainland Southeast Asia: following the paths of Neolithization. *East and West* 57:235-304.

Tsui Yun-chung and W. Meacham. 1978. Pottery. In W. Meacham (ed.), *Sham Wan, Lamma Island,* pp. 127-82. Hong Kong: Hong Kong Archaeological Society Journal Monograph III.

Man Bac pottery from M1-M18 2005 season.

All illustrated vessels are from the 2005 excavation.

2005 M2. Undecorated direct rimmed bowl, no foot. H 5.2, D 11.
2005 M3 (1). Globular pot with everted rim, body cord marked vertically, red-slipped inside and outside rim, H 8.5, D 12.
2005 M3 (2). Globular pot with everted rim, body cord marked vertically, red-slipped inside and outside rim, H 6.5, D 10.5.
2005 M5. Globular pot with everted and angled rim, body decorated with crossed ribbing. H 15.5, D 14.
2005 M8. Globular pot with everted rim, body decorated with crossed ribbing. H 8.5, D 10.
2005 M9 (1). Globular pot with everted and notched rim, body cord marked vertically. H 12, D 14.
2005 M9 (2). Globular pot with everted rim, body decorated with crossed ribbing. H 11, D 11.5.
2005 M10 (1). Globular pot with everted and notched rim, body decorated with parallel ribbing. H 13.5, D 14.
2005 M10 (2). Globular pot with everted and notched rim, body decorated with parallel ribbing, traces of red slip. H 10.5, D 14. (very similar to the previous one – made by the same person?)
2005 M10 (3). Globular pot with everted rim, body decorated with crossed ribbing. Traces of red slip. H 14.5, D 13.
2005 M10 (4). Ring foot with central hole.
2005 M10 (5). Small plain ware cup. H 5, D 8.5.
2005 M10 (6). Globular pot with everted rim, body decorated with parallel ribbing. H 11.5, D 12.
2005 M11 (1). Globular pot with everted rim, body decorated with crossed ribbing. H 12, D 14.
2005 M11 (2). Untanged stone adze, trapezoidal cross-section, rock unidentified, 5.3 cm long, 4.4 wide at bevel.
2005 M12. Globular pot with everted rim, body decorated with crossed ribbing. H 11.5, D 12.
2005 M13 (1). Globular pot with everted red-slipped rim, body decorated with crossed ribbing. H 18, D 14.
2005 M13 (2). Globular pot with everted red-slipped rim, body decorated with parallel ribbing. H 15.5, D 15.
2005 M14. Globular pot with everted red-slipped rim, no decoration, but large flat paddle impressions are visible on the body exterior. H 10, D 12.5.
2005 M15. Direct rimmed bowl with coarse vertical cord marking. H 6.5, D 10.
2005 M17. Three flat-based beakers (one fragmentary) with flaring bodies and lips, decorated with horizontal incised zones filled with punctations and linear impressions that could be from the edge of a bivalve shell (or some form of multi-toothed stamping tool). (1) H 7, D 11; (2) H ?, D 12.5; (3) H 7, D 11. All have local laterite tempers. See M28 and 32 below for parallels.
2005 M18. Globular pot with everted and notched rim, decorated with vertical cord marking. H 10, D 16.

Man Bac pottery from M19-M28 2005 season.

All illustrated vessels are from the 2005 excavation.

2005 M19. Globular pot with everted and notched rim, red-slipped inside, and decorated with an unusual relief net pattern. H 8.5, D 13.5.

2005 M20 (1). Globular pot with everted rim, body decorated with crossed ribbing. H 20.5, D 14.

2005 M20 (2). Globular pot with everted rim, body decorated with crossed ribbing. H 11, D 11.5.

2005 M20 (3). Undecorated open bowl, no pedestal. H not recorded, D 12.

2005 M21. Footed plain vessel, crudely made. H 11, D 9.5.

2005 M23. Globular pot with everted rim, body decorated with crossed ribbing. H 6, D 10.5.

2005 M24 (1). Globular pot with everted and notched rim, body decorated with coarse vertical cord marking. H 7, D 9.5.

2005 M24 (2). Footed dish with notched lip but no visible body decoration. H not measured, D 14.5.

2005 M25. Footed dish with a perforation in the pedestal (photo angled to show decoration). The outside of the dish was first decorated with finely combed lines crossing at an acute vertical angle, overlain by three horizontal burnished bands. Simple curvilinear burnishing occupies the central delineated field. H not measured, D 15 approx.

2005 M26. Globular pot with everted and notched rim, body decorated with crossed ribbing and red-slipped. H 12.5, D 12.5.

2005 M28 (1). Unique barrel-shaped vessel. This has a low ring foot, a high everted rim, and a barrel shaped body decorated with incised zones infilled with the same kind of stamping as the flaring cups from M17 and M28 (above). H 11.

2005 M28 (2). The small pedestalled dish, undecorated. H 5, D 10.

2005 M28 (3). 3 beakers with flaring sides and low ring feet, one fragmentary. These have horizontal zones of decoration formed by incised lines filled with punctations and linear impressions that could be from the edge of a bivalve shell (or some form of multi-toothed stamping tool). In this regard they are very similar to the three beakers in M17, although the latter do not have ring feet. It is very likely that all were made in the same workshop, together with the barrel-shaped vessel described above. Each vessel is about 11 cm high.

2005 M28 (4). Globular pot with everted rim. Very faint traces of ribbing are still visible, but presumably erased before firing. H 11.5, D 12. The many tiny surface holes were perhaps formed by the falling out of laterite sand temper grains.

Man Bac pottery from M29-M35 2005 season.

All illustrated vessels are from the 2005 excavation.

2005 M29. This burial produced five very similar globular everted rimmed vessels with crossed ribbing, one with a notched lip and animal bone contents. (1) H 15.5, D 13.5; (2) H 9, D 10.5; (3) H 12, D 13; (4) with animal bones, H 7.5, D 12.5; (5) H 10, D 10.

2005 M29 (6). T-sectioned green nephrite bracelet, broken in antiquity and mended by the drilling of at least one hole to take a lashing of some kind. External diameter 9.7 (the scale is specific to this item, not to the pots).

2005 M30. Globular pot with everted rim, body decorated with vertical cord marking. H 11, D 13.

2005 M31 (1,2). Two globular pot with everted rims, decorated with crossed ribbing. (1) has a definitely red-slipped interior, visible in the photo. (1) H 11.5, D 13; (2) H 12, D 18.

2005 M31 (3). Open pedestalled dish with carinated profile, decorated below the carination and on the pedestal with fine cord marking. H not measured, D 17.5.

2005 M32 (1). Three footed beakers with flaring sides and complex contours (one carinated), with fragments of a fourth. Decoration is all incised, in various simple horizontal and vertical combinations, often with three or four parallel lines. There is no stamped decoration, and the laterite sand tempers suggest local manufacture. These beakers are similar in concept to those in M17 and M28, but differ in the details. The two larger ones are 12.5 cm high and 9 cm in mouth diameter.

2005 M32 (2). Also shown is a polished and untanged rectangular-sectioned stone adze.

2005 M33. Globular pot with everted rim, body decorated with crossed ribbing, red-slipped. H 14.5, D 13.5.

2005 M35. Globular pot with everted rim, body decorated with crossed ribbing. H 16, D 14.

Man Bac pottery from H1 2007 season.

All illustrated vessels are from the 2007 excavation, H1 (trench 1).

2007 H1 M3. Globular pot with everted rim, body decorated with crossed ribbing. H 20, D 18.

2007 H1 M5. A unique pedestalled dish decorated with horizontal incised zones that contain transverse bands of incision or punctate stamping. It is unclear which is the pedestal and which is the dish, but it is interesting that the photo of this burial in Appendix 1 shows the carinated smaller dish uppermost, with the larger rounded one beneath as a pedestal. H 13, D (at upper rim) 20.5.

2007 H1 M6 Globular pot with everted rim, body decorated with crossed ribbing. H 14.5, D 12.5.

2007 H1 M8. Globular pot with vertical rim, decorated with vertical cord marking. H 10 D 13.

2007 H1 M9 (1). Pedestalled pot with carinated upper profile and fine combing in two directions on the body. H 10.5, D 16.

2007 H1 M9 (2). Globular pot with everted rim, body decorated with mainly parallel ribbing. H 12, D 13.

2007 H1 M10. 6 vessels in total, all shown in the composite photograph, but only four are shown individually. (1) is a tall shouldered pot with everted rim and low ring foot, cross-ribbed, with fine bands of vertical incision on the shoulder. H 23, D 18.

2007 H1 M10 (2). Four unfooted globular pots with everted rims have cross-ribbed decoration in three cases, and parallel and fairly horizontal ribbing in one. All are shown in the composite photograph, and the largest (D 17) and the smallest (H 13, D 15) are shown individually as 10(2).

2007 H1 M10 (3). Presumed pedestal base with the bottom of a narrow diameter cylindrical pedestal still attached. D 21. The convex surface of this pedestal base is decorated with finely combed cross-hatching, over which circular and curvilinear bands of burnishing are delineated by parallel incision. The fine fabric and decoration of this piece suggests importation from a Phung Nguyen source, perhaps further inland from Man Bac.

Man Bac pottery from M2-M18 H2 2007 season.

All illustrated vessels are from the 2007 excavation, H2 (trench 2).

2007 H2 M2. 4 vessels, shown together in one composite photograph. Three are globular pots with everted rims, two cross-ribbed and one horizontally parallel ribbed. The other vessel (1) is a tall footed jar, decorated with cross ribbing (H 43, D 20).

2007 H2 M5. 2 vessels; (1) is a globular pot with cross-ribbed decoration (H 20, D 19), (2) is a very small undecorated cup (H 6, D 10).

2007 H2 M6 (1,2). 2 globular pots with everted rims and vertical cord marked decoration, possibly made by the same potter. M6(1): H 9.5, D 14. M6(2): H 11, D 12.5.

2007 H2 M7 (1,2). 2 globular pots with everted rims. (1) is cross-ribbed, and (2) is vertically cord marked.

2007 H2 M8. Sherd of a possible pedestal or bowl rim decorated with underlying crossed combing beneath horizontal bands and curvilinear motifs formed by smoothing between parallel incised lines. The decoration is very similar to that on the pedestal base in 2007 H1 M10.

2007 H2 M9. Small cup with traces of vertical parallel ribbing in the slightly constricted neck. H 75, D 10.5.

2007 H2 M10 (1). Footed beaker-shaped vessel with flaring upper contours. The decoration comprises zones delineated by parallel incision and filled by parallel lines of rectangular "dentate" stamping. The fine temper suggests that this vessel is an import. H 17.5, D 22.5.

2007 H2 M10 (2). Globular pot with everted rim and notched lip, decorated with what appears to be cord marking applied in two directions. This vessel is unique at Man Bac, both in shape and decoration, and may be an import. H 16, D 14.

2007 H2 M16. Globular pot with cross-ribbed decoration, mostly horizontal in orientation. H 10, D 11.

2007 H2 M17. Globular pot with cross-ribbed decoration, mostly horizontal in orientation. H 14, D 13.5.

Man Bac pottery from M19-M27 H2 2007 season.

All illustrated vessels are from the 2007 excavation, H2 (trench 2).

2007 H2 M19 (1-6). 6 vessels, shown as a group in the composite photograph in the previous figure. Three (1, 2 and 3) are globular pots with everted rims and cross ribbing, and (4) has a notched lip and sloping parallel ribbing.

2007 H2 M19 (5). Globular pot with an unusual inverted rim on a ring foot, with vertical cord marked decoration on the body, and finely combed lines between lines of incision and punctation on the shoulder. H 13.5, D 9.

2007 H2 M19 (6). Footed beaker-shaped vessel with flaring upper contours, like that in H2 M10 (above). But in this case the filling of the horizontal and curvilinear decorative zones is not dentate stamping, but finely crossed incision. The lenticular motifs around the upper part of the pot are filled with lines of punctation. This also appears to be an imported vessel, with fine temper. H 10.5, D 10.5.

2007 H2 M22. Two globular pots with everted rims and cross-ribbed decoration. In (2) the decoration is predominantly horizontal. M22(1): H 17, D 16. M22(2): H 15, D 15.

2007 H2 M24. Globular pot with everted rim and cross-ribbed decoration.

2007 H2 M26. 2 globular pots with everted rims and cross-ribbed decoration. (1) has a notched lip. M26(1): H 13, D 17. M26(2): H17, D 15.

2007 H2 M27. 2 globular pots with everted rims and cross-ribbed decoration. M27(1): H 11.5, D 13.5. M27(2): H 11, D 14.5.

Man Bac pottery from M30-M32 H2 2007 season.

All illustrated vessels are from the 2007 excavation, H2 (trench 2).

2007 H2 M30 (1, 2). 2 globular pots with everted rims and cross-ribbed decoration. M30(1): H 11, D 18.

2007 H2 M30 (3). Dish, once on a pedestal, with very fine Phung Nguyen decoration comprising burnished and incised circular and curvilinear motifs over extremely fine background cord marking. The fabric suggests that this is an import. D 21.

2007 H2 M30 (4). Pedestalled dish with a perforation at the base of the pedestal. The pedestal has incised zonal decoration infilled with regular punctation, whereas the infilling on the dish is of cross-combed lines. H 10, D 16.5.

2007 H2 M30 (5). Tall footed jar with everted rim and very regular cross-ribbed decoration, rather lozenge-like in shape. H 42.5, D 21.5.

2007 H2 M32. This burial, the richest excavated so far at Man Bac, had 8 vessels. Three (1-3) are globular pots with everted rims and cross-ribbed decoration. M32(1): H 13, D 13.5. M32(2): H 14.5, D 13. M32(3): H?, D 12.

2007 H2 M32 (4). Another tall footed jar has similar cross-ribbed decoration to the very similar vessel from burial M30 (item 5) above. Perhaps both came from the same workshop, together with that from 2007 M2. H 25.5, D 13.

2007 H2 M32 (5). Pedestalled dish similar to that from M30 (item 3, above). This pedestal also has incised zonal decoration infilled with punctation, while the infilling on the dish is of cross-combed lines. Both of these dishes-on-stands could have come from the same workshop, and the fabrics appear to be local. H 10, D 17.

2007 H2 M32 (6). Another pedestalled dish, rather roughly decorated with incised and combed motifs similar to the above, but with no curvilinear designs. Again, the fabric appears to be local. H 12.5, D 18.5.

2007 H2 M32 (7,8). Two footed and carinated jars with everted rims. These have cross-ribbed decoration below the carination in each case. Above the carination, the larger jar (7) has non-filled zonal decoration outlined by incised lines, over fine crossed or parallel combing that extends upwards to cover the outside of the rim. The smaller jar (8) has a cruder version of the same idea, with much coarser parallel combing and no rim decoration. Again, the fabrics appear to be local.
M32(7): H 21.5, D 22. M32(8) H 15.5, D 15.

Appendix 3
Individual Descriptions of Human Skeletal Remains at Man Bac: 2005 and 2007 Series

Hirofumi Matsumura[1], Nguyen Lan Cuong[2] and Damien G. Huffer[3]

[1]*Department of Anatomy, Sapporo Medical University, Japan*
[2]*The Vietnamese Institute of Archaeology*
[3]*The Australian National University*

In Appendix 1 sex, age-at-death and mortuary variables were described for each burial. In this chapter the preservation of each burial from the 2004/5 and 2007 seasons are summarised (with photographs provided for the best preserved sets of remains). For adult specimens the cranial morphology is also described and when sexually dimorphic characters are discussed, Acsádi and Nemeskéri's (1970) scoring system is used. Dental occlusal wear was also recorded: Smith's (1984) system for anterior teeth (incisors, canines and premolars); and Scott's (1979) system for molars. For the dentition, tooth presence and condition is recorded via standard recording protocols, as per the following example.

0	M2	M1	P2	P1	C	X	I1	I1	X	C	P1	P2	M1	M2	M3
M3	M2	M1	P2	P1	C	/	X	X	/	C	P1	P2	M1	M2	Δ

I: incisor, C: canine, P: premolar, M: molar
0: postmortem tooth loss, Δ: only tooth root remaining
X: antemortem tooth loss,
/: neither alveolus nor tooth assessable/present

HUMAN REMAINS FROM THE 2005 SERIES

MB05M1:

This burial consists of numerous small fragments of infracranial bones (ribs, the distal ends and epiphyses of both femora and humeri, etc.), as well as the bifurcated skull of one individual. The deciduous teeth are present and completely erupted, except for the second deciduous molars with developing roots.

MB05M2:

The nearly complete skeleton of a neonate. The skeleton is somewhat disarticulated, the cranium is very thin, and no dentition has erupted. Most of the post cranial bones are well preserved. The skull includes a section of the right frontal, both parietals, and the mandible and maxilla. The deciduous mandibular first molars and central incisors are present within the alveolus.

MB05M3:

The cranium is in a moderate state of preservation. The mandible and a portion of maxillary alveoli are present and nearly complete. The fragile facial skeleton is highly fragmented. Almost the entire post cranial skeleton is present without severe damage. All deciduous tooth crowns were completely formed but still within the alveoli.

MB05M4:

Only several large fragments of the calvaria and a diaphyseal section of one humeral shaft remain, together with two small, worn deciduous teeth; the lower canine and first molar. The dental roots are developing, and the eruption of the deciduous molar was incomplete.

MB05M5:

Most of the post-cranial skeleton, including the cranium, is nearly complete apart from the fragmented zygomatico-facial region, the left femur and both feet. The second deciduous molars have begun to erupt.

MB05M6:

This burial consists solely of the right forearm, both os coxae, both femora, tibiae and fibulae.

MB05M7:

A partially preserved skeleton with only the cranium, right arm and leg elements, right os coxa, a few ribs and vertebrae preserved. None of the deciduous teeth have erupted, but are partially formed within their alveoli. The cranium was fragmented into small pieces.

MB05M8:

A partially preserved skeleton. The maxillary region, all dentition, the feet, the entire vertebral column, all ribs, the left arm elements, left clavicle, and the right arm elements (except for the distal humerus) are all missing.

MB05M9:

Skull: almost complete. The forehead of the frontal bone is narrow and the frontal tubercle is clear. The parietal bones are not angled at the sagittal suture, and the frontal slopes backwards steeply. The supranasal suture is absent. The glabella region, superciliary arches and nasal root are flat. The orbital shape is rather round and supraorbital foramina exist on both sides. The zygomatic bones are laterally projecting. The temporal lines are not distinct on the parietal bones. The size of the mastoid process is moderate (score: 3). The external occipital protuberance is weakly protruding, while both superior and inferior nuchal lines are distinctive, and the nuchal plane is moderately rugged. The coronal, sagittal and lambdoidal sutures are not fused ectocranially. Internal suture synostosis is unknown due to soil filling the endocranium. The cranial index is 80.5 (Martin No. 8:1), indicating the cranium is brachyocephalic. The upper facial index is 51.9 (Martin No. 48:45), categorising the proportion of facial height to breadth as medium. The mandibular body is relatively robust, with well reflecting angles and developed pterygoid muscle attachment surfaces. The ramus has a strong posterior inclination. The mental eminence is moderately protruding (score: 3). Dental preservation:

0	M2	M1	P2	P1	C	X	I1	I1	Δ	C	P1	P2	M1	M2	X
M3	M2	M1	P2	P1	C	I2	I1	I1	I2	C	P1	P2	M1	M2	M3

There is incisor occlusal over-bite.

Postcranial skeleton: the remains are in a good state of preservation. Paired upper limb bones include nearly complete scapulae, humeri, radii and ulnae. The lower limb bones of the os coxae, femora, tibiae, fibulae and patellae of both sides are also well preserved. The deltoid muscle attachment area of the humerus is robust. The leg muscles are also well developed, determined from the pilastric form of the linea aspera of the femora, and the high degree of curvature of the femoral shafts. The vertebrae, sacrum and ribs also remain in good condition.

MB05M10:

Very good preservation; only missing the right os coxa, distal femur, hand, wrist and radius, all removed by an intrusive pit or post-hole. The deciduous molars show heavy wear, whereas the permanent teeth are unworn (score: 0-1). Dental preservation:

X	M2	M1	dm2	dm1	dc	0	I1	0	0	dc	dm1	dm2	M1	M2	X
X	M2	M1	dm2	dm1	dc	I2	I1	I1	I2	dc	dm1	dm2	M1	M2	X

MB05M11:

Skull: complete, with only the lateral portion of the right zygomatic bone being damaged. The width of the forehead is moderate and the frontal tubercle is relatively prominent. The parietal bones are slightly angled at the sagittal suture. The frontal is perpendicularly elevated at the forehead. Both supranasal sutures are absent. The glabella region and nasal root are flat, while the superciliary arches are prominently ridged. The cranial index is 78.6 (Martin No. 8:1), indicating a mesocephalic cranium. The facial skeleton is high (upper facial height: 70mm) and relatively narrow in proportions (the upper facial index: 53.8). Orbital shape is rather round with no supraorbital foramina on either side. The zygomatic bones are prominently projecting laterally. The temporal lines are distinct on the parietal bones. The mastoid process is of moderate size (score: 3). The external occipital protuberance is moderate and the crest is not marked (score: 3), while both the superior and inferior nuchal lines are distinct. The nuchal plane is very rugged. The coronal, sagittal and lambdoidal sutures are open ectocranially, while the state of endocranial synostosis is not assessable due to soil filling the endocranium. The mandibular body is rather robust, with well developed pterygoid muscle attachment surfaces. The angle is slightly reflected and the ramus is narrow with a marked posterior inclination. The mental eminence is moderate (score: 3). Dental preservation:

M3	M2	M1	P2	0	C	X	0	0	X	C	P1	P2	M1	M2	0
X	M2	M1	P2	P1	C	I2	I1	I1	I2	C	P1	P2	M1	M2	X

The tooth crown surfaces are flat due to attrition (anterior teeth: score 3, molars: score 4-5), except for the third molars (score: 0).

Postcranial skeleton: the post-cranial remains are in a good state of preservation, except for the lower leg and foot bones. Paired upper limb bones include the scapulae, humeri, radii and ulnae, which are all nearly complete. The humeral and femoral shafts are slender, and most muscle attachment areas are weakly developed. As for the lower limb bones, only the os coxae and femora are preserved well. The tibiae, fibulae, and feet are missing, likely removed by the later burial of MB05M18. The vertebrae, sacrum, and ribs are in good condition.

MB05M12:

Bone preservation is relatively good, with the legs below the patellae missing, also due to the later intrusion of 2005M18. The skull is complete and the metopic suture has not yet fused. All the deciduous teeth are fully erupted, except for the second molars which are partially erupted.

MB05M13:

Skull: good state of preservation and gracile overall. The forehead is relatively broad. The parietal bones are not angled at the sagittal suture. The frontal is perpendicularly elevated at the forehead, and the supranasal suture is absent. The glabella region, superciliary arches and nasal root are flat, while the orbital shape is rather square. Supraorbital foramina are not present on either side. The zygomatic bones display minimal lateral projection. The temporal lines are distinct on the parietal bones. The mastoid process is of a moderate size (score: 3). The external occipital protuberance is weak (score: 2), but both the superior and inferior nuchal lines are distinct, with a smooth nuchal plane. The coronal, sagittal and lambdoidal sutures are open ectocranially, but unassessed internally due to soil filling the endocranium. The mandibular body is also gracile, and the mental eminence is weak (score 2), although the chin protrudes prominently. The gonial region is not everted and the pterygoid muscle attachment is weakly developed. The rami slope backwards steeply. Dental preservation:

X	M2	M1	P2	P1	C	I2	I1	I1	I2	C	P1	P2	M1	M2	M3
M3	M2	M1	P2	P1	C	I2	I1	I1	I2	C	P1	P2	M1	M2	M3

There is incisor occlusal over-bite. The tooth crown surfaces are slightly worn (anterior teeth: score 2, molars: score 2), except for the third molars (score: 0), which were just beginning to erupt.

Postcranial skeleton: all the post cranial elements are in a good state of preservation. The upper limb bones include both scapulae, humeri, radii and ulnae and are nearly complete. The os coxae, femora and tibiae are nearly complete, but the fibulae are poorly preserved. The vertebrae, sacrum, and ribs are also well preserved.

MB05M14:

This burial consists of a near complete skeleton only missing the feet. Bone preservation is good. The skull was damaged only at the cranial base and full set of deciduous dentition has erupted.

MB05M15:

Skull: incomplete due to missing a major part of the frontal bone. The calvarium is dolichocephalic (cranial index: 70.5) and the forehead is wide. Metopism is absent, but the supranasal suture is present. The glabella region, superciliary arches and nasal root are flat. The facial skeleton is high (upper facial height: 71mm), and the upper facial index is 53.4 (Martin No. 48:45), categorising this cranium as moderate, but somewhat narrower than average. Orbital shape is round and supraorbital foramina are bilaterally absent. The zygomatic bones display minimal lateral projection, while the temporal lines are not distinct on the parietal bones. The size of the mastoid process is moderate (score: 3), the external occipital protuberance is weakly developed (score: 2), but both the superior and inferior nuchal lines are distinct, although the nuchal plane is relatively smooth. The sagittal and lambdoidal sutures are not synostosed ectocranially. The coronal suture is not assessable due to postmortem damage. Endocranial synostosis is not assessable due to soil filling the endocranium. The mandibular body is gracile and compact. The pterygoid muscle attachment surfaces are smooth and the angles are not sharp. The rami slope steeply backwards, and the mental eminence is weak (score: 1). Dental preservation:

0	M2	M1	0	0	0	0	0	0	0	C	P1	0	M1	M2	X
M3	M2	M1	P2	P1	C	I2	I1	I1	0	C	P1	P2	M1	M2	M3

The incisor occlusal pattern is unknown due to lack of the upper incisors. The tooth crown surfaces are moderately worn (anterior teeth: score 2, molars: score 3-4), except for the lower third molars (score: 0).

Postcranial skeleton: only the right clavicle and scapula, the upper right ribs and cervical vertebrae, a part of the right ulna, the right hand, the right femur, tibia and fibula and part of one foot are well preserved. Other bones are missing due to post-burial disturbance. The femoral and tibial shafts and are not robust.

MB05M16a:

Skull: partially preserved. The missing regions include the left parietal, cranial base, mandible and maxilla. This skull is characterised as possessing a robust, rugged and compact face. The calvarium is brachycephalic (cranial index: 85.6) and the forehead is narrow and slopes back steeply. The parietal bones are sharply angled at the sagittal suture. The supranasal suture is absent, the glabella region and medial portion of the superciliary arches prominently project and the nasal root is deeply concave. The orbital shape is square, with relatively straight margins and angled corners. The supraorbital foramen was absent on the right side at least; the left is unknown due to post-mortem damage. The zygomatic bones project prominently laterally and the temporal lines are distinct on the parietal bones. The size of the mastoid process is undetermined due to this region being absent. The external occipital protuberance is well developed, but not greatly protruding (score: 3). Both the superior and inferior nuchal lines are quite distinct, but the nuchal plane is relatively smooth. The coronal, sagittal and lambdoidal sutures are un-synostosed ectocranially. Dental preservation:

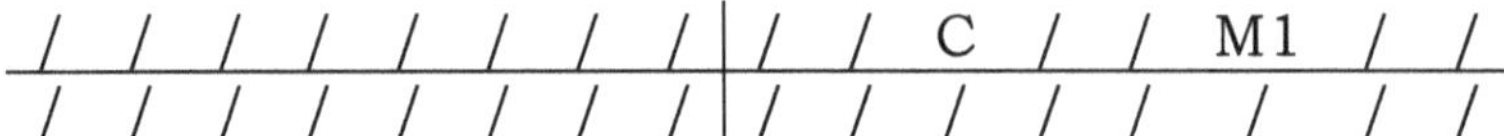

The tooth crown surfaces are lightly worn (score: 2-3).

Postcranial skeleton: the lower leg below the knee was missing due to later disturbance of this burial. Apart from these missing portions, preservation is moderate. The upper limbs and femora are almost complete. The os coxae and vertebrae also remain intact. The dimensions of the limbs are moderate with well developed muscle attachment areas, including the linea aspera of the femora, and the deltoid tuberosity of the humerus.

MB05M16b:

See entry in Appendix 1: 1 x right maxillary di^2 unerupted (crown only formed); 1 x right tibial diaphysis; 1 x fibula diaphysis, 1 x rib fragment, 2 x vertebral arches.

MB05M17:

A very poorly preserved assemblage of post-cranial fragments.

MB05M18:

A nearly complete skeleton missing feet and hands (as well as right radius and ulna). All deciduous teeth except for the second deciduous molars are fully erupted.

MB05M19:

A very poorly preserved skeleton. Only shaft segments of the femur, tibia, and fibula survived.

MB05M20:

Skull: very poor bone quality (extremely friable bone) meant that much of the cranium did not survive removal. Post-excavation the skull lacks its basicranium, including the occipital, sphenoid and temporal bones. The paired malar, zygomatic and nasal bones were also lost, but the cranium could be reconstructed. The cranium is gracile and the forehead is narrow. The parietal bones are weakly angled at the sagittal suture. The frontal has a slight posterior slope. The glabella region and superciliary arches are flat, and the orbital shape is rounded. Supraorbital foramina were absent bilaterally. The temporal lines are slightly distinct on the parietal bones. The mastoid process shows moderate size (score: 3) and the external occipital protuberance is moderately developed (score: 3). Although both the superior and inferior nuchal lines are distinct, the nuchal plane is smooth. The coronal, sagittal and lambdoidal sutures are un-synostosed both ecto and endocranially. The cranial index is 77.8 (Martin No. 8:1), indicating a mesocephalic cranium. The mandibular body is also gracile. The mental eminence is less than moderate (score 2), and the chin is pointed. The gonial region is not everted and the pterygoid muscle attachment surfaces are minimally developed. The wide rami is perpendicularly positioned. Dental preservation:

X	M2	M1	P2	0	C	X	0	I1	X	C	0	P2	M1	M2	M3
0	M2	M1	P2	0	C	I2	I1	I1	I2	C	P1	P2	M1	M2	M3

There is incisor occlusal edge-to-edge bite. The tooth crown surfaces are moderately worn (anterior teeth: score 4, molars: score 4-5), except for the third molars (score: 1).

Postcranial skeleton: the majority of postcranial remains are preserved and bone preservation is good, but condition is fragile, due to interment within the sand/shell matrix.

MB05M21:

The burial consists of a nearly complete skeleton. The long bones are nearly complete, with only the feet and right hand missing. The cranium was crushed laterally, but some portions of the calvarium have been reconstructed. The mandibular incisors are not yet fully erupted, while the maxillary central incisor crowns and deciduous molar crowns were calcified.

MB05M22:

Represented by fragments of a tibia and fibula, parts of an ulna, radius and hand, a femoral head, part of a patella and a few rib sections.

MB05M23:

This individual is represented by only the postcranial skeleton of a young infant. Bone condition is damp and fragile, but generally good. However, the skeleton was missing its head (due to intrusion of the grave shaft of MB05M24), hands, feet (except for a right tarsal), and the right humerus. No dentition was recovered.

MB05M24:

The bone condition is friable, but otherwise good. The skull and dentition are almost complete, with notable features being the eruption of the permanent maxillary and mandibular incisors and first molars, and the retention of the deciduous maxillary and mandibular molars and canines.

MB05M25:

An almost complete skeleton with good to excellent bone preservation. A full set of deciduous teeth has completely erupted and the permanent first molars are visible within their alveoli.

MB05M26:

Represented by a few cranial fragments and several deciduous teeth: left maxillary c, right maxillary i^1, right maxillary i^2 (see notes in Appendix 1).

MB05M27:

The right patella and fragments of the tibia and fibula of an otherwise unexcavated burial.

MB05M28:

A partially preserved (1/2 to 3/4 of elements present in situ). The remains are in poor condition, extremely friable and excavated preserved sections included the right and left humeral shafts and the left femoral shaft. Some fragments of the left maxilla and mandible were recovered, including a few heavily worn teeth.

MB05M29:

Skull: almost complete, although the nasal bones were missing. The cranium is very robust, which differentiates it from most of the other skulls in the Man Bac assemblage. The glabella region and superciliary arches are remarkably prominent and every muscle attachment area is strongly developed. The forehead is narrow and prominently slopes backwards. The parietal bones are not angled at the sagittal suture and the zygomatic bones are large and prominently projecting laterally. The facial region is low and wide and the orbital margins are square, while the nasal root is deeply concave. The degree of dolichocephalism (cranial index: 74.2) and alveolar prognathism are remarkable. Supraorbital foramina are present bilaterally and the temporal lines are distinct on the parietal bones. The mastoid process is quite large (score: 5) and the external occipital protuberance is massive and protruding (score: 4). Both the superior and inferior nuchal lines are well-defined, with the nuchal plane being very rugged. The coronal, sagittal and lambdoidal sutures are not ectocranially synostosed (score: 0), but were not assessable endocranially. The upper facial index is 48.6 (Martin No. 48:45), categorising facial proportions into the low and broad category. The mandibular body is very robust, but the mental eminence is moderate (score 3). The gonial region is strongly everted bilaterally, with well developed attachment surfaces for the pterygoid muscle. The rami are very wide and gently slope backwards. Dental preservation:

M3	M2	M1	P2	P1	C	I2	I1	I1	I2	C	P1	P2	M1	M2	M3
M3	M2	M1	P2	P1	C	I2	I1	I1	I2	C	P1	P2	M1	M2	M3

The incisor occlusal pattern is edge-to-edge type, and the tooth crown surfaces are remarkably worn (anterior teeth: score 5, molars: score 6-8).

Postcranial skeleton: very good preservation with only the distal half of the right tibia and fibula, and the right foot missing. The upper limb bones are robust, with well developed muscle attachment entheses.

MB05M30:

A nearly complete skeleton. The cranium was completely reconstructed, except for the parietal region. Full sets of long bones were recovered, but were damaged at the proximal and/or distal ends. Only the deciduous incisors have erupted to the level of half crown height.

MB05M31:

Skull: cranium (except for the base) and mandible are well preserved. The calvarium is brachyocephalic (cranial index: 88.6), with the forehead being somewhat narrow and vertical. The parietal bones are not angled at the sagittal suture and the supranasal suture is absent. The glabella region and superciliary arches are weakly ridged and the nasal root is rather flat. The facial region is very high (upper facial height: 74mm), but the facial index (50.7) is within the medium range. The orbital shape is round and supraorbital foramina are absent bilaterally. The zygomatic bones strongly project laterally. The temporal lines are not distinct on the parietal bones, the mastoid process is large (score: 4), and the external occipital protuberance is massive and prominent (score: 4). The superior and inferior nuchal lines are well-defined and the nuchal plane is rugged. The coronal and lambdoidal sutures have not synostosed either ecto or endocranially, while the sagittal suture is partially synostosed ecto and endocranially. The mandibular body is robust, but the mental eminence is weak (score 2). The gonial angles are weakly everted and the pterygoid muscle attachment surfaces are well developed. The rami slightly slope posteriorly. Dental preservation:

X	M2	M1	P2	P1	C	X	0	0	X	C	P1	P2	M1	M2	M3
M3	M2	M1	P2	P1	C	0	/	0	0	C	P1	P2	M1	M2	M3

The incisor occlusal pattern was indeterminate and the tooth crown surfaces are remarkably worn (anterior teeth: score 5, molars: score 5, except for the third molars (score: 3).

Postcranial skeleton: preservation condition is good. The upper and lower appendicular elements on the left side are in a better state of preservation than those on the right. The humeri, radii, ulnae, os coxae, femora, and tibiae are all almost complete. The vertebrae, sacrum and ribs are also well preserved.

MB05M32:

Only a relatively thick calvarium with a large mastoid process, a robust mandible and some fragments of the humeral shaft represent this burial. Dental preservation:

/	/	M1	P2	P1	/	/	I1	/	/	/	/	/	/	/	/
/	/	/	/	/	/	/	/	0	0	0	P1	P2	M1	M2	M3

The tooth crown surfaces are moderately worn (score: 3-4, except the third molars (score: 1).

MB05M33:

Unused burial identification number.

MB05M34:

Skull: the zygomatico-facial region was heavily damaged, but the calvarium and the mandible are in relatively good condition. The forehead is relatively broad and only slightly sloped. The parietal bones are smoothly rounded at the sagittal suture and the supranasal suture is absent. The glabella region and superciliary arches are flat. The superior orbital margin at the frontal bone is quite round and a supraorbital foramen is present on the right side alone. The temporal lines are not visible at all on the parietal surface. The mastoid process is small (score: 2) and the external occipital protuberance is flat (score: 1). Both superior nuchal lines are definable. The inferior one is not observable due to the fragmentation of this region, but the nuchal plane is quite smooth. The coronal, sagittal and lambdoidal sutures are not synostosed either ecto or endocranially. The mandibular body is gracile and small, with a weak mental eminence (score 2). The gonial angles are not sharply everted and the pterygoid muscle attachment surfaces are smooth. The rami are narrow and moderately sloping posteriorly. The base line of the mandibular body is slightly curved, indicating the 'rocker jaw' variant, despite being a compact mandible. Dental preservation:

/	/	M1	P2	P1	C	X	I1	I1	X	C	P1	P2	M1	/	/
X	M2	X	P2	P1	C	I2	I1	I1	I2	C	P1	P2	X	X	X

The incisor occlusal pattern is indeterminate, while the tooth crown surfaces are remarkably worn (score: 5), for both the anterior and posterior teeth.

Postcranial skeleton: only the left hand is missing. Most limb bones are in a good state of preservation, although some sections were damaged. All upper limb elements, and those of the right leg, were complete. Their size and robustness, including development of muscle attachment entheses, are within the average range for females. Although preservation of the vertebrae and ribs is poor, the sacrum and lumbar vertebrae are in relatively good condition.

MB05M35:

Unexcavated burial. No data available.

MB05M36:

A partially preserved skeleton. The cranium and mandible are fragmentary, the left leg is missing below the femoral diaphysis and the right leg is missing below the tibial and fibular diaphyses. The mandibular body and part of the maxilla are intact, with a full set of completely erupted deciduous teeth.

HUMAN REMAINS FROM THE 2007 SERIES

2007H1

MB07H1M1:

Skull: moderately well preserved, although it has been deformed by soil pressure over time. The fragile basicranium and the right side of the face were fragmented. Slight dental attrition is observed in the permanent central incisors (score: 1), while the other permanent teeth are newly erupted. The left maxilla retains the deciduous second molar, without the roots, situated between the first premolar and the first molar, under which the second premolar is emerging. Dental preservation:

X	M2	M1	P2	P1	C	I2	I1	I1	I2	C	P1	P2	M1	M2	M3
X	M2	M1	P2	P1	C	I2	I1	I1	I2	C	P1	P2	M1	M2	M3

Postcranial skeleton: The postcranial skeleton is in relatively good condition, but the upper limbs, vertebrae and ribs were highly degraded. The entire left arm is missing, as is the right hand and proximal ulna/radius.

MB07H1M2:

Only the proximal half of a tibia and associated fibula fragment represent this burial. No dentition is present.

MB07H1M3:

Skull: The cranium was damaged only at the maxillo-frontal processes. Dental preservation:

X	M2	M1	P2	P1	C	I2	I1	I1	I2	C	P1	P2	M1	M2	M3
X	M2	M1	P2	P1	C	I2	I1	I1	I2	C	P1	P2	M1	M2	M3

The central incisors and first molars are slightly worn (score: 2), while no wear facets are visible for the other permanent teeth.

Postcranial skeleton: Most of the postcranial bones remain and only a few cranial fragments and distal tarsal and carpal phalanges are missing.

MB07H1M4:

Skull: well preserved, but the facial region is highly fragmented. Only the calvarium and the mandible are complete. The calvarium is brachycephalic (cranial index: 81.3), the forehead is narrow and slopes posteriorly. The parietal bones are minimally angled at the sagittal suture. The supranasal suture is absent. The glabella region is moderately protruding, but the superciliary arches are flat. The nasal root is concave and the orbital shape is square with relatively angular corners. Supraorbital foramina are absent bilaterally. The zygomatic bones project prominently laterally and the temporal lines are not distinct on the parietal bones. The size of the mastoid process is rather small (score: 2). The external occipital protuberance is weakly protruding (score: 2), while both superior and inferior nuchal lines are distinct, and the nuchal plane is moderately rugged. The coronal, sagittal and lambdoidal sutures are not synostosed either ecto or endocranially. The mandibular body is relatively robust, with sharply everted gonial angles, developed pterygoid muscle attachment surfaces, and weakly sloping rami. The mental eminence is moderately protruding (score: 3). Dental preservation:

X	X	M1	P2	/	C	X	I1	I1	X	C	P1	X	/	/	/
M3	0	M1	P2	P1	C	X	X	X	X	C	P1	P2	M1	M2	M3

The tooth crown surfaces are heavily worn overall (anterior teeth: score 5, molars: score 6-7).

Postcranial skeleton: In situ preservation was good, although bone quality was poor leading to fragmentation on removal. Limb bone sizes are moderate and display well-developed muscle attachment entheses, including the deltoid area of the humerus and the femoral linea aspera. Squatting facets are present on the talus.

MB07H1M5:

Skull: almost complete, and the overall dimensions are robust. The shape of the calvaria is classified as brachyocephalic (cranial index: 80.1). The parietal bones are weakly angled at the sagittal suture and the frontal displays a strong posterior slope. The supranasal suture is absent, the glabella region and superciliary arches are moderately ridged and the nasal root is slightly concave. The orbital shape is rather square with straight frontal margins. Supraorbital foramina are absent bilaterally and the zygomatic bones project prominently laterally. The facial skeleton is relatively low and wide, but the index is within the range of moderate facial proportions (upper facial index: 51.9), while the temporal lines are not distinct on the parietal bones. The mastoid process is quite large (score: 4), while the external occipital protuberance is weakly protruding. The superior and inferior nuchal lines are distinctive and the nuchal plane is moderately ridged. The coronal, sagittal and lambdoidal sutures are not synostosed ectocranially externally, while the endocranium could not be assessed. The mandible is robust, displaying sharply everted gonial angles and well-developed pterygoid muscle attachment surfaces. The rami are minimally sloping, and the baseline of the mandibular body is rounded, typical of the 'rocker jaw' variant. The mental eminence is moderately protruding (score: 3). Dental preservation:

0	M2	M1	P2	P1	C	X	I1	I1	X	C	P1	P2	M1	M2	M3
M3	M2	M1	P2	P1	C	X	X	X	X	C	P1	P2	M1	M2	X

All the crown surfaces are heavily worn (anterior teeth: score 6, molars: score 7).

Postcranial skeleton: all elements are in a good state of preservation. Paired upper limb bones include the scapulae, humeri, radii and ulnae, and all are nearly complete. The os coxae, femora, tibiae, fibulae and patellae of both sides are also almost complete. The deltoid muscle attachment area of humerus is large and robust. Well developed leg muscles are presumed from the overall femoral morphology, specifically the highly curved shaft and strongly protruding linea aspera. The vertebrae, sacrum and ribs also remain in good condition.

MB07H1M6:

The cranium and mandible are well preserved, except for the fragmentary facial region. Most of the postcranial skeleton was recovered in good condition as well, although both feet and hands are missing (except for a few disarticulated proximal phalanges), as are many ribs. All deciduous tooth crowns were completely formed within the alveoli, but not yet erupted.

MB07H1M7:

The preservation of the cranium and mandible are very poor. Only the left maxilla is nearly complete. The upper first deciduous molars are visible at 2/3 crown height within the alveolus. Only the lumbar vertebrae, os coxae and lower limbs (feet missing) are preserved due to the disturbance caused by the later burial of MB07H1M3.

MB07H1M8:

Skull: nearly complete, with robust cranial morphology. The parietal bones are weakly angled at the sagittal suture, and the frontal slopes steeply. The glabella region and superciliary arches are moderately ridged, while the orbital shape is squared with straight orbital margins. Supraorbital foramina were absent bilaterally. The temporal lines are slightly visible on the parietal bones. The mastoid processes are prominent (score: 5). The external occipital protuberance is also prominently projecting (score: 4). The superior nuchal line is distinct, but the inferior one is not and the nuchal plane is smooth. The coronal, sagittal and lambdoidal sutures are not synostosed either ecto or endocranially. The mandibular body is of moderate size and robusticity, while the mental eminence is moderately expressed (score 3) and the tri-angular region is clear. The gonial angles are not everted and the pterygoid muscle attachment surfaces are weakly developed. The wide rami are perpendicularly elevated and the baseline of the mandibular body is strongly curved, exhibiting the 'rocker jaw' variant. Dental preservation:

M3	M2	M1	P2	P1	C	X	I1	I1	I2	C	P1	P2	M1	M2	M3
M3	M2	M1	P2	P1	C	I2	I1	I1	I2	C	P1	P2	M1	M2	M3

There is an edge-to-edge incisor occlusal bite. The tooth crown surfaces are moderately worn (anterior teeth score: 4, molars score: 5 except for the third molars score: 3).

Postcranial skeleton: the limb bones are in a good state of preservation. The humeri, radii and ulnae are complete, but the scapulae are only partially preserved. The femora, tibiae and fibulae are also complete, while the os coxae were fragmentary. The dimensions of these limb bones are moderate and relatively slender. All the muscle attachment areas are smooth and minimally developed. Much of the vertebrae are in good condition. A clear squatting facet is present on the talus.

MB07H1M9:

Skull: cranium and mandible are well preserved, except for the basicranium and a section of the zygomatico-facial skeleton. The calvarium is quite long and narrow (cranial index: 72.9) or dolichocephalic. The forehead is also narrow and minimally sloping. The parietal bones are moderately angled at the sagittal, while the supranasal suture is absent. The glabella region is flat, while the superciliary arches are weakly ridged at their medial extent. The orbital shape is slightly rounded. A supraorbital foramen is present on the left side alone. The zygomatic bones are moderately projecting laterally. The temporal lines are not distinct on the parietal bones. The mastoid process is large (score: 5), and the external occipital protuberance is massive and prominent (score: 5). The superior nuchal lines are well-defined, but the inferior is not as ridged as the superior and the nuchal plane is smooth. The sagittal, coronal, and lambdoidal sutures have not synostosed either ecto or endocranially. A large Inca bone subdivides the occipital region. The mandibular body is large and robust, with a well developed mental eminence (score 4). However, the gonial angles are not everted, but the pterygoid muscle attachment surfaces are well developed. The rami do not slope backwards and the mandibular body exhibits the 'rocker jaw' variant with a strongly curved baseline. Dental preservation:

M3	M2	M1	P2	P1	C	I2	I1	I1	0	C	P1	P2	M1	M2	M3
/	M2	M1	P2	P1	C	/	/	0	I2	C	P1	P2	M1	M2	M3

The incisor occlusal bite is edge to edge, while the occlusal surfaces were not severely worn (score: 4 anterior teeth, score 4-5 first and second molars, score 2 third molars).

Postcranial skeleton: unfortunately, the lower thoracic region, including most ribs and the lumbar vertebrae did not survive. The limbs, with the exception of the left arm, are variously preserved. A range of pathological conditions are apparent, including extensive lower limb atrophy and a completely ankylosed cervical spine (see Oxenham et al. 2009).

MB07H1M10

Skull: damaged at the zygomatico-facial region and basicranium, although the calvaria and mandible are well preserved. The forehead of the frontal bone is narrow and the cranial vault is mesocephalic (cranial index: 77.4). The parietal bones are not angled at the sagittal suture and the superciliary arches are moderately ridged anteriorly. The frontal bone has a slight slope, while the supranasal suture is absent. The glabella region and the nasal root are flat, the orbital margins are round and supraorbital foramina are absent bilaterally. The temporal lines are not distinct on the parietal bones. The mastoid processes are of moderate size (score: 3), while the external occipital protuberance is not protruding. The degree of development of the superior and inferior nuchal lines are unknown due to postmortem damage. The coronal, sagittal and lambdoidal sutures display nearly complete synostosis both ecto and endocranially. The mandibular body is large but gracile, with minimally developed pterygoid muscle attachment surfaces and rami that have a marked posterior slope. The mental eminence is moderately protruding (score: 3). Dental preservation:

X	X	M1	P2	P1	C	X	I1	I1	X	C	P1	P2	M1	M2	X
X	0	M1	P2	P1	C	I2	I1	I1	I2	C	P1	P2	M1	M2	0

The occlusal surfaces are heavily worn (anterior teeth: score 5, molars: score 5-6).

Postcranial skeleton: moderately well preserved. The humeri, radii and ulnae are nearly complete, except for the humeral heads; however, the scapulae were quite fragmentary. The femora, tibiae, fibulae and patellae on both sides are also well preserved, except for the distal condyles of the femora. The muscle attachment areas of the lower limbs, including the pilastric form of the linea aspera of the femora, are moderate, whereas the deltoid tuberosity of the humerus is well developed. The vertebrae are highly fragmented.

MB07H1M11:

Skull: nearly complete and intact, only small sections of the zygomaxillary and basicranial regions were damaged. The calvaria is mesocephalic (cranial index: 76.1), with a wide forehead. The supranasal suture is absent and the glabella region, superciliary arches and nasal root are all remarkably flat, with minimally elevated nasal bones. The orbital shape is slightly round and a supraorbital foramen is present on the left side only. The zygomatic bones are moderately projecting laterally. The temporal lines are not distinct on the parietal bones and the size of the mastoid process is moderate (score: 3). The external occipital protuberance is slightly ridged (score: 2) and the superior nuchal lines are indistinct, with the inferior line being unobservable due to postmortem damage. The nuchal plane is quite smooth. The coronal, sagittal and lambdoidal sutures are not synostosed either ecto or endocranially. The width/height ratio of the facial region is moderate as indicated by a facial index of 52.2. The mandibular body is relatively gracile and compact. The pterygoid muscle attachment surfaces are smooth, the gonial angles are not everted and the rami are minimally sloping. The mental eminence is moderate (score: 3), despite the small mandibular body. The baseline of the mandibular body is straight. Dental preservation:

/	/	/	0	0	0	X	0	I1	X	0	P1	0	M1	M2	M3
X	X	M1	P2	P1	C	X	X	X	X	0	P1	P2	M1	X	X

Incisor occlusal pattern not assessable. The tooth crown surfaces are heavily worn (anterior teeth score: 5, molars score: 5-7), with the left mandibular molars especially reaching a severe stage of attrition (score: 7).

Postcranial skeleton: most postcranial elements are nearly complete. The size and development of muscle attachment regions are moderate. Clear squatting facets are visible on both tali.

MB07H1M12:

The cranium is crushed and mostly missing, and most preserved elements (ribs, left os coxa, clavicles, scapulae, right humerus, left proximal femur, cervical and thoracic vertebrae) are disarticulated.

MB07H1M13a:

Skull: the cranium is absent, but a mandible of appropriate age and dimensions was recovered in level 15(III), square A1H2. The mandibular body is relatively large and robust. The base line is prominently curved, indicating a clear 'rocker jaw' variant. The pterygoid muscle attachment surfaces are well developed, despite the minimally everted gonial angles. The rami are wide and minimally angled. The mental eminence is prominent (score: 4). Dental preservation:

/	/	/	/	/	/	/	/	/	/	/	/	/	/	/	/
X	X	M1	P2	P1	C	I2	I1	I1	I2	C	P1	P2	M1	M2	M3

All elements above the radius/ulna are either missing or disarticulated, with the exception of the middle thoracic vertebrae and four ribs. The cranium is absent, but a mandible of appropriate age and dimensions was recovered in level 15(III), square A1H2. Both hands and feet are intact and in situ.

The attrition of the occlusal surfaces is severe (anterior teeth score: 5, first and second molars score: 5-7, third molars score 5).

Postcranial skeleton: Only the appendicular bones are well preserved, though generally incomplete. The os coxae, most vertebrae, and the ribs were recovered highly fragmented. The diaphyses that remain are slender, suggestive of weakly developed muscle attachment areas, although the linea aspera are prominently ridged. Clear squatting facets are present on the talus.

MB07H1M13b:

See Appendix 1 entry.

MB07H1M14a

See Appendix 1 entry.

2007H2

MB07H2M1 Age: middle-aged adult. Sex: male.

Skull: nearly complete, with a robust cranial morphology overall. The cranial index is 85.2 (Martin No. 8:1), indicating a brachyocephalic cranium. The upper facial index is 48.4 (Martin No. 48:45), indicating low and wide facial proportions. The parietal bones are moderately angled at the sagittal suture and the frontal slopes steeply. The glabella region and superciliary arches are strongly ridged (score: 4), the orbital shape is square and the orbital margins are straight. A supraorbital foramen is present on the right orbit only. The temporal lines are clearly visible on the parietal bones. The size of the mastoid processes are moderate (score: 3). The external occipital protuberance is extremely prominent (score: 5), while both the superior and inferior nuchal lines are distinct, associated with a rugged nuchal plane. The coronal sutures are not synostosed either ecto or endocranially, while the sagittal and lambdoidal sutures are partially synostosed both ecto and endocranially. The mandibular body is large and robust. The mental eminence is moderate expressed (score: 3), with a distinctive triangular shape on the superior surface. The gonial angles are sharply everted and the pterygoid muscle attachment surfaces are rugged. The mandibular rami are wide and perpendicularly oriented. The baseline of the mandibular body is moderately curved, exhibiting the 'rocker jaw' variant. Dental preservation:

M3	M2	M1	P2	P1	C	I2	I1	I1	I2	C	P1	P2	M1	M2	M3
X	M2	M1	P2	P1	C	X	X	X	X	C	P1	P2	M1	M2	X

Incisor occlusal bite could not be assessed. The occlusal surfaces were moderately worn (anterior teeth score: 4-5, molars score: 6).

Postcranial skeleton: all postcranial elements were recovered in good condition. The humeri, radii and ulnae are complete, but the scapulae are fragmentary. The os coxae, femora, tibiae and fibulae are complete, except for postmortem damage around the knee joint. The dimensions of these limb bones are moderate and relatively slender, with all major muscle attachment areas being relatively smooth. Almost all vertebrae are in good condition. The tali exhibit clear squatting facets.

MB07H2M2:

Skull: well-preserved and gracile, with a perpendicularly elevated forehead and parietal bones not angled at the sagittal suture. The supranasal suture is absent and the glabella region, superciliary arches and nasal root are very flat. The orbital shape is more square than round and supraorbital foramina are present bilaterally. The zygomatic bones are relatively forward projecting, forming a generally flat face, together with a quite flat nasal root. Dental preservation:

X	M2	M1	P2	P1	C	I2	I1	I1	I2	C	P1	P2	M1	M2	X
M3	M2	M1	P2	P1	C	I2	I1	I1	I2	C	P1	P2	M1	M2	M3

The occlusal surfaces are slightly worn (anterior teeth: score 1, molars: score 1), except for the third molars (score: 2).

Postcranial skeleton: all postcranial elements are in good condition. The scapulae, humeri, radii and ulnae are nearly complete, as are the os coxae, femora and tibiae, whereas the fibulae are incomplete.

MB07H2M3:

Poorly preserved in general. All of the upper body is missing, almost all of the skull, the right tibia/fibula, most of both os coxae, both hands and feet are missing.

MB07H2M4:

Only the limb bones were recovered and these of right side alone.

MB07H2M5:

Skull: Except for a high degree of damage to some sections of the parietal and frontal bones, the skull was relatively well preserved, including a nearly complete zygomatico-facial region. The cranial index is 76.4 (Martin No. 8:1), indicating a mesocephalic cranium. The forehead is relatively broad, flat and elevated perpendicularly. The parietal bones are not angled and are smoothly rounded at the sagittal suture. A metopic suture is present. The glabella region, superciliary arches, nasal root and nasal bones are very flat. The superior orbital margin of the frontal bone is weakly rounded, while a supraorbital foramen of the 'notch' type is present on the right side; the left side being unknown due to post-mortem damage. The temporal lines are not visible at all on the parietal bones and the mastoid processes are quite small (score: 1). The external occipital protuberance is very flat (score: 1), and the nuchal plane, including both the superior and inferior nuchal lines, are weak and smooth. The coronal, sagittal and lambdoidal sutures are not synostosed ectocranially, and are not assessable endocranially. The facial skeleton is quite low and wide (upper facial height: 64mm, the upper facial index: 48.0). The mandibular body is gracile and relatively small, but the mental eminence is prominent (score: 2). The gonial angles are not everted, the pterygoid muscle attachment surfaces are smooth and the rami are narrow and steeply sloping. The baseline of the mandibular body is remarkably curved, exhibiting pronounced 'rocker jaw,' despite the compact mandible. Preserved dentition:

0	M2	M1	P2	P1	C	I2	I1	I1	0	C	P1	P2	M1	M2	0
0	M2	M1	P2	P1	C	I2	I1	I1	I2	C	P1	P2	M1	0	X

The incisor occlusal pattern displays an over-bite. The occlusal surfaces are moderately worn (score: 5 for both the anterior and posterior teeth, except for the first molars, scored as 7).

Postcranial skeleton: A nearly complete skeleton postcranial skeleton. The distal 4/5ths of the right and left tibia and left fibula, distal 1/3rd of the right fibula, all of the left tarsals and tarsal phalanges, and the right calcaneus and distal tarsal phalanges are missing. The left hand is disarticulated and scattered underneath the os coxae. Weak development of the muscle attachment regions generally.

MB07H2M6:

The cranium and most postcranial elements are well preserved, except for the fragmentary facial region. All deciduous tooth crowns are completely formed.

MB07H2M7:

The entire skeleton is in good condition, although the cranium was deformed by soil pressure and the facial skeleton was crushed into small fragments. The postcranial skeleton, including all four limbs, is nearly complete. Full eruption of the deciduous teeth was observed.

MB07H2M8:

Good skeletal preservation. Only the distal left and right tibiae and fibulae, both feet, and distal phalanges from both hands are missing. All teeth are deciduous, and right deciduous m^1 was observed to be erupting.

MB07H2M9:

Poorly preserved skeleton. Only the left half of the cranium, the right ulna, distal radius, os coxae, one half of an un-sided femur, and a few rib fragments were recovered.

MB07H2M10

Skull: The skull is nearly complete, except for the heavily damaged left side of the face. This cranium is relatively large and robust, but the glabella and nasal region are quite flat. The cranial index is 77.9 (Martin No. 8:1), or mesocephalic. The upper facial index is 54.1 (Martin No. 48:45), categorizing the proportion of facial height to breadth as medium. The forehead is broad and steeply sloping. The parietal bones are sharply angled at the sagittal suture. The supranasal suture is absent, the superciliary arches are not protruding and the orbital shape is slightly rounded. Supraorbital foramina ('notch' variant) are present bilaterally. The zygomatic bones project prominently laterally and the temporal lines are distinct on the parietal bones. The size of the mastoid process is large (score: 4), the nuchal crest is weakly protruding (score: 5) and both the superior and inferior nuchal lines are distinct, on a moderately rugged nuchal plane. The coronal, sagittal and lambdoidal sutures are partially synostosed ectocranially, while they cannot be assessed endocranially. The mandibular body is large and robust, exhibiting strongly everted gonial angles and well developed pterygoid muscle attachment surfaces. The rami are rather narrow and moderately sloping. The mental eminence is moderately protruding (score: 3). Dental preservation:

M3	M2	M1	P2	P1	C	I2	0	/	/	/	/	/	M1	M2	0
M3	M2	M1	P2	P1	C	I2	I1	I1	0	C	0	P2	M1	M2	0

The incisor occlusal pattern is an edge to edge-bite. Occlusal wear is severe (anterior teeth: score 5, molars: score 6-7).

Postcranial skeleton: all postcranial elements are in good condition. The scapulae, humeri, radii and ulnae are nearly complete, as are the os coxae, femora, tibiae, fibulae and patellae. The deltoid muscle enthesis of the humerus is robust, as are the major leg muscle entheses, associated with the well developed pilastric form of the femoral linea asperae and prominently curved femoral shafts. The vertebrae, sacrum and ribs also remain in good condition.

MB07H2M11:

See Appendix 1 entry.

MB07H2M12:

Skull: The cranium and mandible are nearly complete, except for the basicranial region. The overall view of this specimen is of a small and gracile individual. The calvarium is dolichocephalic (cranial index: 72.2), the forehead is narrow and sloped and the parietal bones are moderately angled. Presence of the supranasal suture is unknown, due to post-mortem damage to the supranasal region. The glabella, supraorbital ridges and nasal bones, including the root, are very flat (supraorbital ridge score: 2), while the superior orbital margins are slightly round. A supraorbital foramen is present bilaterally and the temporal lines are faintly traceable on the parietal surfaces. The mastoid processes are small (score: 1), and the external occipital protuberance is weakly ridged (score: 2). Both the superior and inferior nuchal lines are definable, although the nuchal plane is smooth. The coronal, sagittal and lambdoidal sutures exhibit minimal synostosis ectocranially. Endocranial synostosis is minimal at the coronal suture, but almost complete at sagittal and lambdoidal sutures. The upper facial height is relatively high (the height is 68 mm). The mandibular body is gracile and small, with a weak mental eminence (score 2). The gonial angles are not everted and the pterygoid muscle attachment surfaces are smooth. The rami are narrow and moderately sloping. The baseline of the mandibular body is slightly curved, exhibiting the 'rocker jaw' variant, despite the compact mandible. Dental preservation:

X	X	X	P2	P1	C	X	X	X	0	0	X	/	/	/	/
X	X	X	X	P1	C	X	X	X	X	C	P1	P2	M1	X	X

The incisor occlusal pattern is not assessable. The occlusal surfaces are severely worn for the anterior teeth (score: 6), while wear of the left premolars and first molar are minimal (score: 3).

Postcranial skeleton: All postcranial elements are in good condition, although some sections suffered minimal damage. The limbs are moderate in length but slender, with muscle attachment sites being relatively smooth. The preservation level of the vertebrae and ribs is high.

MB07H2M13:

Preservation condition is good, with much of the skeleton (except the right os coxae, femur, tibia, fibula, tarsals, metatarsals and phalanges) remaining, including the skull and the complete facial region. All deciduous molars have erupted and show some wear.

MB07H2M14:

A partially preserved, fragile skeleton. The entire right arm, right tibia and both feet and hands are missing, and those elements present are moderately disarticulated The cranium is fragmented and no teeth have yet erupted in the near complete maxilla and mandible.

MB07H2M15:

A nearly complete and well preserved skeleton with only the proximal left ulna/radius, right tibial diaphysis and a few carpal phalanges missing. A full set of deciduous molars has erupted, with faintly worn crowns.

MB07H2M16:

A nearly complete skeleton. The right hand, femur, tibia and fibula (except for the distal ends of the latter), as well as the left foot, are missing. The deciduous teeth are fully erupted, except for the deciduous second molars which are in the process of erupting.

MB07H2M17:

Skull: The cranium is almost completely preserved, including a full set of permanent teeth exhibiting slight occlusal wear (score: 1), with the exception of the third molars which are partially erupted. Dental preservation:

X	M2	M1	P2	P1	C	I2	I1	I1	I2	C	P1	P2	M1	M2	X
X	M2	M1	P2	P1	C	I2	I1	I1	I2	C	P1	P2	M1	M2	X

Postcranial skeleton: Intrusive destruction post-interment has removed the right distal humerus, ulna, radius, most of the hand, ilium and ischium, most of the sacrum, vertebrae below T9/10, and ribs below the 5th; otherwise well preserved.

MB07H2M18:

Skull: the cranium is in good condition. The glabella region, superciliary arches, nasal root and nasal bones are quite flat, and the orbital shape is round rather than square. A supraorbital foramen was found on the right side alone. Dental preservation:

X	M2	M1	P2	P1	C	I2	I1	I1	I2	C	P1	P2	M1	M2	X
X	M2	M1	P2	P1	C	I2	I1	I1	I2	C	P1	P2	M1	M2	X

The incisor occlusal pattern forms an over-bite. The occlusal surfaces are slightly worn (anterior teeth: score 2, molars: score 2).

Postcranial skeleton: All postcranial elements are in good condition. The scapulae, humeri, radii and ulnae are nearly complete, as are the os coxae, femora and tibiae, but the fibulae are incomplete. The vertebrae, sacrum, and ribs are also in good condition.

MB07H2M19:

Skull: severely damaged with much of the calvarium and face missing. Only a part of the right parietal bone, maxilla and mandible were excavated in situ. The mastoid processes are large (score: 4), as is the mandibular body, which exhibits the 'rocker jaw' variant. The gonial angles are moderately everted, with well developed pterygoid muscle attachment surfaces. The mandibular rami are minimally sloping and the mental eminence is weakly protruding (score: 1). Dental preservation:

0	M2	M1	P2	P1	C	X	I1	I1	X	C	P1	P2	M1	/	/
0	X	M1	P2	P1	C	X	X	X	X	C	P1	P2	M1	M2	M3

The incisor occlusal pattern is not assessable. All occlusal surfaces are heavily worn (anterior teeth: score 6, molars: score 4-6).

Postcranial skeleton: All postcranial remains are in a good state of preservation. The humeri, radii and ulnae are nearly complete, as are the femora, tibiae, fibulae and patellae. The muscle attachment areas represented by the deltoid tuberosity of the humerus and the pilastric form of the femoral linea aspera, were well developed. Most of the vertebrae were preserved.

MB07H2M20:

A fragmentary and fragile skeleton only missing the hands.

MB07H2M21:

Reasonable condition, albeit fragmentary. Missing the left half of the cranium, right ribs, humerus, most foot phalanges and the sacrum.

MB07H2M22

Skull: perfectly preserved. The calvarium is brachycephalic shape (cranial index: 80.1) and the glabella region, superciliary arches and nasal root are quite flat. The nasal bones are slightly elevated. The orbital shape is round rather than squared and a supraorbital foramen is present on the left side only. The zygomatic bones are moderately projecting laterally, the forehead is of moderate width, the supranasal suture is faintly present, and the temporal lines are distinct on the parietal bones. The size of the mastoid processes is small (score: 2). The external occipital protuberance is not ridged (score: 1), the superior and inferior nuchal lines are indistinct and the nuchal plane is quite smooth. The coronal and lambdoidal sutures are minimally fused synostosed and the sagittal suture is partially synostosed ectocranially. Endocranial synostosis could not be assessed. The width/height proportion of facial skeleton is moderate, as indicated by a facial index of 54.3. The mandibular body is gracile, but the chin is relatively high. The pterygoid muscle attachment surfaces are smooth, and the gonial angles are slightly everted. The narrow rami have a slight slope. The mental eminence is weak (score: 2), but the baseline of the body is slightly curved, exhibiting the 'rocker jaw' variant. Dental preservation:

X	M2	M1	P2	P1	C	I2	I1	I1	X	C	P1	P2	M1	0	X
X	M2	M1	P2	P1	C	I2	I1	I1	I2	C	P1	P2	M1	M2	X

The incisor occlusal pattern is an over-bite. The tooth crown surfaces are moderately worn (anterior teeth score: 4, molars score: 4-5).

Postcranial skeleton: most elements are nearly complete, but most tarsals, all metatarsals and foot phalanges, most elements of the left hand, and a few distal phalanges from the right hand are missing. The long bones are relatively slender, and the development of muscle attachment sites is minimal. Clear squatting facets are present on the tali.

MB07H2M23:

See entry in Appendix 1.

MB07H2M24:

Skull: The facial region was heavily damaged postmortem, but the calvarium and mandible are in relatively good condition. The forehead is broad and relatively vertical. The parietal bones are not smoothly rounded at the sagittal suture. The supranasal suture is absent and the glabella region and superciliary arches are very flat. The superior orbital margin of the frontal bone is weakly rounded and supraorbital foramina are bilaterally absent. The temporal lines are not distinguishable at all on the parietal bones. The mastoid processes are small (score: 2), and the external occipital protuberance is flat (score: 1). Both the superior and inferior nuchal lines are definable, although the nuchal plane is quite smooth. The coronal, sagittal and lambdoidal sutures are not synostosed either ecto or endocranially. The mandibular body is relatively thick and robust and the mental eminence is distinctive (score 3). The status of the gonial angles, ramus angle and pterygoid muscle attachment site rugosity are all unobservable due to postmortem damage. Dental preservation:

X	X	M1	P2	P1	C	X	/	/	/	/	/	/	/	/	/
X	M2	M1	P2	P1	C	X	X	X	X	C	P1	P2	M1	M2	0

The incisor occlusal pattern is not assessable. The tooth crown surfaces were worn (score: 5 for both the anterior and posterior teeth).

Postcranial skeleton: most postcranial elements are in a moderate state of preservation, with some areas damaged. Both upper limbs and the right lower limbs are preserved. The long bones are relatively slender and their robustness, including the development of muscle attachment sites, is moderate. Although preservation of the vertebrae and ribs is poor, the sacrum and the lumbar vertebrae are in relatively good condition.

MB07H2M25:

See entry in Appendix 1.

MB07H2M26:

Well preserved, only missing the hands and feet, sacrum, and distal right tibia and fibula. The deciduous teeth are fully erupted, with the exception of the deciduous second molars.

MB07H2M27:

Skull: nearly intact. The cranial index is 70.3 (Martin No. 8:1), indicating a dolichocephalic cranium. The upper facial index is 55.6 (Martin No. 48:45), categorizing the facial region as high and narrow. The forehead is narrow, with a prominent frontal tubercle. The parietal bones are moderately angled at the sagittal suture and the frontal bone slopes steeply. The supranasal suture is present, while the glabella region, superciliary arches and nasal root are relatively flat. The orbital shape is slightly round, and supraorbital foramina exist bilaterally. The zygomatic bones strongly project laterally, while the temporal lines are distinct on the parietal bones. The size of the mastoid process is moderate (score: 3) and the external occipital protuberance is moderately protruding (score: 3). The superior and inferior nuchal lines are distinct upon a rugged nuchal plane. The coronal, sagittal and lambdoidal sutures are not synostosed ectocranially and are not assessable endocranially. The mandibular body is tall, thick and robust, with a perpendicular gonial region. The pterygoid muscle attachment surfaces are smooth, and the rami are moderately sloping. The mental eminence is distinctively protruding (score: 3), but the baseline of the mandibular body does not exhibit the 'rocker jaw' variant. Dental preservation:

M3	M2	M1	P2	P1	C	X	X	I1	X	C	P1	P2	M1	M2	M3
M3	M2	M1	P2	P1	C	X	X	X	X	C	P1	P2	M1	M2	M3

The incisor occlusion pattern was originally an edge-to-edge bite. The occlusal surfaces are severely worn (anterior and posterior teeth: score 5).

Postcranial skeleton: all postcranial elements are in good condition. The scapulae, humeri, radii and ulnae are nearly complete, as are the os coxae, femora, tibiae, fibulae and patellae. The deltoid muscle attachment areas of the humeri are robust, as are the leg muscles, presumed from the pilastric form of the linea aspera of the femora and the highly curved femoral shafts. The vertebrae, sacrum and ribs remain in good condition as well. Squatting facets are present on both tali.

MB07H2M28:

Reasonable preservation, although both hands and feet, most epiphyses, left fibula, distal right fibula, and distal right radius and ulna are missing. No dentition observable or recovered.

MB07H2M29:

The preservation is reasonable but the entire cranium is missing, and the mandible is broken in two, with both halves facing outwards. The entire right hand, most of the left hand, and most of both feet (except for the left talus, cuboid and medial cuneiform and right talus and cuboid) are missing. The deciduous molars show heavy attrition, but only the permanent incisors and first molars have erupted.

MB07H2M30:

Skull: good condition, although the neuro-cranium was obliquely deformed to a certain extent, probably due to subterranean soil pressure. Overall, this skull is large in size and robust in morphology. The upper facial index is 44.9 (Martin No. 48:45), describing the facial height to breadth ratio as low and broad. The forehead is wide and the frontal tubercle is very clear, while the parietal bones are not well angled at the sagittal suture and the frontal itself is minimally sloped. The supranasal suture is present but faint, and the glabella region and superciliary arches are moderately protruding (supra orbital ridge, score: 3). Orbital shape is square rather than round and supraorbital foramina are present bilaterally. The lateral projection of the zygomatic bones is strong, forming a very broad facial profile and the temporal lines are distinct on the parietal bones. The size of the mastoid processes are also very large (score: 5), while the external occipital protuberance is minimally protruding (score: 2). Both the superior and inferior nuchal lines are distinctive, with a very rugged nuchal plane. The coronal, sagittal and lambdoidal sutures are slightly synostosed ectocranially, while endocranial sagittal and lambdoidal suture synostosis is considerable. The mandibular body is tall, thick and robust. Although the gonial angles are minimally everted, the pterygoid muscle attachment surfaces are well developed. The mandibular rami are wide and slightly sloping. The mental eminence is moderately protruding (score: 3). The baseline of the mandibular body exhibits a faint 'rocker jaw' shape. Dental preservation:

X	X	X	0	P1	C	X	I1	I1	X	C	P1	P2	M1	X	X
M3	M2	M1	P2	P1	C	I2	I1	I1	I2	C	P1	P2	0	M2	M3

Incisor occlusion was an edge-to-edge bite. Occlusal wear was heavy attrition (anterior teeth, score: 5-6, molar's score: 5-7).

Postcranial skeleton: good condition. The scapulae, humeri, radii and ulnae are nearly complete, as are the os coxae, femora, tibiae, fibulae and patellae. The deltoid muscle attachment areas of the humeri are quite robust and the leg muscles are also well developed, indicated by the pilastric form of the femoral linea aspera and the highly curved shafts. The vertebrae, sacrum and ribs are also in good condition. Squatting facets are present on the tali.

MB07H2M31:

A well preserved skeleton only missing most of the right hand, the left carpals and distal phalanges, distal phalanges from both feet and most of the sacrum. All deciduous teeth are fully erupted.

MB07H2M32:

Skull: excellent preservation. The cranial index is 74.4 (Martin No. 8:1), indicating a dolichocephalic shape. The upper facial index is 50.7 (Martin No. 48:45), classifying the facial height to breadth ratio as medium. The forehead of the frontal bone is broad and the frontal tubercle is sharp. The frontal bone is steeply sloping and the parietal bones are moderately angled at the sagittal suture. The supranasal suture is present but faint, while the glabella region and superciliary arches are moderately protruding (supra orbital ridge, score: 3) and the nasal root is very concave. The orbital shape is square rather than round and supraorbital foramina are present bilaterally. The lateral projections of the zygomatic bones are pronounced, forming a very broad facial profile. The wide piriformis of the nasal opening is another peculiarity of this specimen. The temporal lines are visible on the parietal bones, the size of the mastoid processes are quite large (score: 5), while the external occipital protuberance is minimally protruding (score: 2). Both the superior and inferior nuchal lines are distinguishable within a very rugged nuchal plane. The coronal, sagittal and lambdoidal sutures are not synostosed wither ecto or endocranially. The size and robustness of the mandibular body is moderate when compared to the general Man Bac male subsample. The gonial angles are minimally everted, but the pterygoid muscle attachment surfaces are well developed. Ramus width and angle are also moderate, but the baseline of the mandibular body exhibits a typical 'rocker jaw'. The mental eminence is moderately protruding (score: 3). Dental preservation:

M3	M2	M1	P2	P1	C	X	I1	I1	X	C	P1	P2	M1	M2	M3
M3	M2	M1	P2	P1	C	X	X	X	X	C	P1	P2	M1	M2	M3

Incisor occlusion cannot be assessed. Tooth wear is minimal (anterior and posterior teeth, score: 3-4).

Postcranial skeleton: all postcranial remains are in good condition. The scapulae, humeri, radii and ulnae are almost complete, as are the os coxae, femora, tibiae, fibulae and patellae. The deltoid muscle attachment areas of the humeri are robust, while the lower limb bones suggest that some of the leg muscles were well developed as well. The linea aspera of the femora are pilastric in form and the femoral shafts have pronounced curvatures. The vertebrae, sacrum and ribs are also in good condition. Squatting facets are present on the tali.

LITERATURE CITED

Acsádi G, Nemeskéri J. 1970. History of Human Life Span and Mortality. Budapest: Akadémiai Kiadó.

Oxenham MF, Tilley L, Matsumura H, Cuong NL. Thuy NK, Dung NK, Domett K, Huffer D. 2009. Paralysis and severe disability requiring intensive care in Neolithic Asia. Anthropological Science 2: 107-112.

Scott EC. 1979. Dental wear scoring technique. Am J Phys Anthropol 51:213-218.

Smith BH. 1984. Patterns of molar wear in hunter gatherers and agriculturalists. Am J Phys Anthropol 63: 39-56.

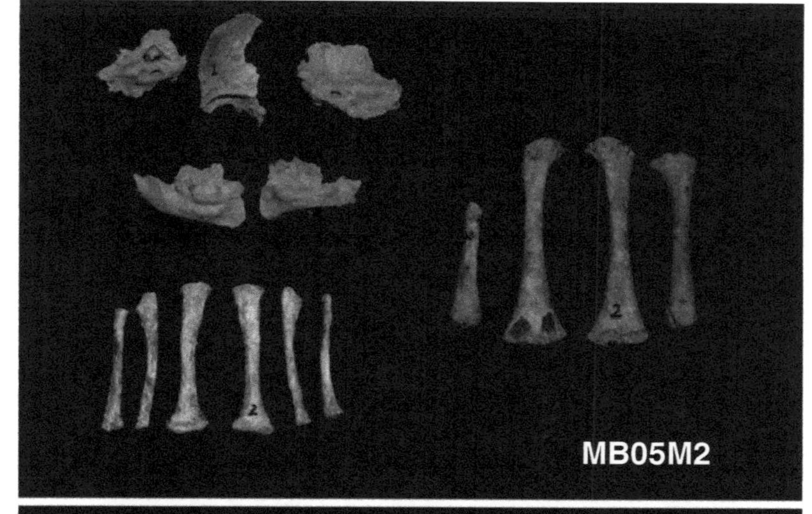

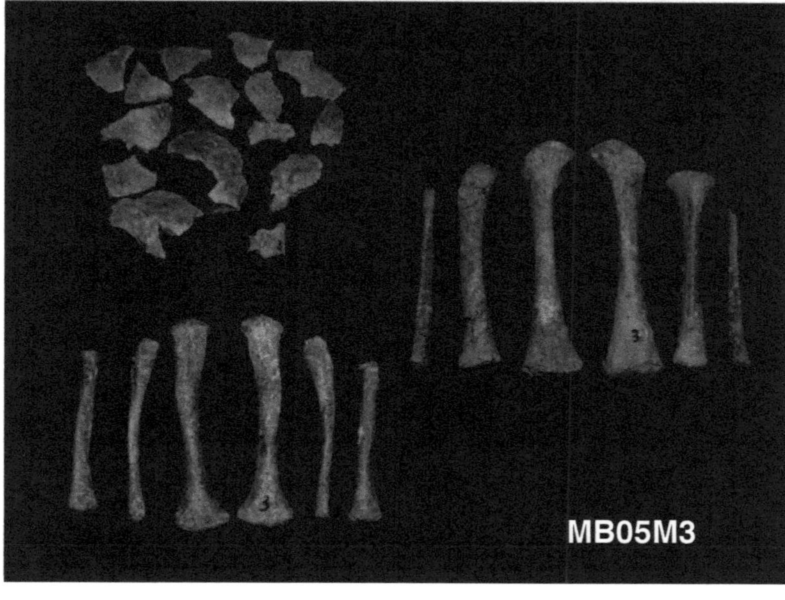

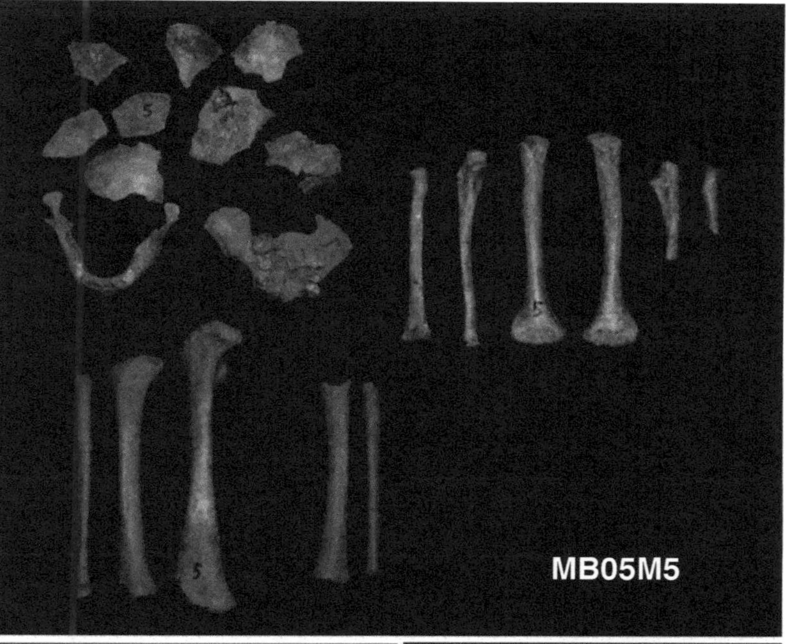

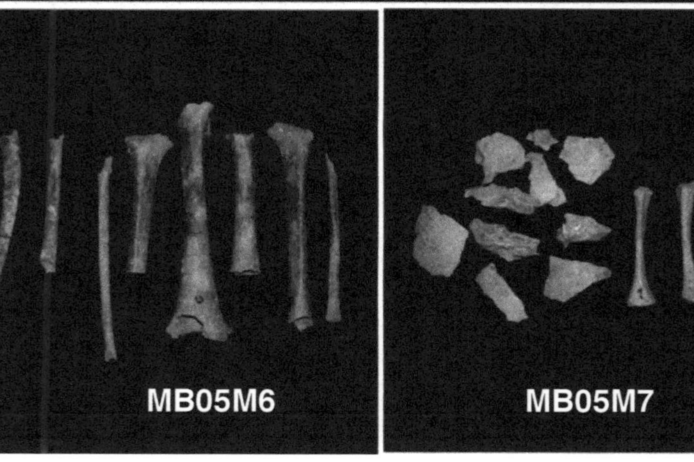

Human skeletal remains from the Mac Bac site

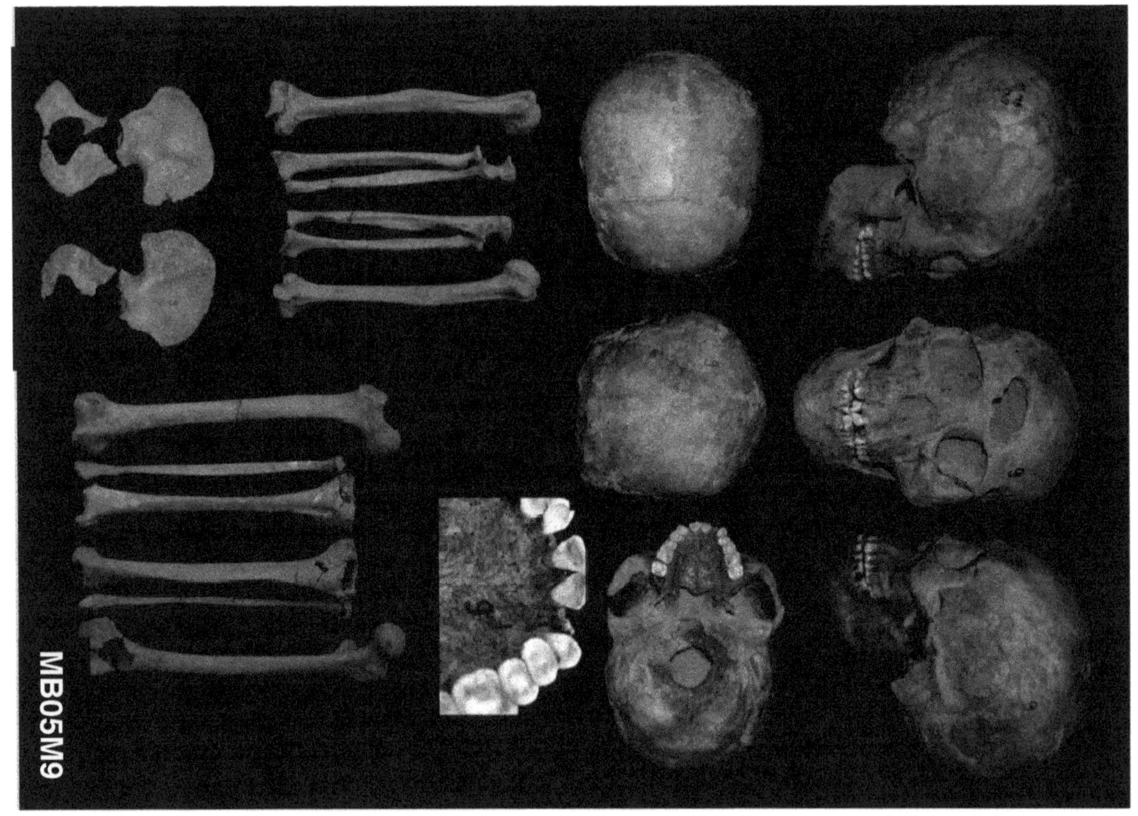

Human skeletal remains from the Mac Bac site

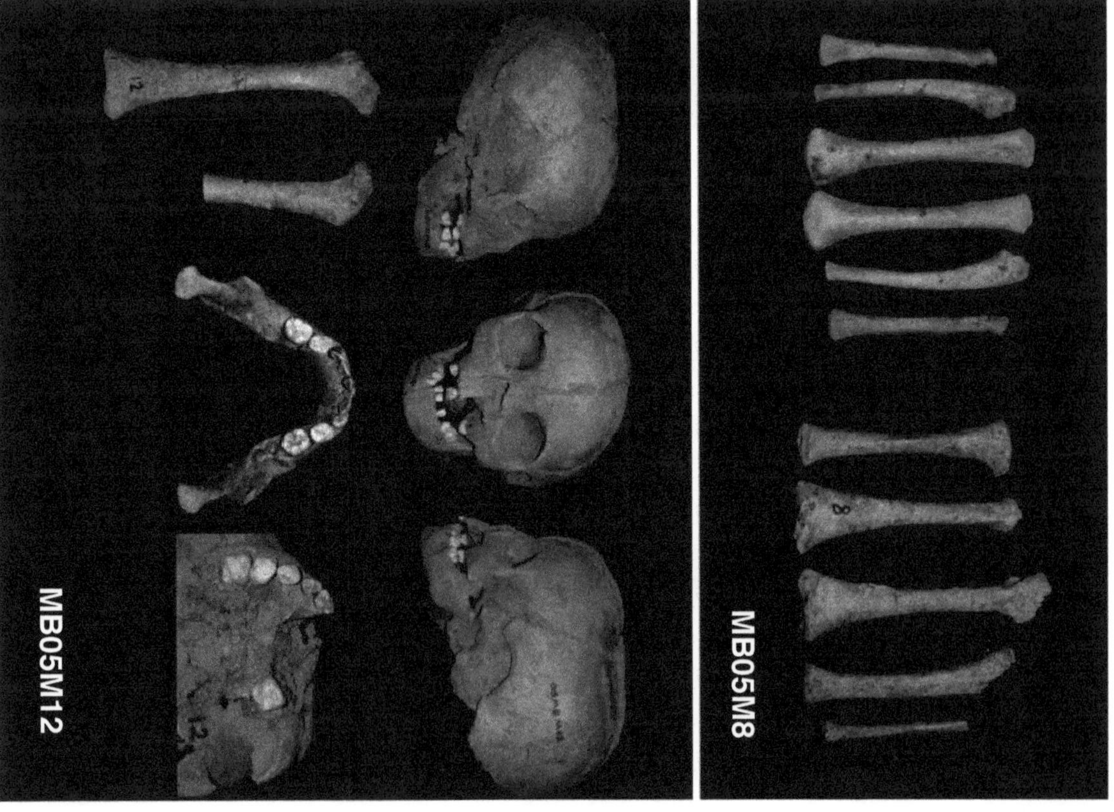

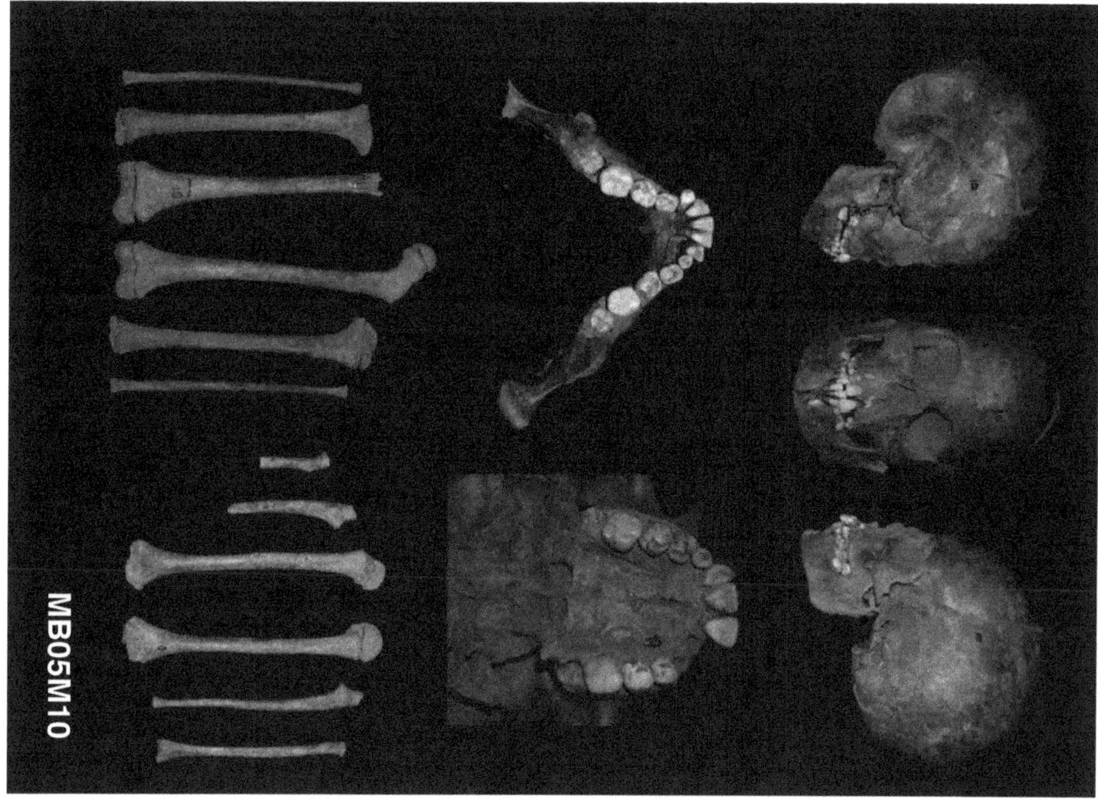

Human skeletal remains from the Mac Bac site

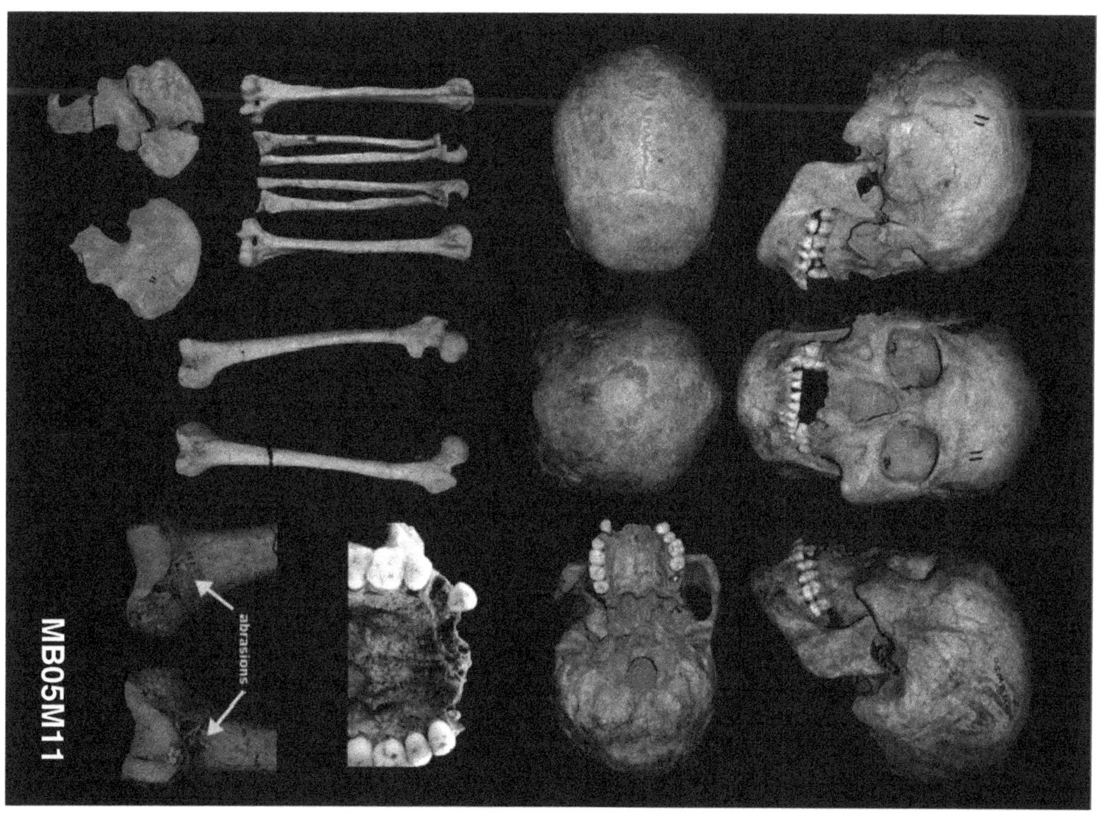

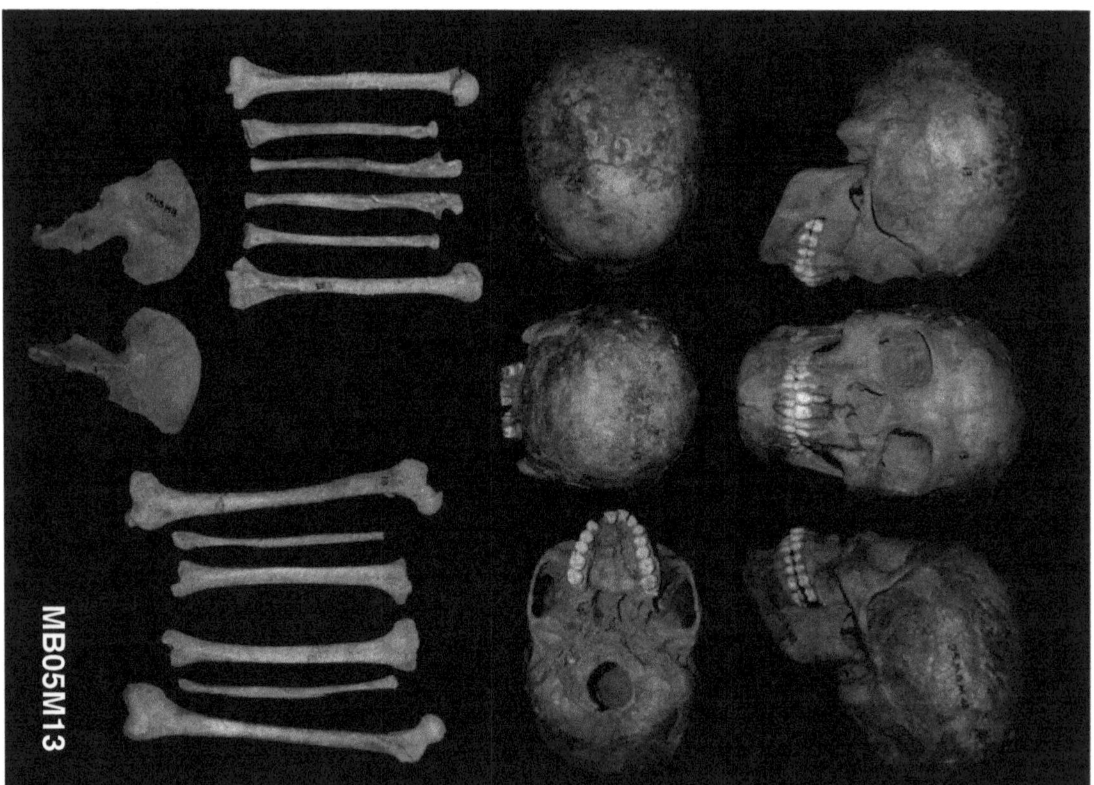

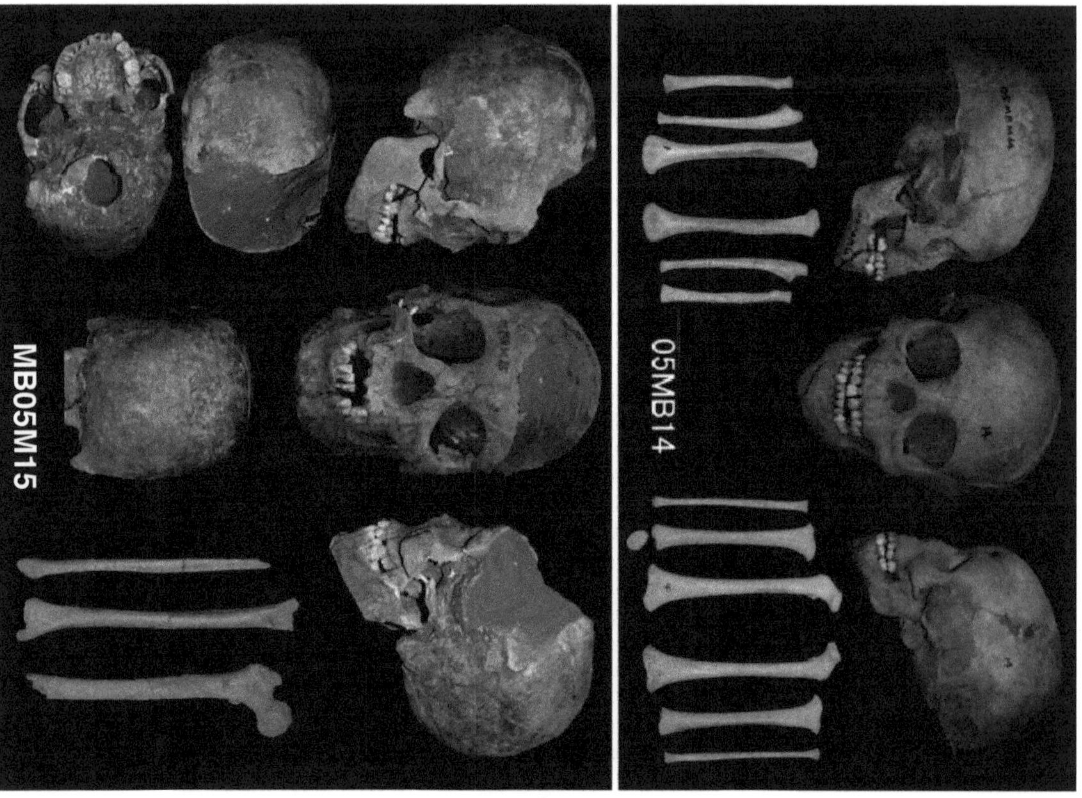

Human skeletal remains from the Mac Bac site

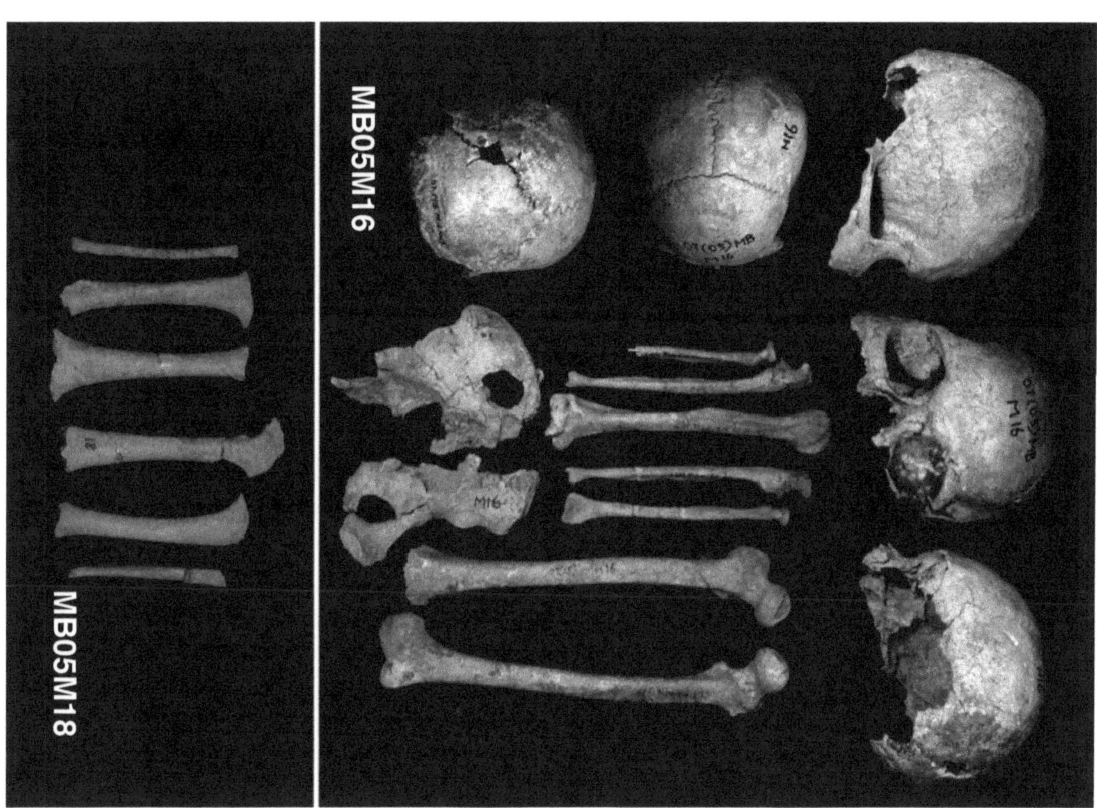

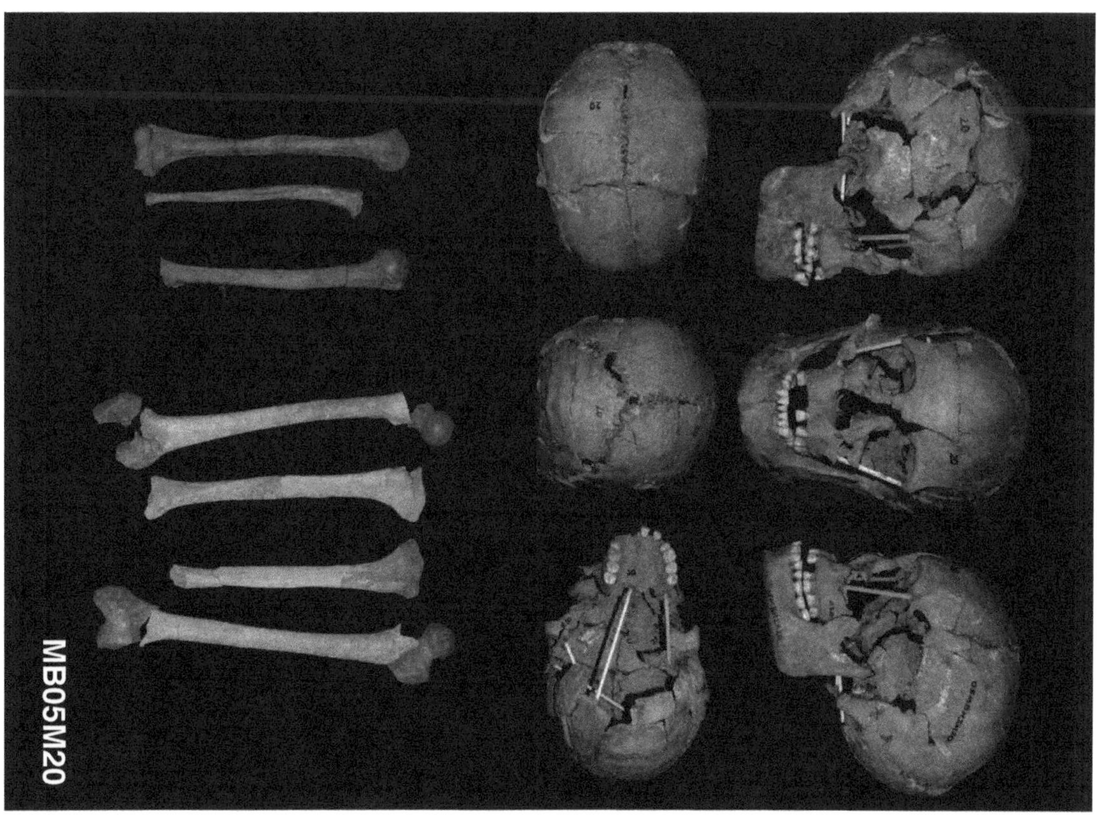

Human skeletal remains from the Mac Bac site

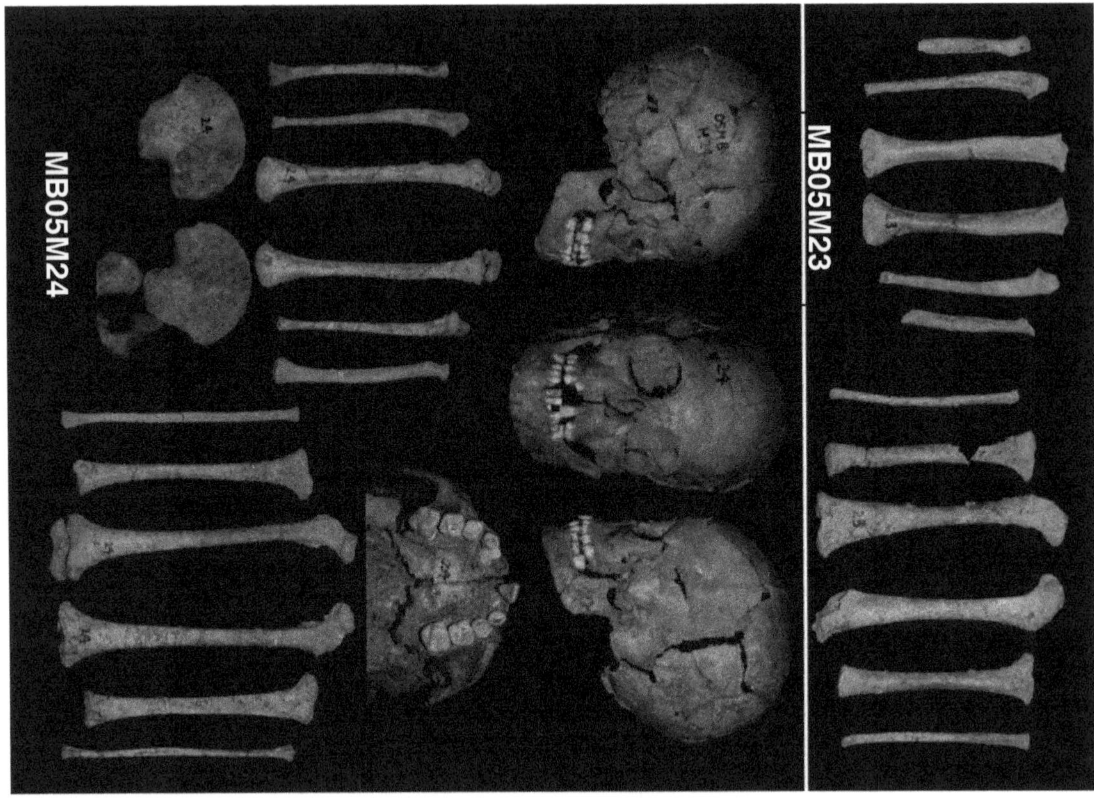

Human skeletal remains from the Mac Bac site

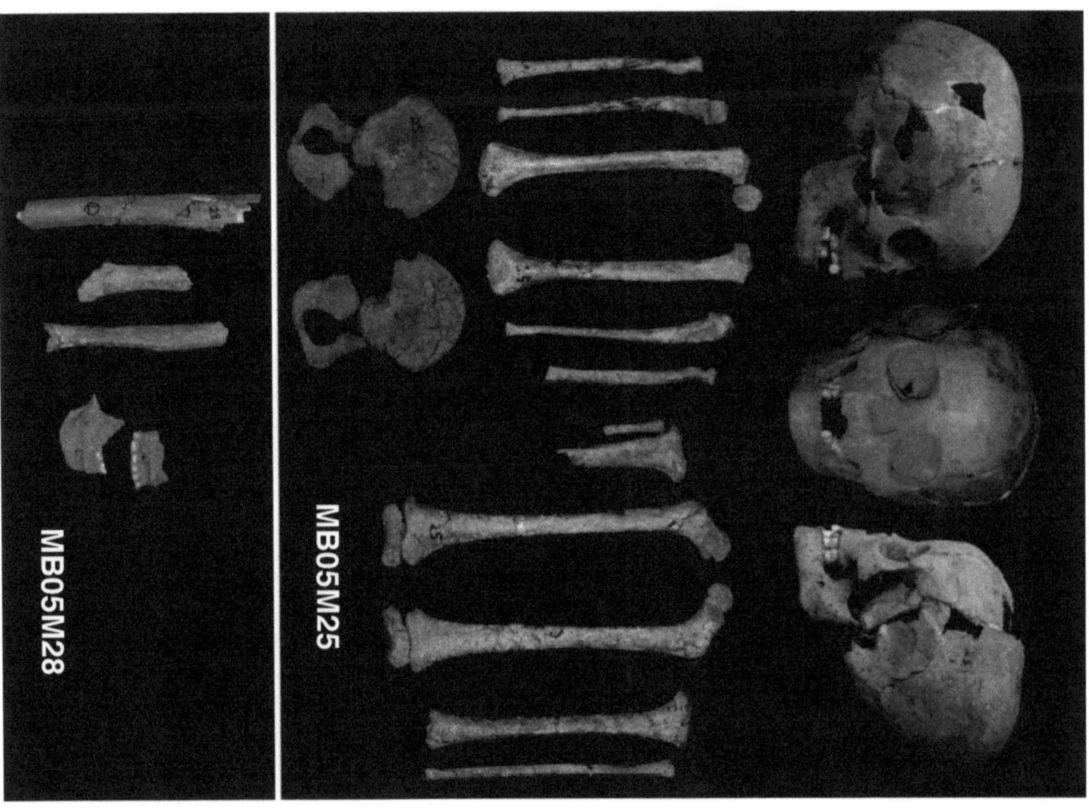

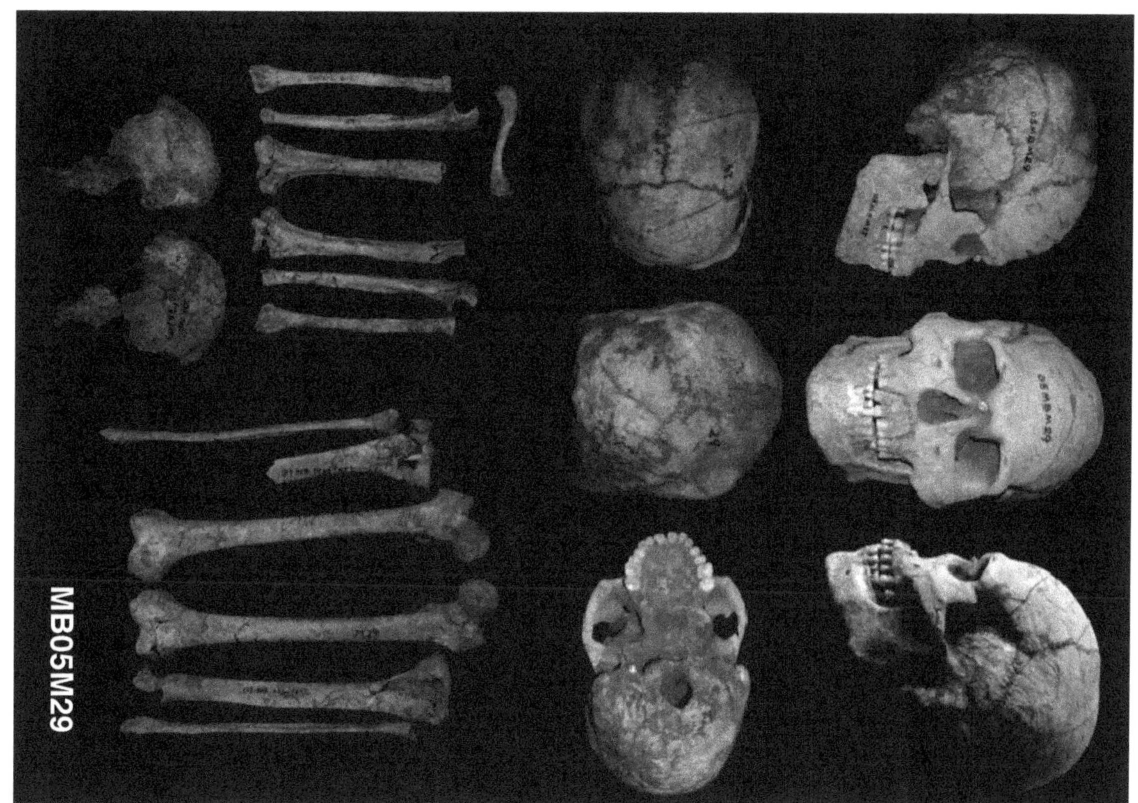

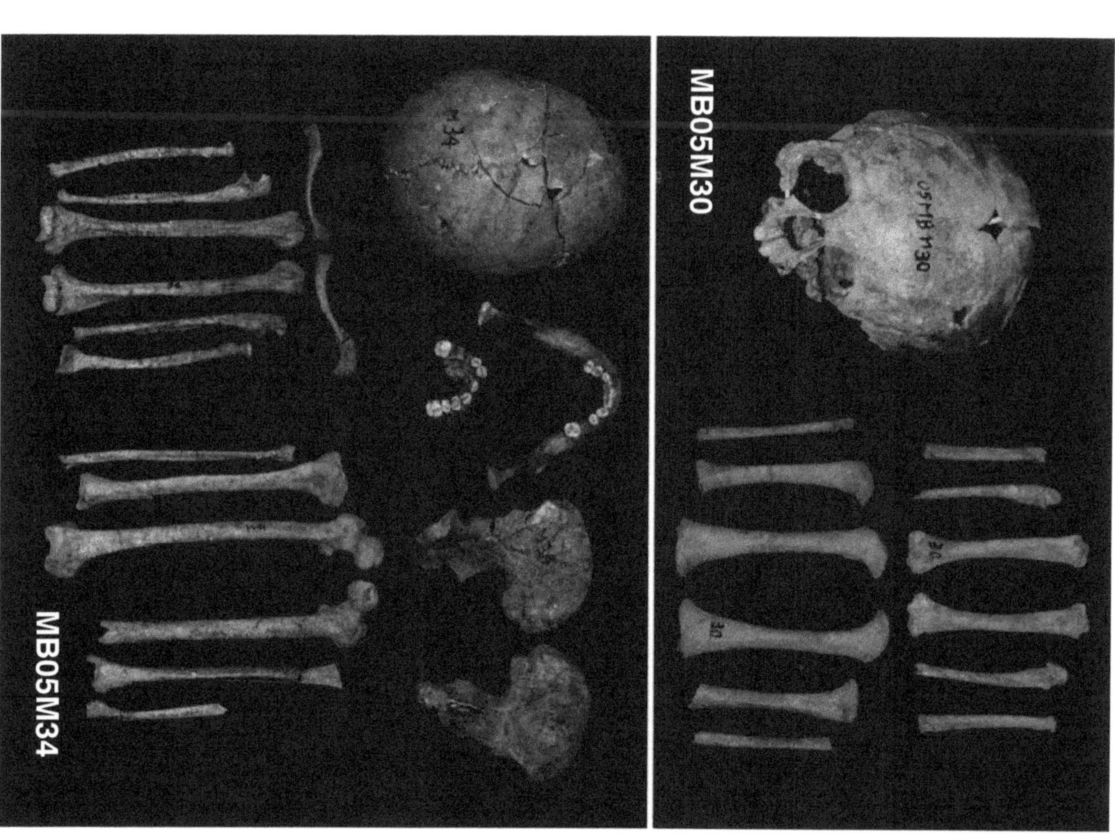

Human skeletal remains from the Mac Bac site

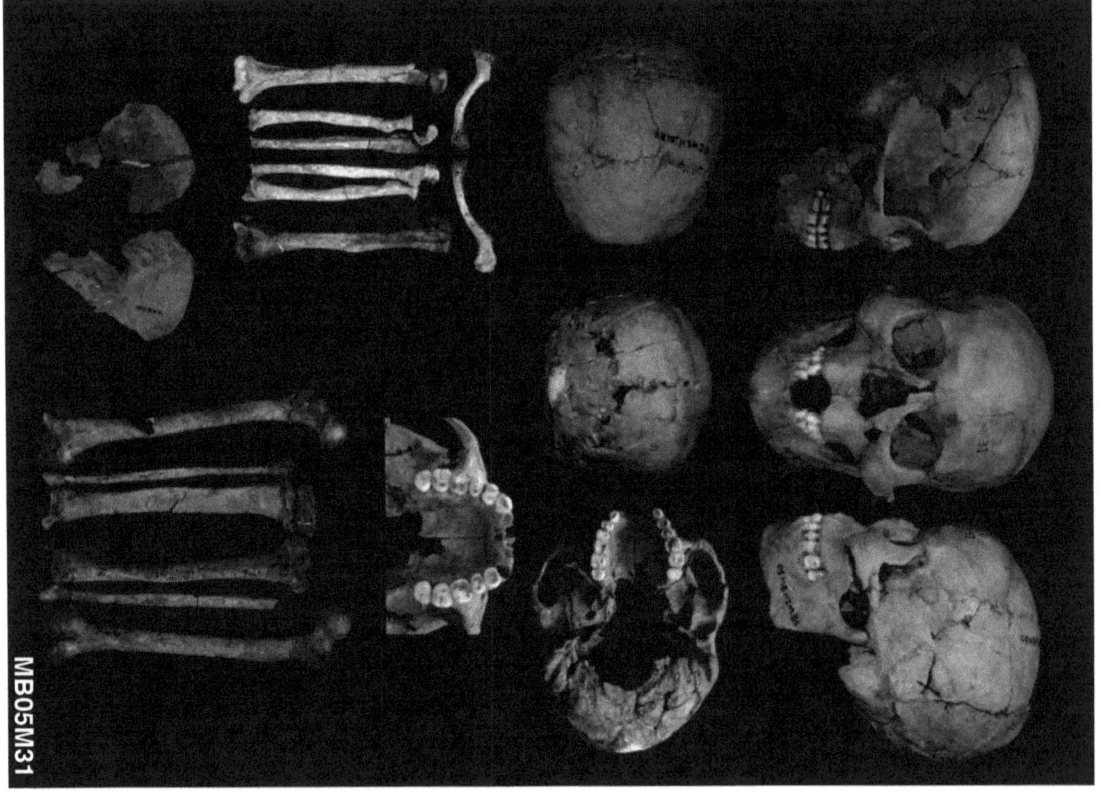

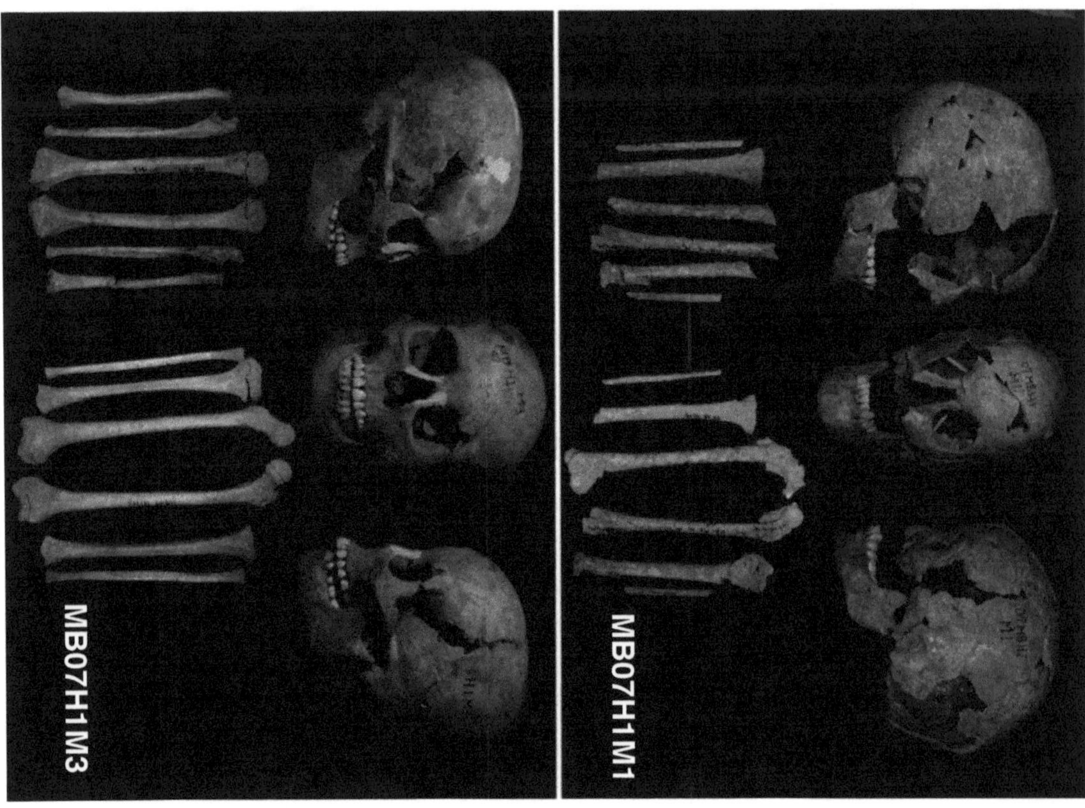

Human skeletal remains from the Mac Bac site

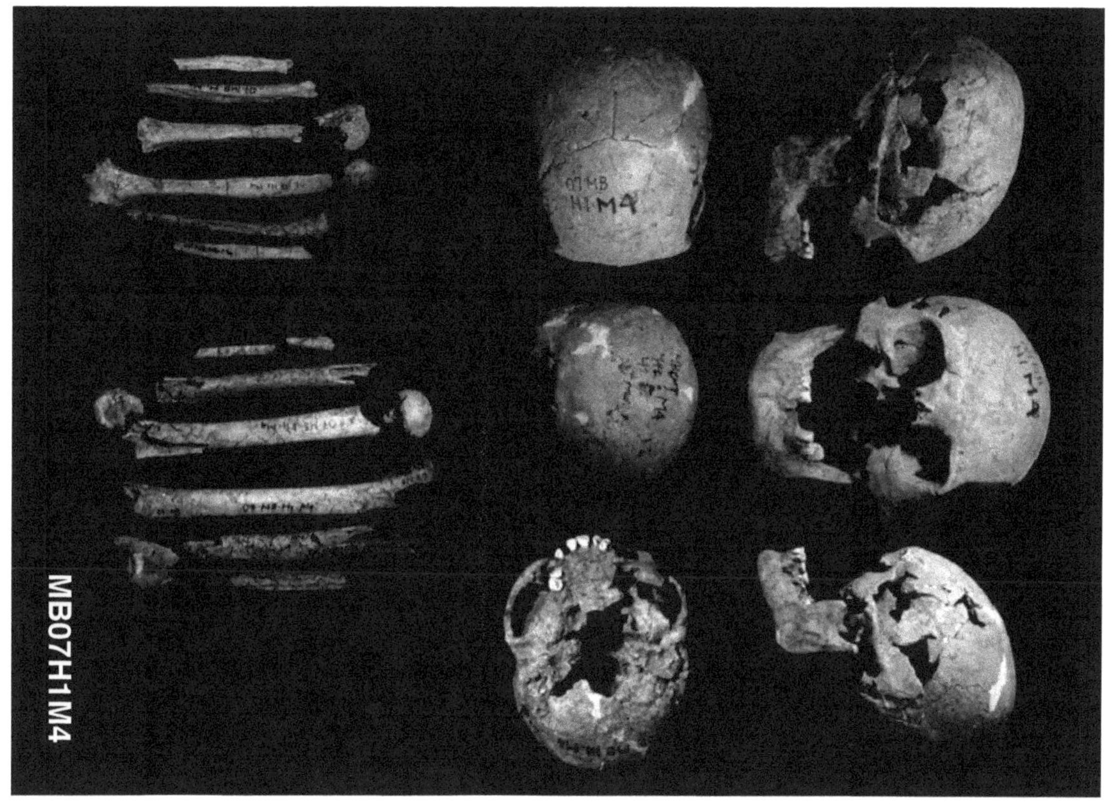

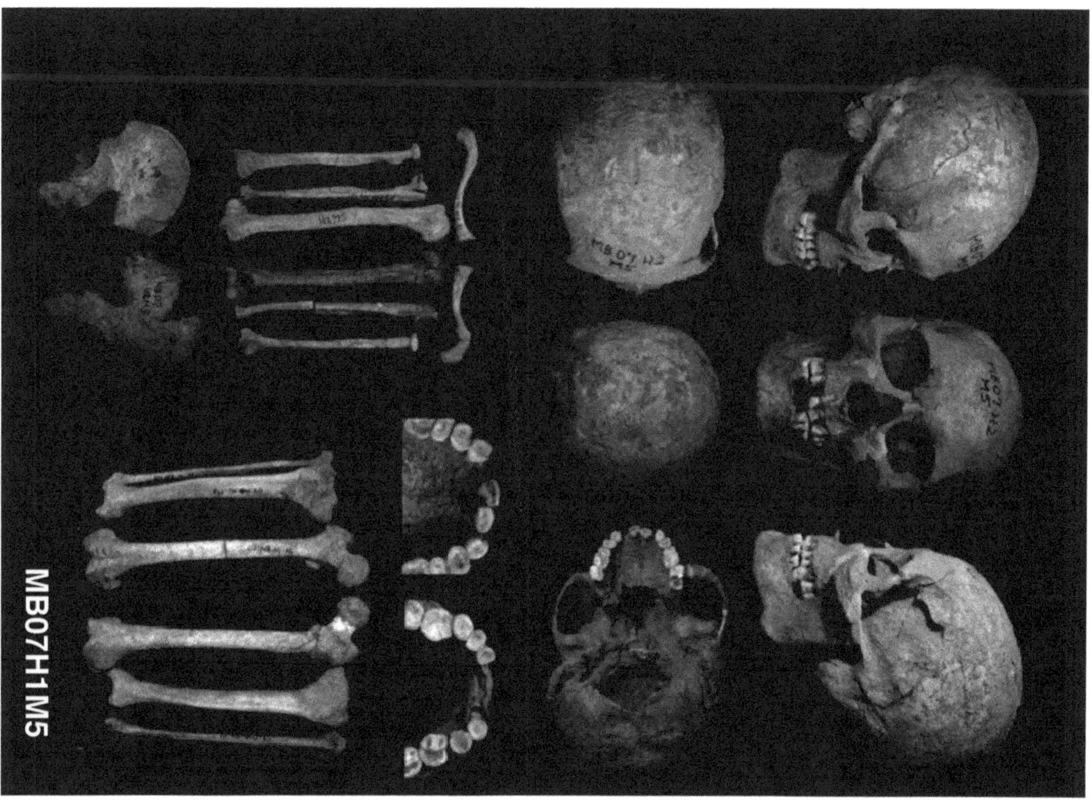

Human skeletal remains from the Mac Bac site

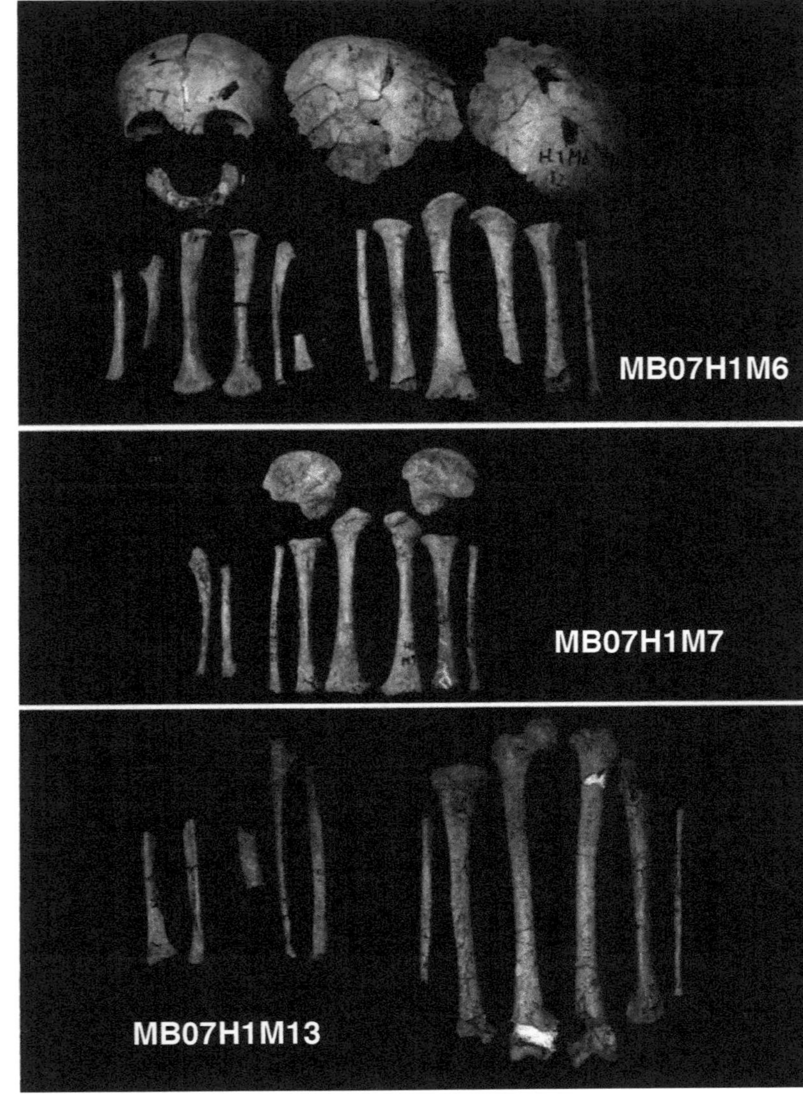

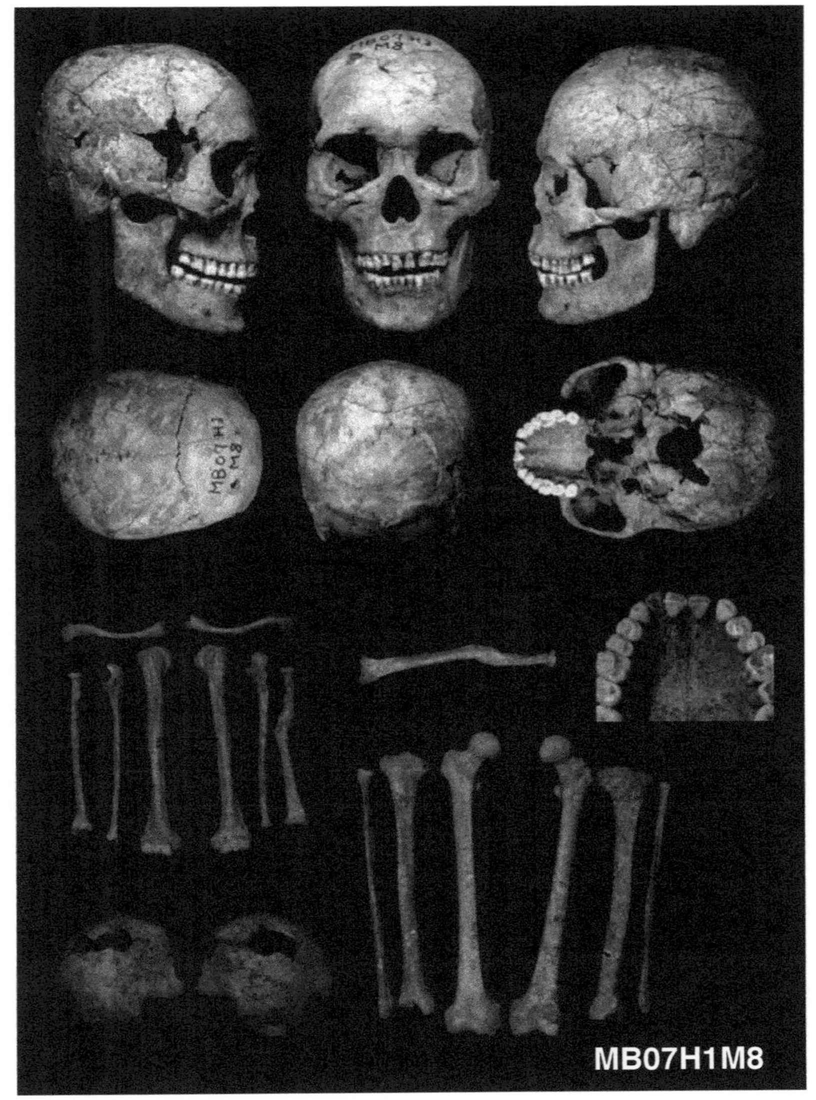

Human skeletal remains from the Mac Bac site

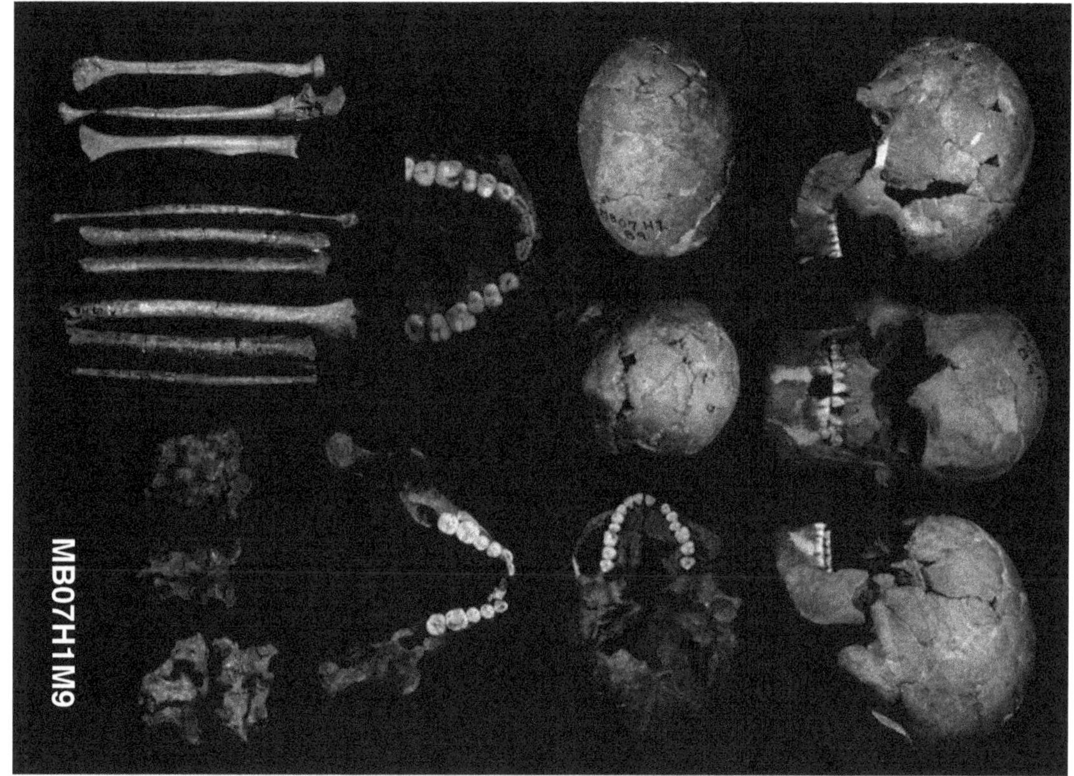

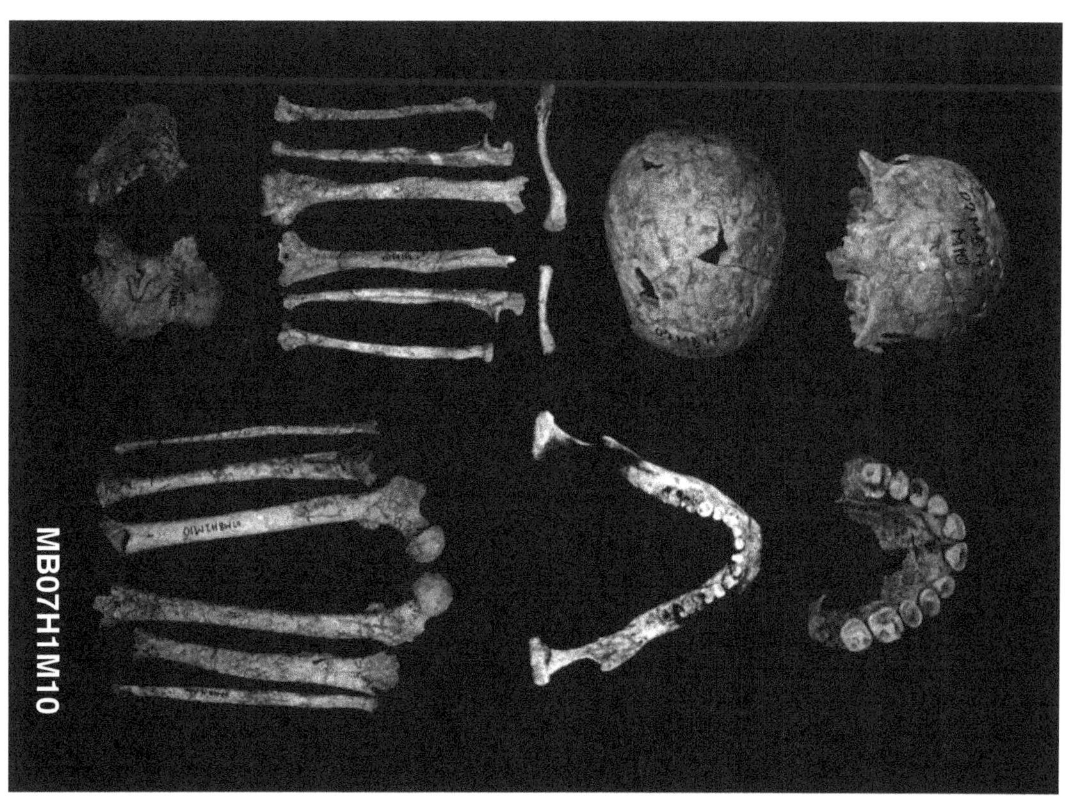

Human skeletal remains from the Mac Bac site

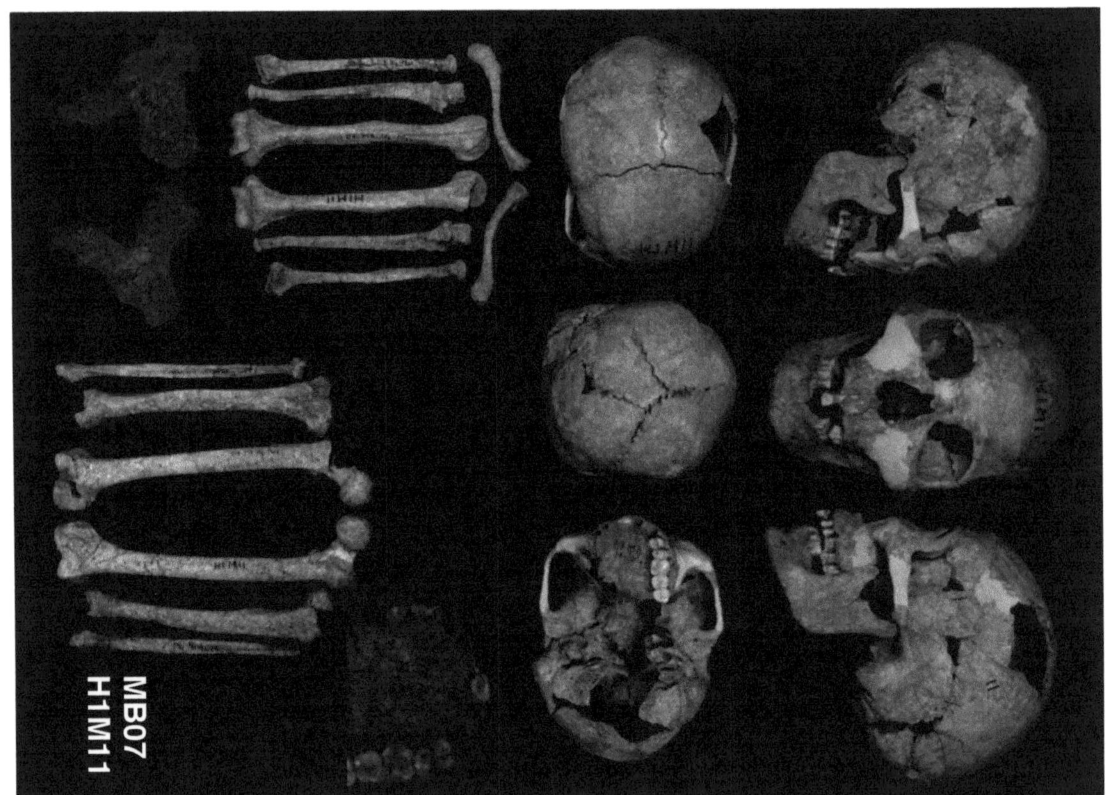

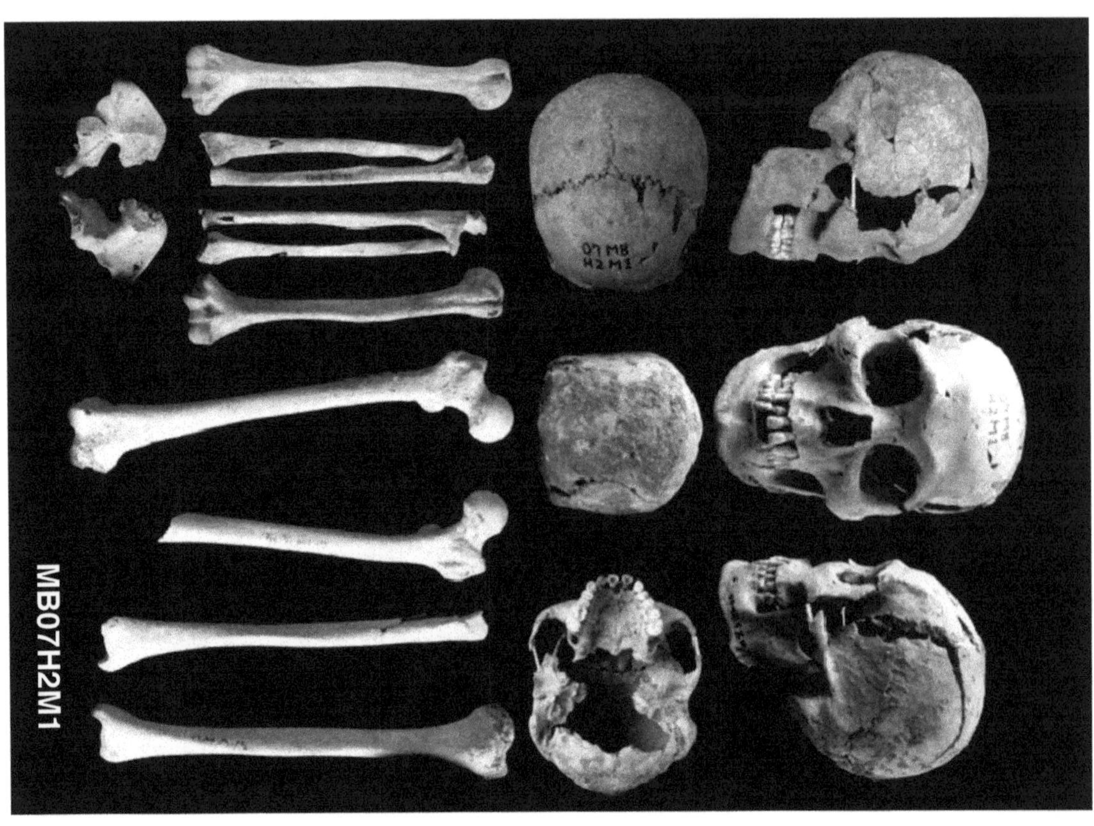

Human skeletal remains from the Mac Bac site

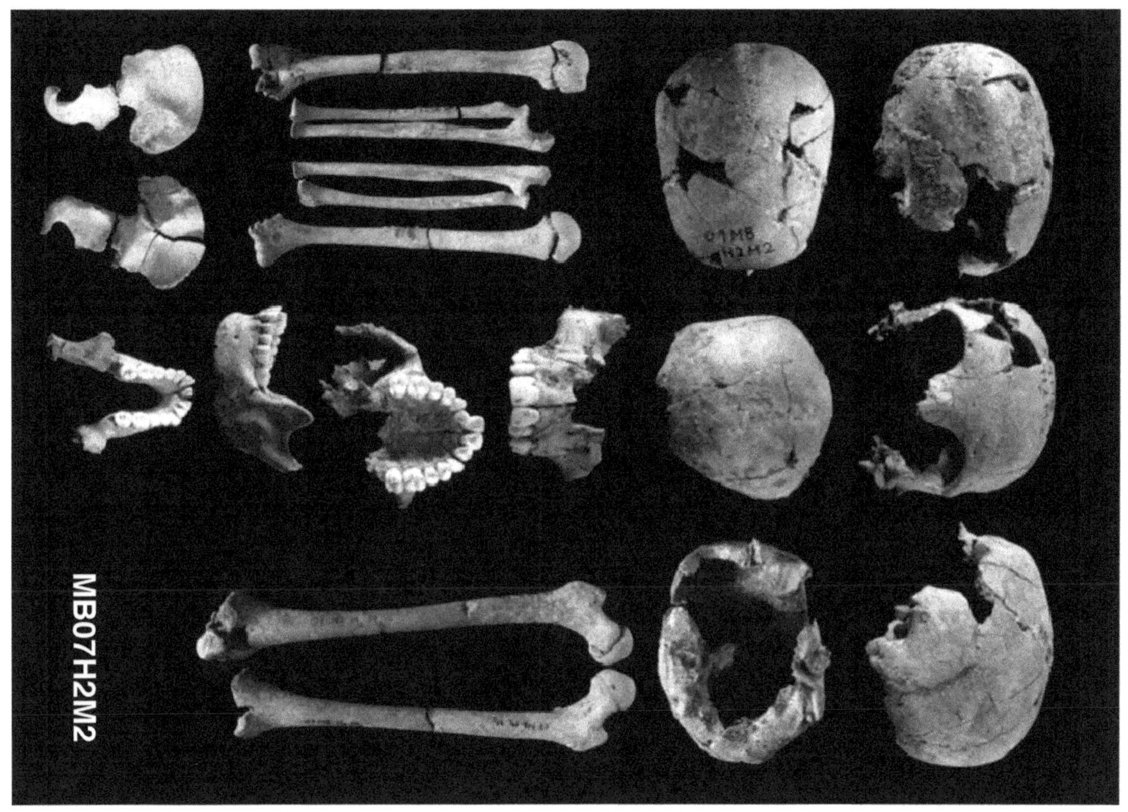

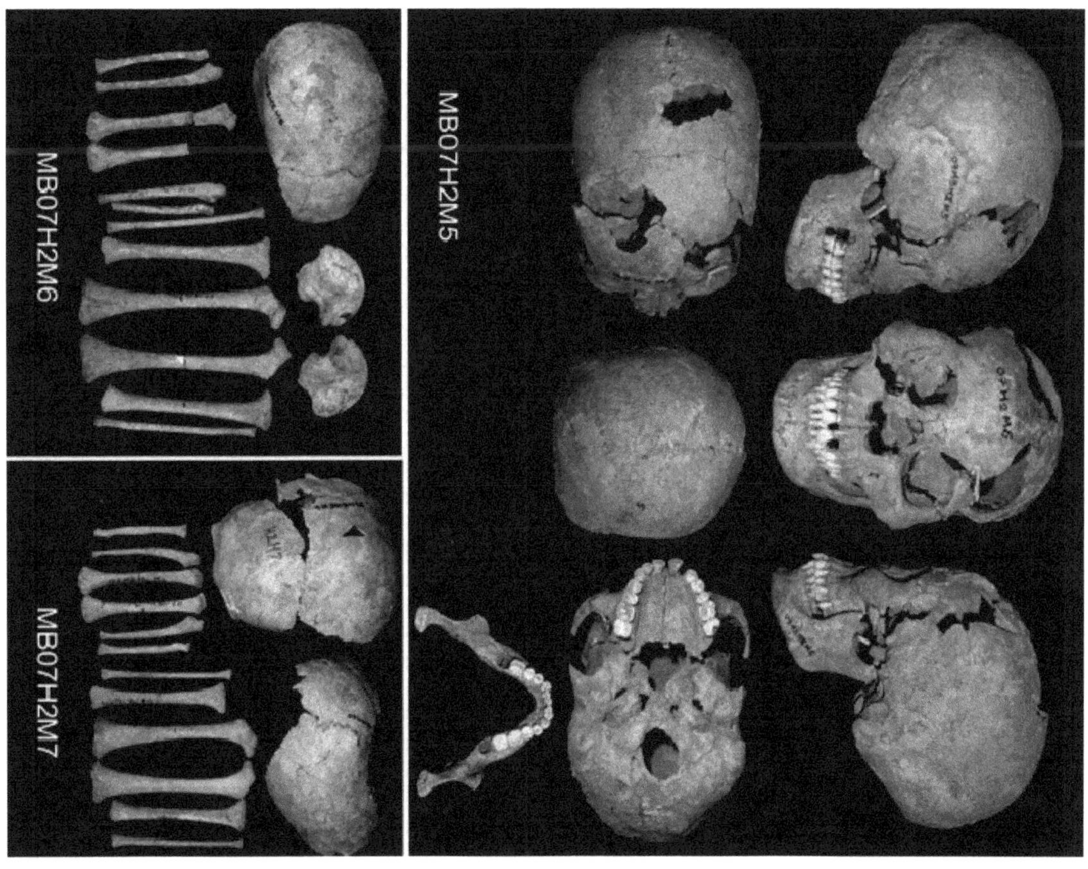

Human skeletal remains from the Mac Bac site

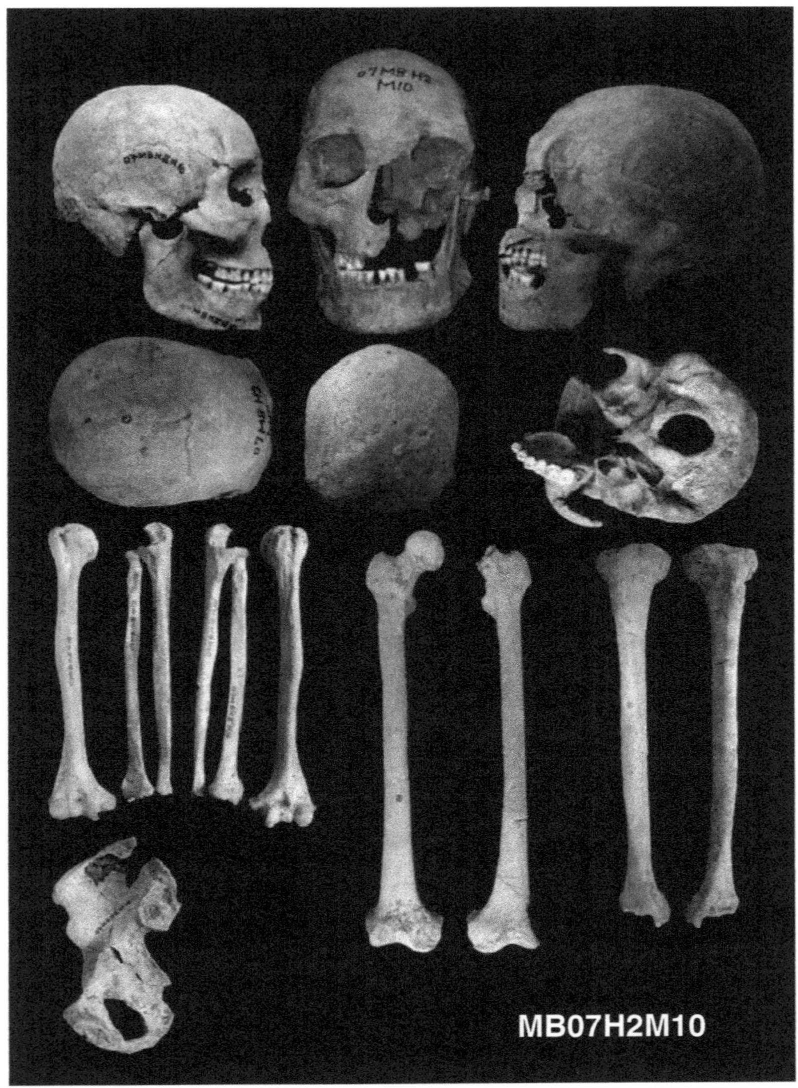

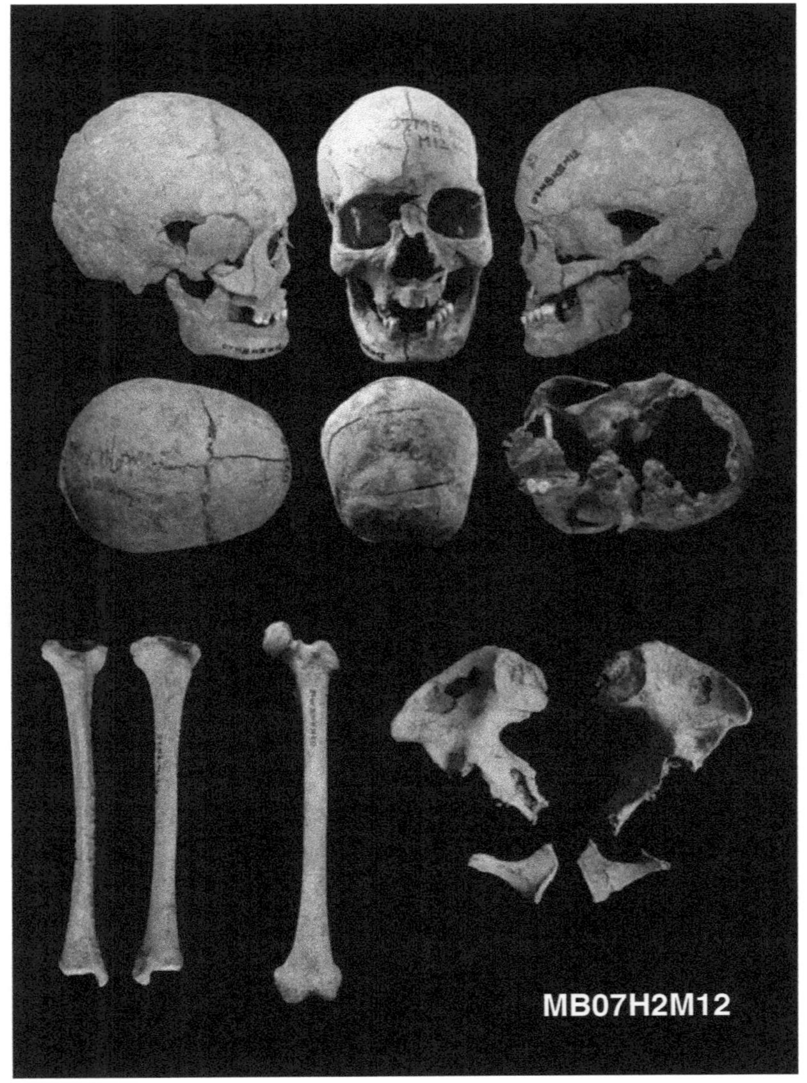

Human skeletal remains from the Mac Bac site

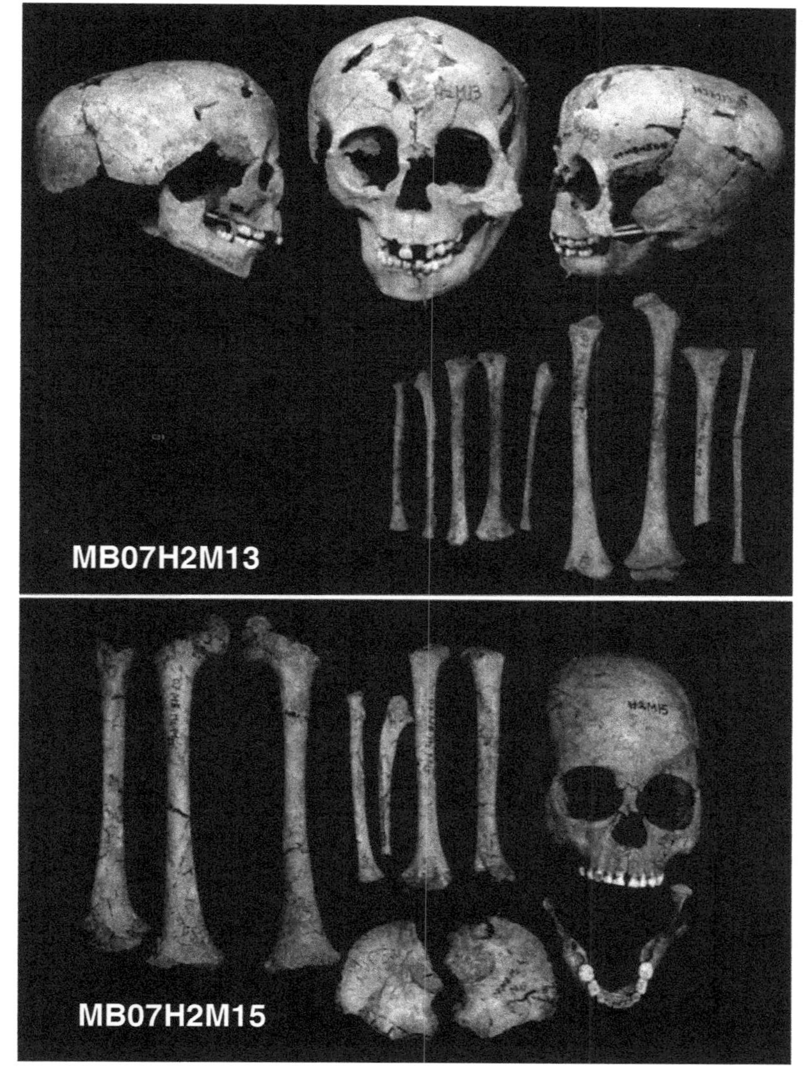

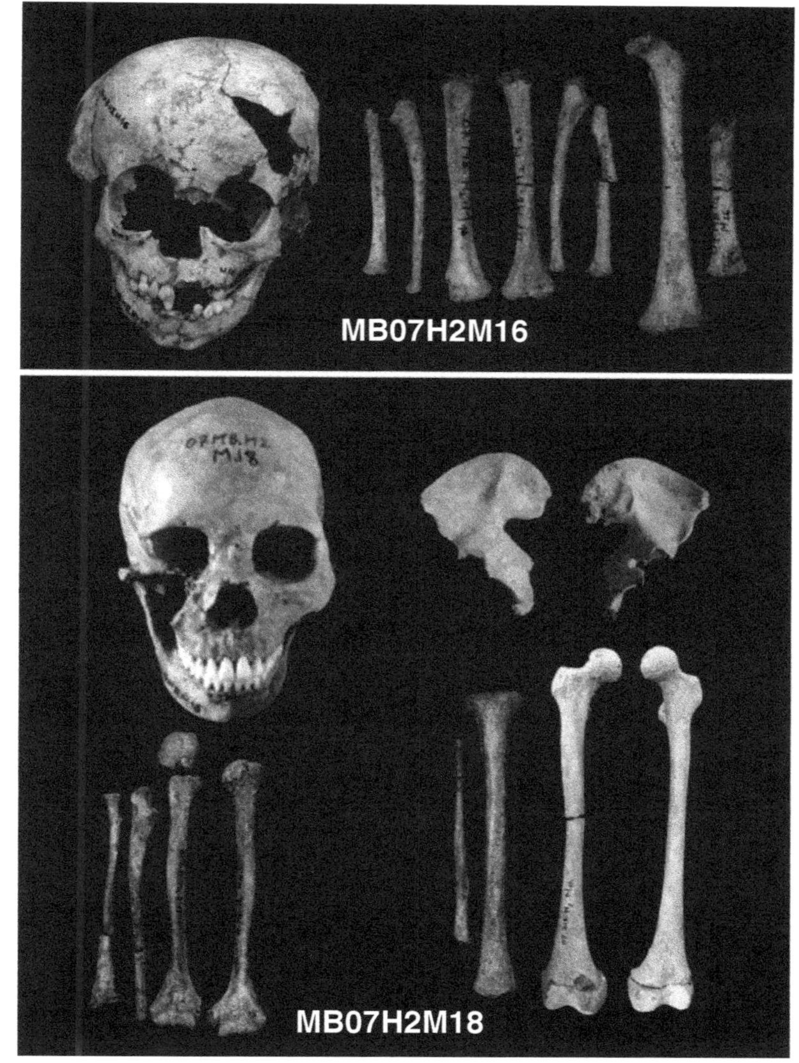

Human skeletal remains from the Mac Bac site

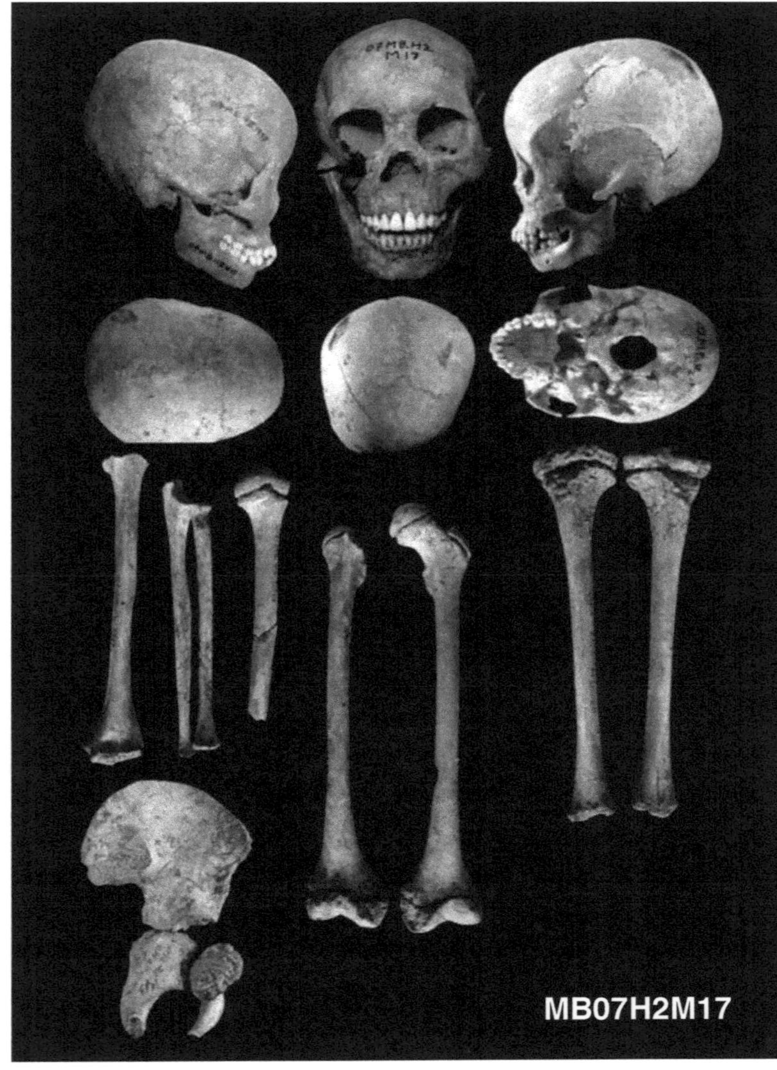

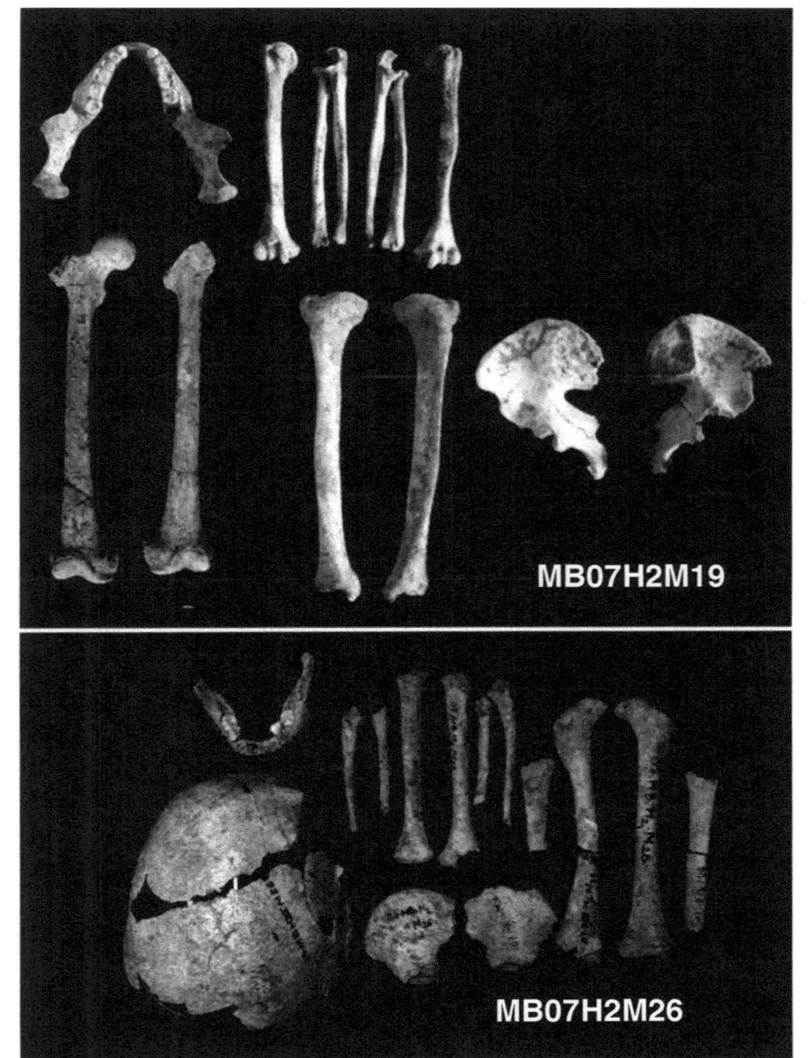

Human skeletal remains from the Mac Bac site

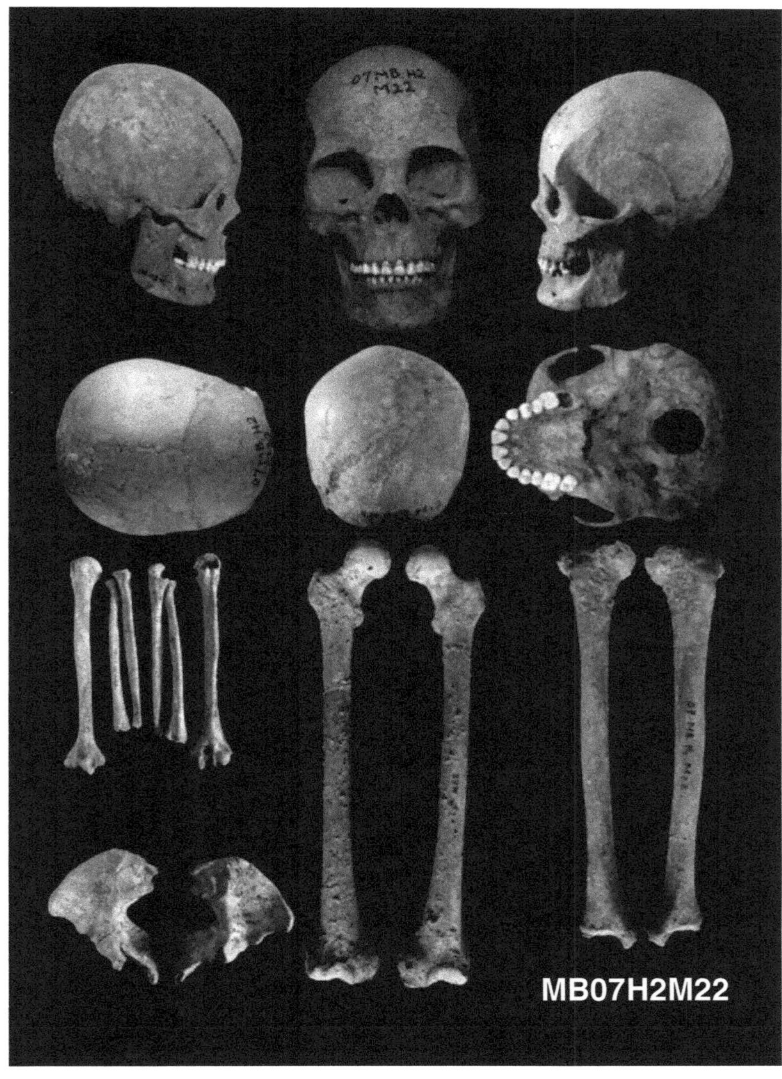

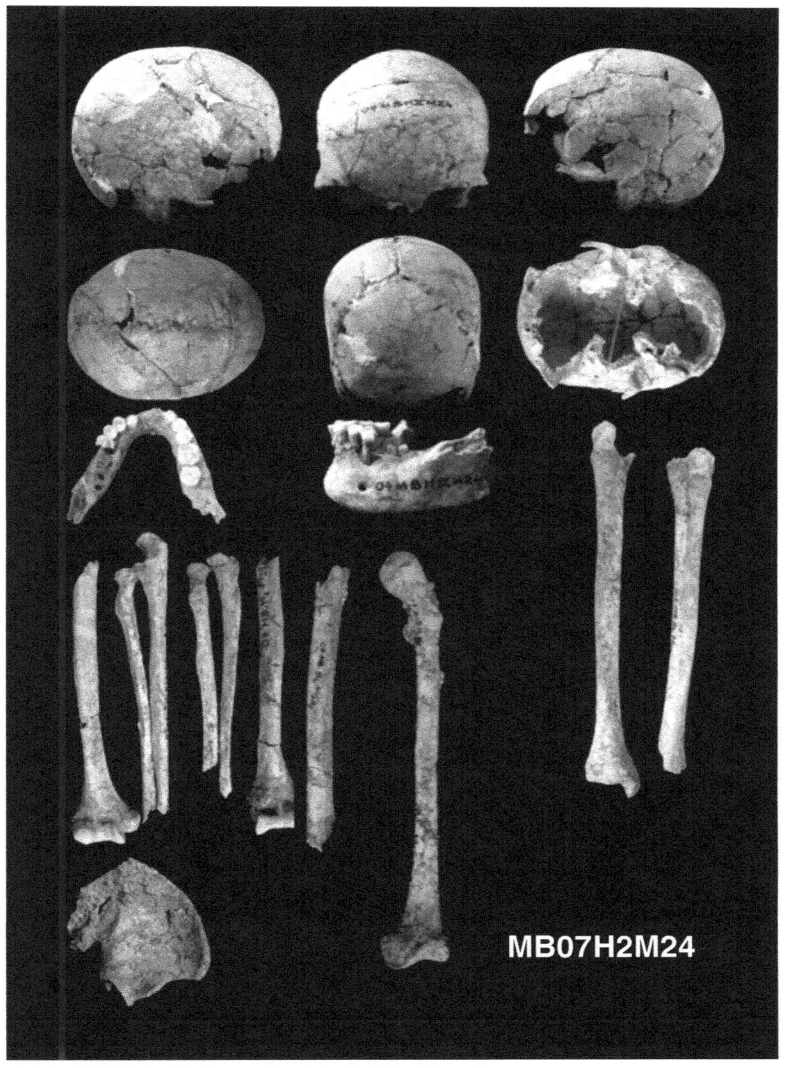

Human skeletal remains from the Mac Bac site

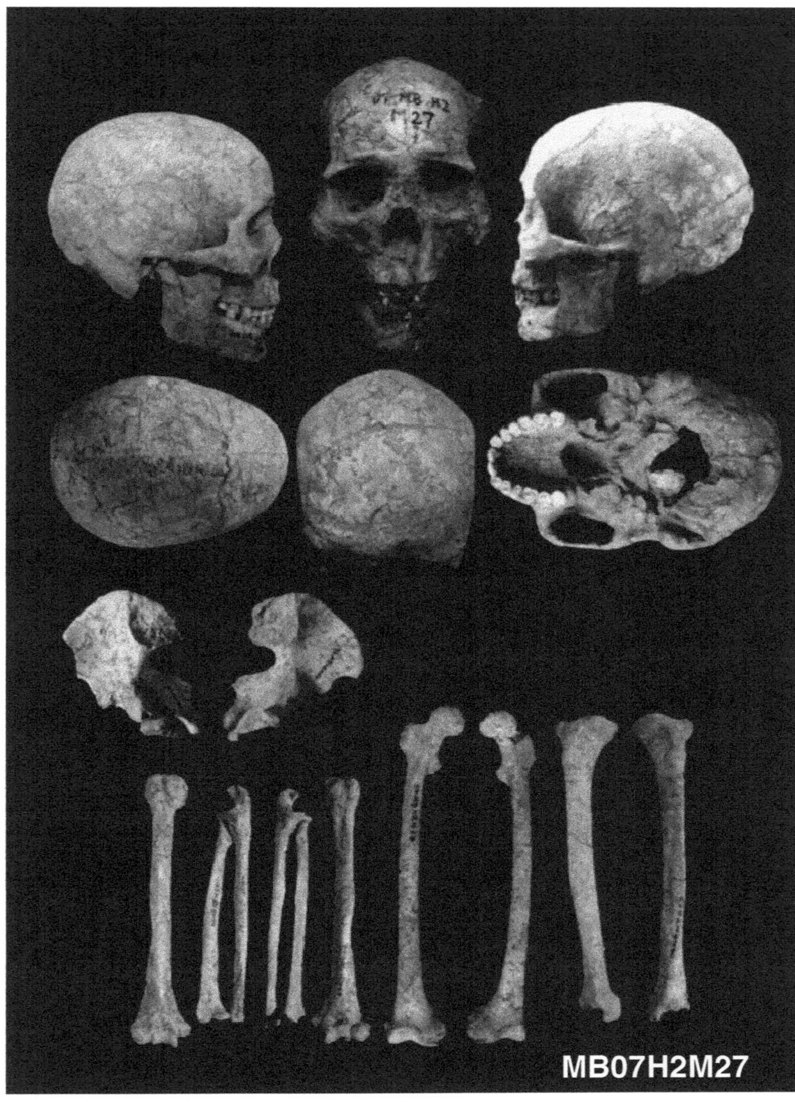

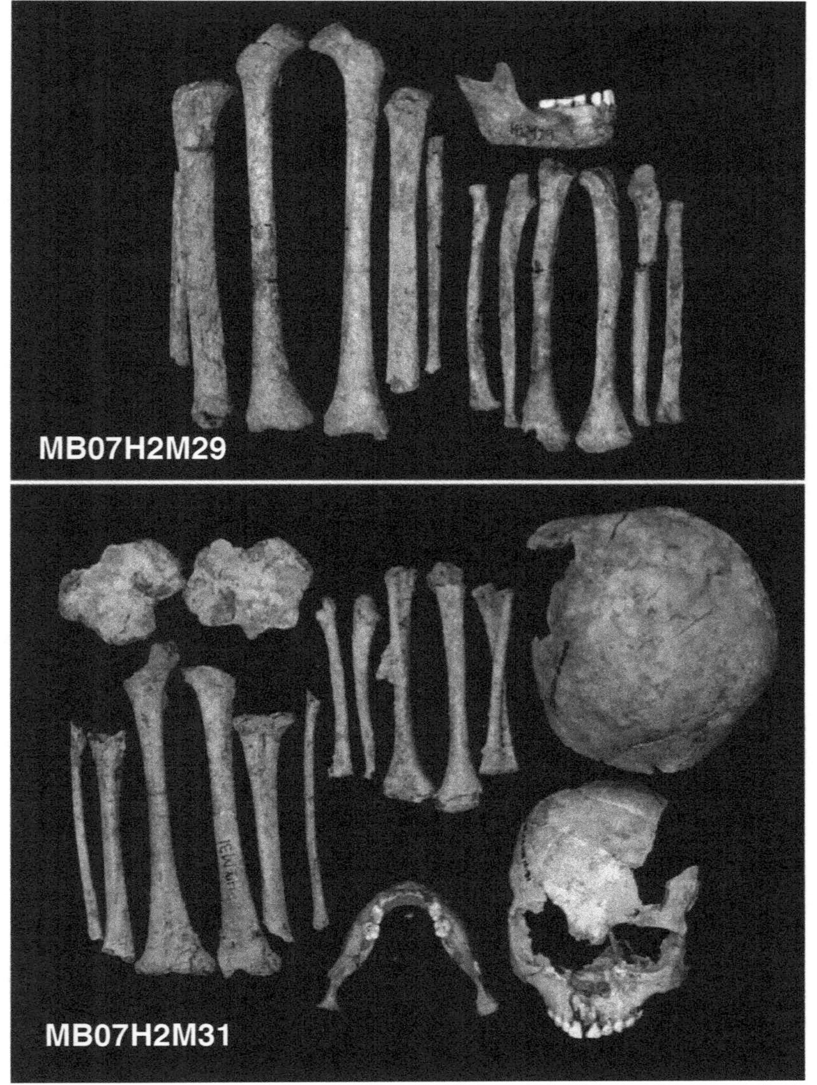

Human skeletal remains from the Mac Bac site

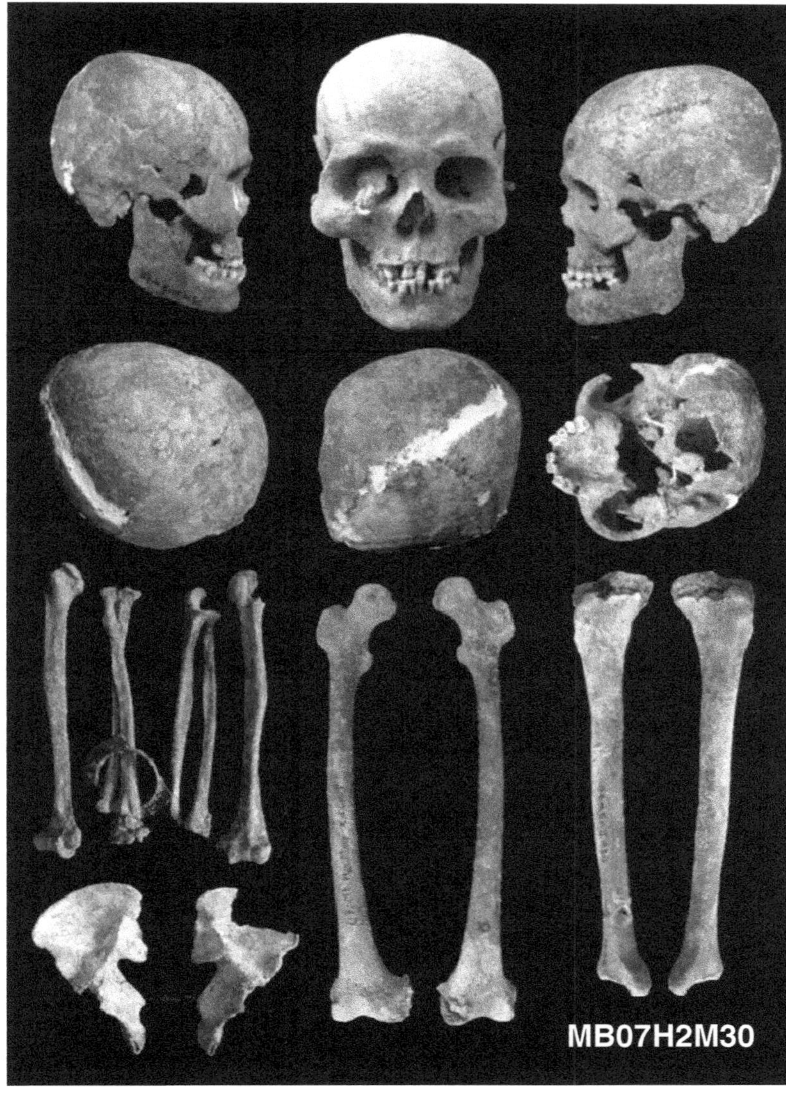

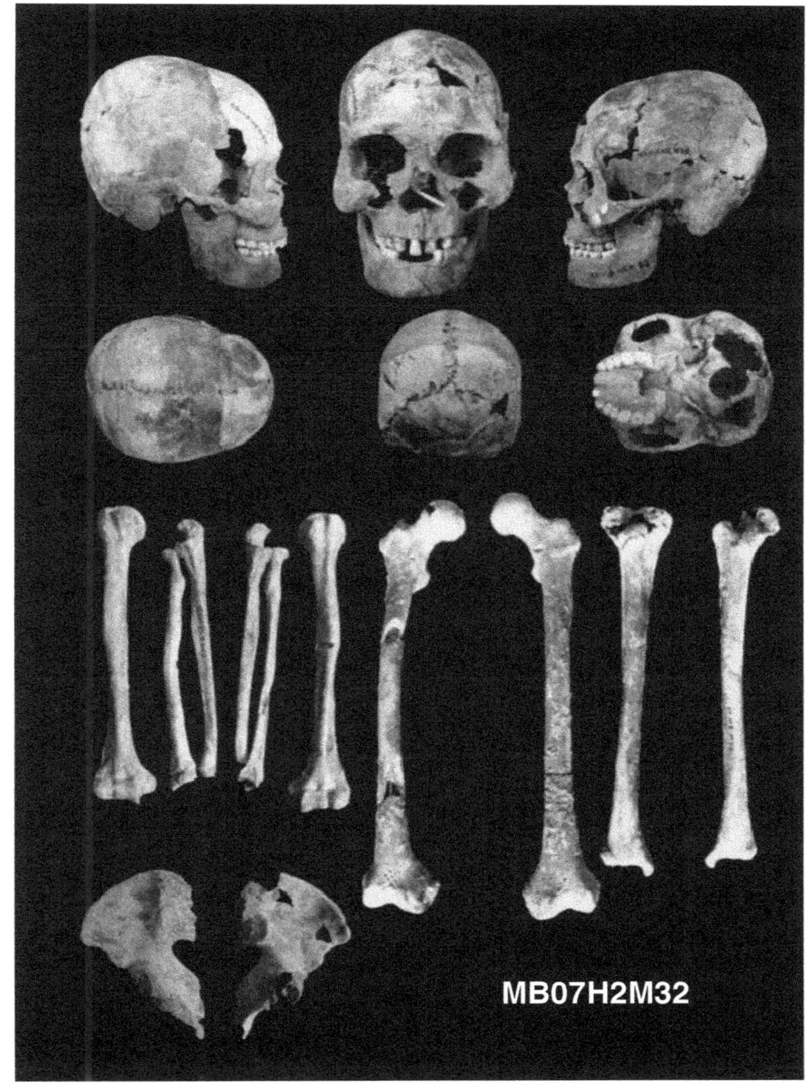

Human skeletal remains from the Mac Bac site

www.ingramcontent.com/pod-product-compliance
Lightning Source LLC
Chambersburg PA
CBHW051309270326
41929CB00029B/3461
* 9 7 8 1 9 2 1 8 6 2 2 2 9 *